Hans Bühlmann
Alois Gisler

A Course in Credibility Theory and its Applications

Professor Hans Bühlmann

Swiss Federal Institute
of Technology Zurich
ETH Zentrum, HG
Rämistrasse 101
8092 Zürich
Switzerland
E-mail: hbuhl@math.ethz.ch

Professor Alois Gisler

Winterthur Insurance Company
WSAN
Römerstrasse 17
8401 Winterthur
Switzerland
E-mail: alois.gisler@winterthur.ch

Mathematics Subject Classification: 91B30, 62P05, 62J05, 62C10, 62C12

Library of Congress Control Number: 2005930887

ISBN-10 3-540-25753-5 Springer Berlin Heidelberg New York
ISBN-13 978-3-540-25753-0 Springer Berlin Heidelberg New York

Springer is a part of Springer Science+Business Media
springeronline.com

Printed in The Netherlands

Typesetting: by the authors and TechBooks using a Springer LaTeX macro package

Cover design: *design & production* GmbH, Heidelberg

Printed on acid-free paper SPIN: 11320753 41/TechBooks 5 4 3 2 1 0

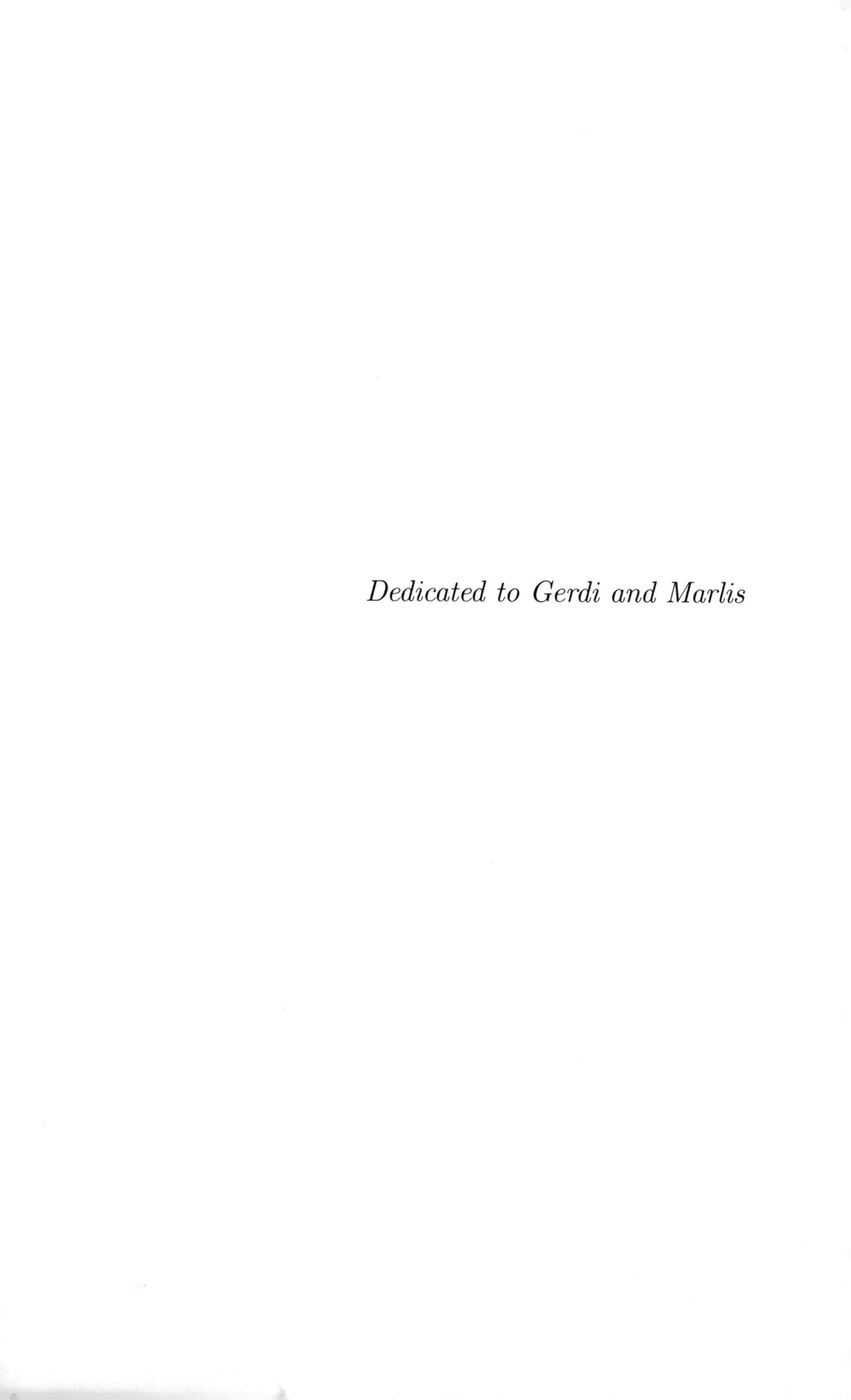

Dedicated to Gerdi and Marlis

Preface

The topic of credibility theory has been for many years – and still is – one of our major interests. This interest has led us not only to many publications, but also has been the motivation for teaching many courses on this topic over more than 20 years. These courses have undergone considerable changes over time. What we present here, "A Course in Credibility Theory and its Applications", is the final product of this evolution.

Credibility theory can be seen as the basic paradigm underlying the pricing of insurance products. It resides on the two fundamental concepts "individual risk" and "collective" and solves in a rigorous way the problem of how to analyse the information obtained from these sources to arrive at the "insurance premium". The expression "credibility" was originally coined for the weight given to the experience from the "individual risk".

Credibility theory as a mathematical discipline borrows its methods from many fields of mathematics, e.g. Bayesian statistics, $\mathcal{L}^2$ Hilbert space techniques, least squares, and state space modelling to mention only the most important ones. However, credibility theory remains a lifeless topic if it is not linked closely with its applications. Only through these applications has credibility won its status in insurance thinking. The present book aims to convey this dual aspect of credibility and to transmit the flavour of the insurance applications also to those readers who are not directly involved in insurance activities. In particular we are convinced that many insurance applications have the potential to be powerful tools that could also be used in the wider area of finance.

The present text would have never been completed without the great help we received from many colleagues, friends and coworkers. In the first place we want to mention Paul Embrechts, who encouraged us to write this book. Without his constant support and advice we probably would never have succeeded in finishing this text. During the time of writing and rewriting our (several) manuscripts we profited from discussions with colleagues in Winterthur and Zürich, especially Anne Sheehy, Mario Wüthrich, Peter Eberle, Graham Masters, Werner Stahel and Peter Bühlmann. Anne translated a first

German version into English, she and Mario assisted us in the summer school where the text was tested, Peter E. was a big support for conceiving the exercises, Graham has checked the English language in the final manuscript, and Werner and Peter B. provided advice on statistical issues whenever needed. Also the referees who judged an earlier version, have taken great care in reading our text and making valuable suggestions. A great "thank you" goes also to Howard Waters and to Heriot-Watt University. It provided the good atmosphere for a two-month sabbatical for one of the authors to write part of the book.

Our book would not have been possible without the great community of researchers who have worked in the area of credibility. Some names of these researchers appear in the bibliography of our book, but many do not, since we have only listed those publications which are related to our text. We want to remember explicitly those researchers who are already dead but have had a decisive influence in the field: Charles A. Hachemeister, William S. Jewell, Etienne de Vylder and Erwin Straub.

The physical writing of this book was done by Elisabeth Werner. She typed and retyped all our versions and was a wonderful support for improving the script quality of the presentation. We owe her a very special "thank you".

We also profited from the generosity of our employers, ETH Zürich and Winterthur Insurance Company. These two institutions have provided us with the logistics necessary for the whole book project. The sustained support given by Winterthur to the actuarial programme at ETH is an excellent example of a fruitful cooperation between academia and practice.

Writing this book turned out to be much more time consuming than anticipated. This time was mainly taken to the detriment of our families. As a sign of gratitude we therefore want to dedicate this book to our wives Gerdi and Marlis. Their support goes, of course, far beyond this project. They have given to us lasting companionship, sharing with us happy and difficult times.

Hans Bühlmann and Alois Gisler
August 2005

Notes and Comments

Notes on Chapter 1 and Chapter 2

These two chapters give the basic framework for the whole book: Chapter 1 does so using common language, Chapter 2 uses mathematical modelling.

The basic understanding of insurance risk is built upon two fundamental concepts, the individual risk and the collective. The insurer has typically little knowledge about the individual risk, but quite extensive statistical information about the collective and possibly about subcollectives. The theoretical concept of a *homogeneous* collective is a fiction. Credibility theory relies on the more realistic concept of a *heterogeneous* collective.

The fundamental mathematical paradigm underlying credibility theory is the "two-urn model": first the characteristics of the individual model (parameter ϑ) are drawn using a distribution U from the collective population. The claims produced by the individual risk are then generated by a claims distribution F_ϑ. This paradigm coincides – in abstract analogy – with the Bayes model in mathematical statistics.

It is interesting to note that the Poisson–Gamma model appeared as early as 1929 in [Kef29], where it is used for pricing of group life insurance. Ove Lundberg [Lun40] and Arthur Bailey [Bai50] seem to be the first actuaries to recognize the general Bayesian structure. It is not clear how they interpreted the role of the prior distribution U. But it is obvious that in insurance applications the prior distribution must be understood as the description of the collective. The method of how to infer the distribution U from collective data has – in the abstract context – been advocated by Herbert Robbins [Rob55] under the heading "empirical Bayes".

Whoever is looking for a concise description of credibility theory should read Ragnar Norberg's article in the actuarial encyclopedia [Nor04]. This article reduces credibility to its essential mathematical structure, hence giving an excellent insight into the subject from a theoretical point of view.

As a textbook on Bayesian statistics we recommend DeGroot [DeG70]. The explicit calculation of the posterior mean is one of the standard tasks in Bayesian statistics. In an actuarial context such calculations date back to Ove Lundberg [Lun40], Arthur Bailey [Bai50], Bruno De Finetti [DF64] and for the general formulation in the case of conjugate exponential families to William S. Jewell [Jew74a].

In the early sixties, the Poisson–Gamma model served as the mathematical basis for the development of Bonus–Malus systems in motor liability insurance (see [Bic64], [Del65]), which form one of the most prominent applications of Bayesian statistics in the insurance context.

Notes on Chapter 3 and Chapter 4

For practical use it is convenient to require that the estimates for the individual premium be linear in the data. There is a considerable freedom of choice for this linearity, as it may refer to transformations of any kind of the original data.

In a simple context (Bühlmann model) this linear estimate turns out to be a weighted average between the individual and the collective mean. Such weighted means (credibility formulae) were already used by American actuaries at the beginning of the 19th century (Whitney [Whi18], Perryman [Per32]) in the context of workmen's compensation insurance. The link with Bayesian statistics was made by Arthur Bailey [Bai50] (see also notes on Chapters 1 and 2).

Mowbray [Mow14], [Mow27] tried to derive credibility formulae on purely classical statistics arguments using confidence bounds to arrive at "full credibility" (i.e. giving weight 1 to the individual estimate). If n_0 observations give you full credibility, the weight $\sqrt{n/n_0}$ is given to the individual estimator if you have only $n < n_0$ observations. This rather heuristic approach is called "limited fluctuation theory". According to our view it is only of historical interest. See also Longley and Cook [LC62] for a description of this method.

The model treated in Chapter 4 (Bühlmann–Straub) is widely used in practice. The reason is that it combines credibility thinking with the concept of the loss ratio, which is one of the most frequently used statistics in insurance.

For surveys on credibility theory we recommend W.S. Jewell [Jew76b], and M.J. Goovaerts and W.J. Hoogstad [GH87]. An extensive bibliography is given by DePril, D'Hooge and Goovaerts [DPDG76] and in De Wit [Wi86]. Recent textbooks on credibility theory are [DKG96] and [Her99].

Notes on Chapter 5

As credibility theory deals with estimators which are linear in the data or in transformed data, the question arises of how to choose such transformations. This chapter studies the transformation by truncation. Truncating is very useful in the presence of large claims. The general idea of using transformations goes back to de Vylder [DV76b], [DVB79]. His goal was the search for transformations which are optimal within the least squares framework. Theoretically this goal can be achieved by solving an integral equation of the Fredholm type. Gisler's [Gis80] approach to restrict the transformations to simple truncations turns out to be more fruitful in practice. In this approach the truncation point is the only one-dimensional parameter to be optimized. From a pragmatic point of view even this optimization may often be left aside. The truncation point is then chosen by pure a priori judgement. The point is that truncation avoids considerable premium jumps due to one or a few large claims. Such premium jumps are very often considered as violating the very purpose of insurance.

Other possibilities in the framework of one-dimensional methods of dealing with large claims such as combining credibility and robust statistics are mentioned at the end of Chapter 5 without going into details and referring to the literature. The topic of treating large claims is resumed under multidimensional credibility in Chapter 7.

Notes on Chapter 6

Hierarchical models generalize the "two-urn model" into a "many-urn model". In such models the risk characteristics of the individual risk are drawn from a collective urn whose risk characteristics are drawn from a further parent urn etc. Such hierarchies appear very naturally in the tariff structure of an insurance company.

Jewell [Jew75c] and Taylor [Tay79] have introduced the concept of hierarchical credibility. Sundt [Sun80] and Norberg [Nor86] have clarified the basic model assumptions and given important insights into the model structure.

In the context of hierarchical credibility the Hilbert space technique proves particularly useful as it replaces tedious calculations by the use of the more intuitive projection operator. The iterativity property of this operator is crucial in this context. This route of description of the hierarchical credibility model is followed in this text. It essentially summarizes Bühlmann and Jewell [BJ87].

Notes on Chapter 7 and Chapter 8

The basic set-up is the idea that we have to estimate a real-valued vector

$$\boldsymbol{\mu}(\Theta) = (\mu_1(\Theta), \ldots, \mu_n(\Theta))'.$$

In the abstract setting of Chapter 7 we assume that we have an observation vector of the same dimension

$$\mathbf{X} = (X_1, \ldots, X_n)'$$

such that

$$E[\mathbf{X}|\Theta] = \boldsymbol{\mu}(\Theta).$$

This vector $\mathbf{X}$ may be obtained by data compression from the raw data. Such an abstract set-up is convenient and covers most practical applications. It does not include the case where we have fewer observations than parameters to estimate. This case is, however, treated in Chapters 9 and 10 and is also intuitively much easier to understand in a recursive estimation procedure as treated there. In the multidimensional philosophy of Chapters 7 and 8 regression can be treated as a special case of data compression.

Multidimensional credibility was introduced by Jewell [Jew73]. By the appropriate use of matrix notation, the Jewell extension inherits the weighted average structure of the credibility estimator from the simple model.

Hachemeister's credibility regression model [Hac75] is an extension of multidimensional regression as treated in classical statistics. The regression parameters become random. In this view it is natural to keep the classical assumption of a design matrix of full rank smaller than the number of observations. Already Jewell [Jew76a] has pointed out that in Bayesian regressions this assumption could be dropped. He then continues: "However, to relate our results to classical theory we shall... [keep the restriction]". As already mentioned, we follow here the route of Jewell, whereas in Chapters 9 and 10 the restriction is dropped.

Although it is well known in classical statistics that one should work with an orthogonal design matrix – whenever feasible – this pragmatic rule was originally disregarded in the actuarial literature. This led the practitioners not to use the credibility regression model as it often produced "strange results". This handicap is overcome by using orthogonal design matrices (see [BG97]). Of course the corresponding orthogonal transformation may in some cases produce parameters that are difficult to interpret.

Finally it should be noted that in our approach (which is essentially Jewell's) the credibility regression model is a special case of multidimensional credibility. One could as well have introduced multidimensional credibility as a special case of the credibility regression model.

Notes on Chapter 9 and Chapter 10

The recursive procedure for credibility estimators is naturally tied to evolutionary models: the parameters to be estimated vary in time. In such evolutionary models our understanding of the collective as an urn, from which the individual risk characteristics are drawn independently, reaches its limits. If changes in risk parameters derive from common causes, the independence assumptions can no longer hold. It is quite natural that the question of how to model the collective recurs throughout these two chapters.

Early actuarial papers devoted to recursive credibility are [GJ75a], [GJ75b], [Kre82], [Sun81], [Sun82]. The Kalman Filter was developed originally in an engineering context [Kal60]. Its applicability in the insurance area has been advocated e.g. by Mehra [Meh75], Zehnwirth [Zeh85], Neuhaus [Neu87]. Most texts start with the most general form of the Kalman Filter and then specialize the assumptions to obtain the credibility type formulae. In these two chapters we go the opposite route. We build successively the understanding of the procedure by starting first in a one-dimensional world and proceed from there to more complicated models. This route shows quite clearly the connection with the static (non-evolutionary) credibility models. It is important to note that not all static credibility models can be evaluated recursively. The condition (in a somewhat different setting) under which this is possible goes back to Gerber and Jones [GJ75b].

Contents

1

Introduction

1.1 Rating of Risks and Claims Experience in Insurance

The basic idea underlying insurance is that individuals, who have the "same" exposure to a particular risk, join together to form together a "community-at-risk" in order to bear this perceived risk. In modern society, this idea is most often realized in the form of insurance. On payment of a premium, the individuals in the "community-at-risk" transfer their risk to an insurance company.

We consider an insurance company with a portfolio consisting of I insured risks numbered $i = 1, 2, \ldots, I$. In a well-defined insurance period, the risk i produces

- a number of claims N_i,
- with claim sizes $Y_i^{(\nu)} (\nu = 1, 2, \ldots, N_i)$,
- which together give the aggregate claim amount $X_i = \sum_{\nu=1}^{N_i} Y_i^{(\nu)}$.

We will refer to the premium payable by the insured to the insurer, for the bearing of the risk, as the gross premium. The premium volume is the sum, over the whole portfolio, of all gross premiums in the insurance period. The basic task underlying the rating of a risk is the determination of the so-called pure risk premium $P_i = E[X_i]$. Often, we use just the term "risk premium". The classical point of view assumes that, on the basis of some objectively quantifiable characteristics, the risks can be classified into homogeneous groups (risk classes), and that statistical data and theory (in particular, the Law of Large Numbers) then allow one to determine the risk premium to a high degree of accuracy.

In reality, of course, it is clear that a huge number of factors contribute to the size of the risk premium. In order to have reasonably homogeneous risk classes, one would have to subdivide the portfolio into a very large number of classes. For example, if we were to use four characteristics to rate the risk, then assuming that each characteristic had 10 possible values, this would lead to 10 000 risk classes. Many of these classes, by their very definition, will contain very few risks and will therefore provide little statistical information for rating future risks. On the other hand, if we subdivide the portfolio into larger classes, the assumption of homogeneous risk profiles within the classes becomes a fiction. In fact, it is clear that no risk is exactly the same as another. For example, every car driver is an individual with his own personal associated risk which is influenced by many factors, not least his particular character and constitution. This observation leads to the following thesis.

Thesis: *There are no homogeneous risk classes in insurance.*

Indeed in practice, relatively few so-called risk characteristics are explicitly used in the rating of risks. The segmentation of risk classes is relatively coarse. There are a number of reasons why some characteristics, which would perhaps be useful in determining the quality of a risk, are not considered.

a) They are not objectively quantifiable: for example, the temperament of a car driver, or the work atmosphere in a firm.
b) They are difficult to check: for example, the total number of kilometres driven by an individual, or his abstinence from alcohol.
c) They are politically delicate: for example, sex, nationality, or mother-tongue.
d) They would lead to excessive administrative work.
e) They would make the rating structure overly complicated and, as discussed above, the accurate estimation of the premium may be impossible due to limited statistical information.

In general, we need to know what one should do, and indeed, what one can do, in order to find "fair" premiums. The following pragmatic considerations can guide us:

- The specific risk exposure of each risk has an impact on the observed individual claim experience of that risk. However, the observed claim experience of an individual is too limited in order to be statistically reliable.
- Every individual risk is part of a risk collective. The collective claim experience, for large collectives, does provide reliable statistical information. This can be used to calculate the expected value of the average (over the risk collective) aggregate claim amount per risk. However, for the assessment of the quality of a given individual risk, this information is only of limited use.
- Intuitively, it seems obvious that both sources of information should be used for a fair rating of risks.

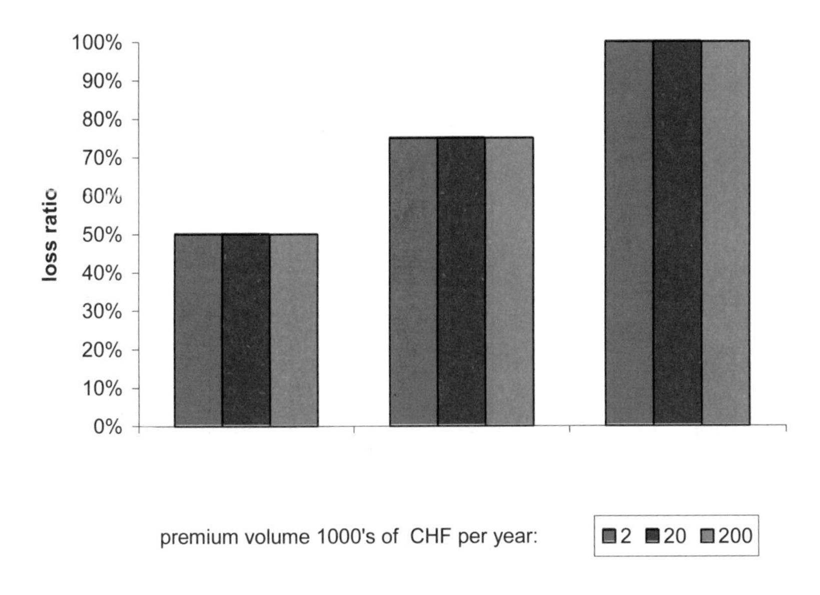

Fig. 1.1. Average loss ratio of a fictitious portfolio of nine contracts

Let us consider the following fictional example (Figure 1.1). We have data on a portfolio consisting of nine group health insurance contracts, giving the claim experience and gross premiums over a period of 5 years. The observed loss ratio over the 5 years (the aggregate claim amount divided by the gross premium for this period) is 50% for contracts 1–3, 75% for contracts 4–6 and 100% for contracts 7–9. The break-even aggregate loss ratio (the percentage of the premium which should be used for covering claim amounts) is 75%, that is, 25% of the premiums are needed for administrative costs and other loadings. New premiums are to be calculated for each of these contracts for the coming period. In particular, the loss ratio for each of these contracts, assuming the existing premiums, needs to be forecast. Note that the premium volume of the contracts also varies from one contract to another. In assessing these contracts on the basis of their individual loss ratios over the last 5-year period, two extreme points of view are possible.

a) The differences in the observed loss ratios between the individual contracts are entirely the result of the inherently random nature of the occurrence of claims. No contract is a better or worse risk than another. With this point of view, the best forecast that we can make of the future loss ratio of a given contract, on the basis of the available data, is equal to the average observed loss ratio over all nine contracts, that is, 75%. Hence, no change to the existing premium level is necessary.

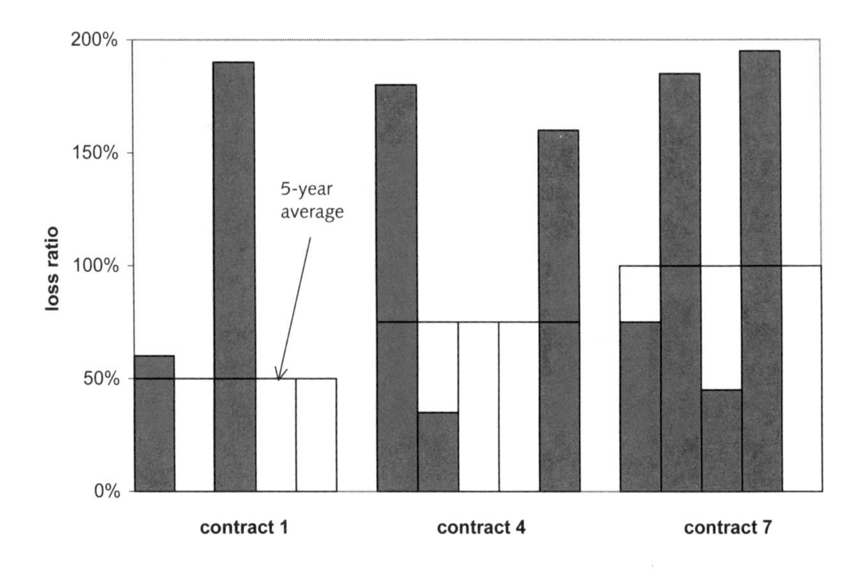

Fig. 1.2. yearly loss ratios of the small contracts

b) The differences in the observed loss ratios are not random, they are systematic. They are due to the varying risk profiles across the contracts. As a consequence, with this point of view, the premiums for the first three contracts should be decreased by 1/3, while those of the last three should be increased by 1/3.

The truth probably lies somewhere between the two extremes.

For a more refined analysis we consider next the yearly loss ratios separately for the small, medium and large contracts.

Figure 1.2 shows the yearly loss ratios of contracts 1, 4 and 7, each of which has a yearly premium volume of CHF 2 000. We see that the variability of the yearly loss ratios is high. One might seriously ask whether the observed differences in the 5-year loss ratios should be attributed to random fluctuations only. This fact decreases our confidence in the accuracy of this average as a reflection of the long-term average.

Figure 1.3 shows the analogous picture for the medium-sized contracts with a yearly premium volume of CHF 20 000. The yearly variability is still considerable, but it appears doubtful that the differences between the observed average loss ratios are purely the result of random events.

Finally, in Figure 1.4, we see displayed the yearly loss ratios for the three largest contracts 3, 6 and 9 (yearly premium volume of CHF 200 000). It seems clear that there are real differences in the risk profiles of the three contracts.

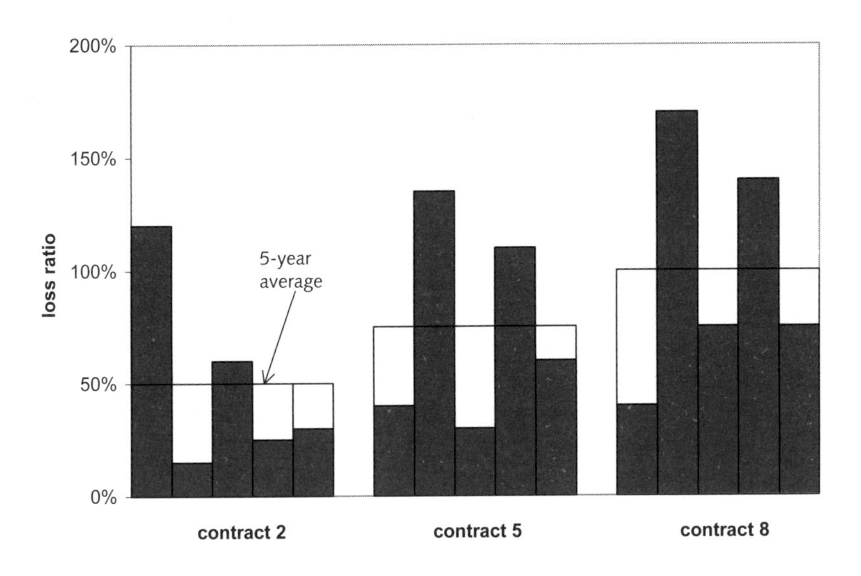

Fig. 1.3. Yearly loss ratios of the medium-sized contracts

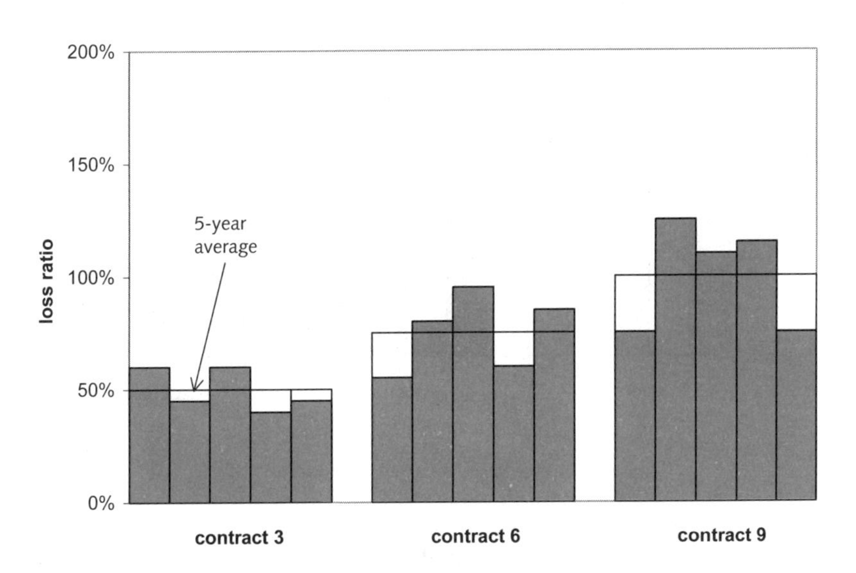

Fig. 1.4. Yearly loss ratios of the large contracts

Let us look again at contracts 1–3 (see Figure 1.5). For all three contracts, the average loss ratio over the past 5 years is 50%. However, at least intuitively, the data suggest that the "best" forecast should not be the same for all three. Should we use a forecast of 50% (the average observed loss ratio of the individual contracts) or one of 75% (the average loss ratio over the whole portfolio)? For contract 1, because of the high volatility in the yearly observed loss ratios, we would not have much confidence in the accuracy of the average of these as a forecast of future loss ratios, and would be inclined to use the average over all nine contracts, 75%, as our forecast. On the other hand, for contract 3, the information from its individual claim experience seems reliable and one would be inclined to use a figure in the neighbourhood of the observed average of 50% as our forecast. For contract 2, our forecast would probably lie somewhere between that for contracts 1 and 3.

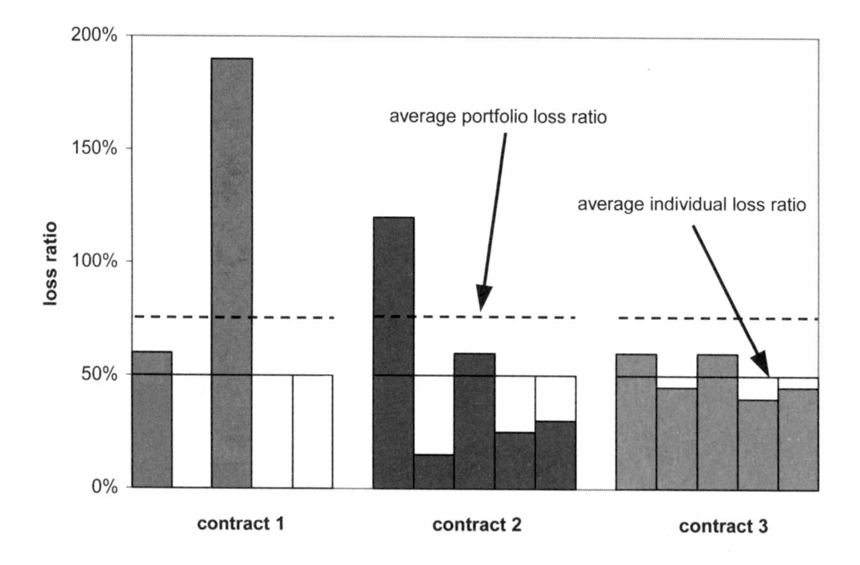

Fig. 1.5. Yearly loss ratios of contracts 1–3

Credibility theory provides a sound mathematical basis for the intuitive ideas and heuristic arguments sketched above.

Credibility theory:

- is the mathematical tool to describe heterogeneous collectives;
- answers the question of how one should combine individual and collective claims experience;
- belongs mathematically to the area of Bayesian statistics;

- is motivated by questions arising in insurance practice.

Credibility theory is one of the basic tools that every modern actuary should know and understand.

1.2 Mathematical Formulation of the Problem of Rating Risks

1.2.1 Individual Risk

The individual risk can be regarded as a black box that produces aggregate claim amounts X_j $(j = 1, 2, \ldots, n)$, where X_j denotes the claim amount in year j (or in some other well-specified time period j). Examples of the "black box" associated with an individual risk include:

- an individual driver in third-party motor liability insurance,
- a group of insured individuals in life insurance,
- all employees of a firm in collective workman's compensation insurance,
- a ceding company in reinsurance.

Observe: X_j $(j = 1, 2, \ldots, n)$ are, mathematically, interpreted as random variables. This interpretation holds also for past periods (the observed values could have been different).

On the basis of the observations in the *previous periods*, $\mathbf{X} = (X_1, \ldots, X_n)'$, we want to determine the risk premium for the aggregate claims in a *future period,* for example, X_{n+1}. In order to do this, we must make certain assumptions about the distribution function of the random variables X_j. The simplest standard assumptions are:

Assumption 1.1

A1: Stationarity: All the X_j's are identically distributed with (conditional) distribution function $F(x)$.

A2: (Conditional) Independence: The random variables X_j, $j = 1, 2, \ldots$, are (conditionally) independent (given the distribution $F(x)$).

Remarks:

- In this text we often use the symbols $g(y)$ when referring to the function g with argument y, e.g. $F(x)$ instead of F in Assumptions 1.1, thus highlighting the argument x.
- The *stationarity* assumption allows one to establish a relationship between the past and the future. A (possibly weaker) assumption of stationarity is always needed for the calculation of insurance premiums on the basis

of historical data. The (strong) stationarity assumption A1 will later be weakened and generalized. In practice, stationarity can often be achieved by making adjustments for inflation, indexing, as-if-statistics, trend elimination, and so on.
- Precisely what we mean in Assumption A2 by (conditional) independence (given the distribution $F(x)$), will be explained shortly (see page 12). We will also encounter models where this independence assumption is weakened.

Typically, in insurance practice, we have to deal with the following situation:

i) F is unknown,
ii) F varies from risk to risk.

In order to formalize i) and ii) more clearly, we use the common notation from mathematical statistics: we index F by a parameter ϑ and write F_ϑ (instead of F) and say

i) ϑ is unknown,
ii) ϑ varies from risk to risk.

Remarks:

- Parameterization is always possible (in the most extreme case, one can choose the distribution function itself as the parameter). In general, we write that ϑ is an element of some abstract space Θ.
- ϑ can be thought of as the "risk profile".

1.2.2 The Correct Individual Premium

By a premium calculation principle $\mathcal{H}$, we mean a function that assigns a real number to a random variable X having distribution function F, i.e.

$$X \longmapsto \mathcal{H}(X),$$
or indicating that the value of $\mathcal{H}$ depends only on F,
$$F \longmapsto \mathcal{H}(F).$$

Some of the more well-known "classical" premium calculation principles are:

Expectation principle:	$X \longmapsto (1+\alpha)\,E[X]$	$\alpha > 0$;
Standard deviation principle:	$X \longmapsto E[X] + \beta\sigma(X)$	$\beta > 0$;
Variance principle:	$X \longmapsto E[X] + \gamma\sigma^2(X)$	$\gamma > 0$;
Exponential principle:	$X \longmapsto \frac{1}{\delta}\ln E\left[e^{\delta X}\right]$	$\delta > 0$.

For each of these principles the premium $\mathcal{H}(X)$ can be decomposed into the *pure risk premium* $E[X]$ and a positive so-called risk loading determined by the parameters α, β, γ, and δ. Note also that the pure risk premium is obtained by the expectation principle with $\alpha = 0$.

Remark:

- The economic necessity of a risk loading is given by the fact that insurance companies need risk-bearing capital to cope with the volatility of the insurance results and in order to be able to fulfil the obligations towards the insured in an unfavourable year. The investors should be compensated for exposing their capital to risk. The risk loading compensates for this risk and can be seen as the cost of risk capital.

By applying the risk calculation principle $\mathcal{H}$ the resulting *correct individual premium* (for a risk with risk profile ϑ) would be $\mathcal{H}(F_\vartheta)$. *In this book we limit ourselves to the discussion of the pure risk premium.* Thus, we arrive at the following definition of the correct individual premium.

Definition 1.2. *The correct individual premium of a risk with risk profile ϑ is*

$$P^{\text{ind}}(\vartheta) = E[X_{n+1} | \vartheta] =: \mu(\vartheta). \tag{1.1}$$

The correct individual premium is also referred to as the fair risk premium. In order to simplify notation we will, in the following, often write $E_\vartheta[\cdot]$ instead of $E[\cdot | F_\vartheta]$ or $E[\cdot | \vartheta]$.

The individual rating problem can then be described as the determination of the quantity $\mu(\vartheta)$. However, in insurance practice, both ϑ and $\mu(\vartheta)$ are unknown. Therefore we have to find an estimator $\widehat{\mu(\vartheta)}$ for $\mu(\vartheta)$.

1.2.3 The Risk Rating Problem in the Collective

An insurance company insures many kinds of risks. For the purpose of rating risks, these risks are grouped into classes of "similar risks", on the basis of so-called "objective" risk characteristics. In motor insurance, examples of such risk characteristics are cylinder capacity, make of car and power/weight ratio, as well as individual characteristics such as the driver's age, sex and region. In industrial fire insurance, important characteristics might be the type of construction of the insured building, the kind of business conducted in the building, or the fire extinguishing facilities in the building. How exactly such groups are constructed is important in practice, but will not be part of our discussion in this book. Important for the credibility theory framework is that we do not consider each risk individually but that we rather consider each risk as being embedded in a group of "similar" risks, called the collective.

In the language of Subsection 1.2.2 we consider the following situation: every risk i in the collective is characterized by its *individual risk profile* ϑ_i.

These parameters ϑ_i are elements of a set Θ, where Θ is the set of all potential and possible values of the (unknown) risk profiles of the risks in the collective. In the special case of a homogeneous collective, Θ consists of just one element. This corresponds to the classical point of view of insurance: every member of the collective has exactly the same risk profile and therefore the same distribution function for its corresponding aggregate claim amount. However, the risk groups or collectives considered in insurance are mostly heterogeneous. In other words, the ϑ-values of the different risks in the collective are not all the same, they are rather samples taken from a set Θ with more than one element. But while the risks in the collective are different, they also have something in common: they all belong to the same collective, i.e. they are drawn from the same set Θ. This situation is referred to when we say that the risks in the collective are similar.

The specific ϑ-values attached to the different risks in the collective are typically unknown to the insurer. But, on the basis of a priori knowledge and statistical information, the insurer does know something about the structure of the collective. He knows, for example, that most car drivers are "good" risks and seldom make a claim, while a small percentage of drivers make frequent claims. Formally (at least intellectually), this information can be summarized by a probability distribution $U(\vartheta)$ over the space Θ.

Definition 1.3. *The probability distribution $U(\vartheta)$ is called the structural function of the collective.*

We can interpret $U(\vartheta)$ in a number of ways:

- In the *frequentist interpretation,* we consider the ϑ's in the collective as being a random sample from some fixed set Θ; then the function $U(\vartheta)$ describes the idealized frequencies of the ϑ's over Θ. This interpretation is called *empirical Bayes* and it is the predominant view taken in this text.
- In the pure *Bayesian interpretation* we consider the distribution function $U(\vartheta)$ as a description of the personal beliefs, a priori knowledge, and experience of the actuary.

In Subsection 1.2.2 we defined the correct individual premium. Now, we will define the collective premium.

Definition 1.4. *The collective premium is defined by*

$$P^{\text{coll}} = \int_{\Theta} \mu(\vartheta)\, dU(\vartheta) =: \mu_0. \tag{1.2}$$

Let us summarize the two kinds of premiums that we have so far considered:

- The "correct" *individual premium* $P^{\text{ind}}(\vartheta) = \mu(\vartheta) = E_\vartheta[X_{n+1}]$. This corresponds to the expected claim amount of the individual risk (given the individual risk profile ϑ) for the rating period $n+1$. Since

ϑ is unknown to the insurer, $\mu(\vartheta)$ is also unknown. In order to estimate this premium, in the best case, the insurer has at his disposal the information about the claim experience of this risk over the past few periods. However, more often than not, this information is very limited and has little predictive value.

- The *collective premium* $P^{\text{coll}} = \mu_0 = \int_{\Theta} \mu(\vartheta)\, dU(\vartheta)$.

  This corresponds to the average, over all risks of the collective, of the expected claim amount per individual risk. In most cases, this quantity is also unknown to the insurer. However, for most reasonably sized collectives, in contrast to the individual premium, this premium can be estimated with considerable accuracy on the basis of observations from the past.

It is of central importance for an insurance company to be able to calculate the collective premium (also known as the tariff level). If the insurer demands the same premium, P^{coll}, from every member of the collective, the books will be balanced, that is, over the whole collective, the premiums will be equal to the aggregate claim amount (in expected value). One could then ask why an insurance company is so interested in the "correct" individual premium. The answer is simple and can be most clearly expressed by the following thesis.

Thesis: *The most competitive rate is the correct individual premium.*

It is competition that forces companies to offer the fairest rates possible. If a company sets rates at the same level for all risks in a heterogeneous collective, the good risks will pay too much and the bad ones too little. If a competing company then offers rates which are more differentiated and fair, then its premiums for the good risks will be cheaper. In comparison to the first company, this competing company will be more attractive to the good risks and less attractive to the bad ones. This can have disastrous consequences for the first company: it loses good risks and has an increase in bad risks. Insurers refer to such an effect as anti-selection. In the language of this section this means that the collective changes, and the structural function of the insurer becomes less favourable.

1.2.4 Formulating the Rating Problem in the Language of Bayesian Statistics

Mathematically, the collective rating problem sketched in Subsection 1.2.3 can be most elegantly described in the language of Bayesian statistics. In the simple case considered so far this can be demonstrated with a two-urn model (see Figure 1.6).

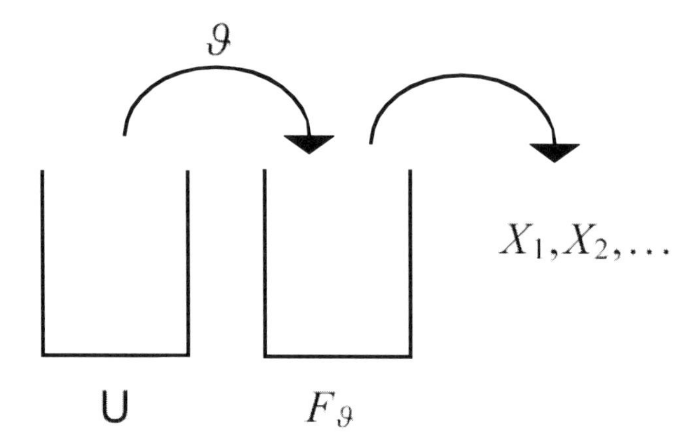

Fig. 1.6. Two-urn model

The first urn represents the urn containing the collective with distribution function U. From this urn, we select the individual risk, or equivalently its risk profile ϑ. This ϑ then determines the content of the second urn, or equivalently the distribution function F_ϑ. From this urn we select the values of the random variables $X_1, X_2, \ldots$, which are thus independent and identically distributed with distribution function F_ϑ. This results in the following mathematical structure.

Every risk is characterized by its individual risk profile ϑ, which is itself the realization of a random variable Θ, and the following holds:

i) Conditional on the event $\Theta = \vartheta$, $X_1, X_2, \ldots$ are i.i.d. (independent and identically distributed) with distribution function F_ϑ.
ii) Θ is itself a random variable with distribution function U.

Remarks:

- This model describes in mathematical language exactly the scenario which we considered in Section 1.1. The risks in the collective differ from each other. Each has its own risk profile ϑ. The risks do, however, have something in common. The risk parameters of the individual risks are themselves realizations of random variables. They are all independently drawn from the same urn with distribution function U.
- In this interpretation, the individual premium itself becomes a random variable $\mu(\Theta)$. We do not know the correct value of the individual premium. We do, however, know something about the possible values of $\mu(\Theta)$ and with what probability these values occur. It is therefore natural to model $\mu(\Theta)$ as a random variable.
- We therefore denote the individual premium by

$$P^{\text{ind}} = \mu(\Theta) := E[X_{n+1} | \Theta],$$

which is a conditional expectation and therefore a random variable. Note the change of notation and interpretation from the correct individual premium as defined in (1.1).

- Notice also, that a priori, all risks are equal. We know that there are better and worse risks in the portfolio. But we cannot, a priori, see to which class a particular risk belongs. Only a posteriori, after observations on the individual risk have been made, we can draw conclusions. This remark formalizes our intuition about the collective as a group of different, but similar risks.
- In contrast, the collective premium is

$$P^{\text{coll}} = \mu_0 = \int_{\Theta} \mu(\vartheta)\, dU(\vartheta) = E\left[X_{n+1}\right],$$

an unconditional expectation, and therefore a fixed number.

- Notice also that $X_1, X_2, \ldots$ are only conditionally independent, given Θ. Unconditionally, they are positively correlated. This is clear from the following:

$$\begin{aligned}\operatorname{Cov}(X_1, X_2) &= E\left[\operatorname{Cov}(X_1, X_2 \,|\Theta)\right] + \operatorname{Cov}\left(E\left[X_1 \,|\Theta\right], E\left[X_2 \,|\Theta\right]\right)\\ &= \operatorname{Cov}(\mu(\Theta), \mu(\Theta))\\ &= \operatorname{Var}\left[\mu(\Theta)\right].\end{aligned}$$

It is now clear what was intended on page 7 when we talked about "conditional independence" in Assumption A2. Within the Bayesian framework, we have now described this in a mathematically exact way.

Remark on the notation:

- Earlier we denoted the *space of the possible* ϑ *values* by Θ, while in this section Θ was used to denote *the random variable* from which ϑ is observed. This imprecision is deliberate. In order to reduce the number of symbols in our notation, we will in general use the same symbol to denote both the random variable and the space of its possible realizations.

Our goal is to estimate, for each risk, the correct premium $\mu(\Theta)$ as precisely as possible. One potential estimator is the collective premium μ_0, i.e. the premium for the considered particular risk is estimated by the "average" expected value over the whole collective. This estimator is appropriate when we are considering a new risk, about which there is no pre-existing claim experience. It takes into account the fact that the risk belongs to the collective and that its risk profile was selected at random from the "collective urn" U. However, this estimator does not take the individual claim experience into account. If we have observed the risk over a period of n years and if $\mathbf{X}$ denotes the vector of aggregate claim amounts associated with this period, then this information should contribute to the estimation process. This brings us to the concept of experience rating. The best experience premium depending on the individual claim experience vector $\mathbf{X}$ is called the Bayes premium, which we will now define. The sense in which it is "best" and the reason why it is called the Bayes premium will be explained in Chapter 2.

Definition 1.5. *The Bayes premium (= best experience premium) is defined by*

$$P^{\text{Bayes}} = \widetilde{\mu(\Theta)} := E\left[\mu(\Theta)|\,\mathbf{X}\right]. \tag{1.3}$$

1.2.5 Summary

We have so far encountered the following premium types:

- The *individual premium*

$$P^{\text{ind}} = \mu(\Theta) = E\left[X_{n+1}|\,\Theta\right].$$

  It should be noted here, that in this interpretation P^{ind}, contrary to $P^{\text{ind}}(\vartheta)$, is a random variable and not a "true" premium, i.e. P^{ind} is not a real number, that assigns a fixed cost for assuming a given risk. It is more a sort of "fictional" premium, from which we could derive the size of the "true" premium if we knew the value ϑ of the random variable Θ.
- The collective premium

$$P^{\text{coll}} = \mu_0 = \int_{\Theta} \mu(\vartheta)\, dU(\vartheta) = E\left[X_{n+1}\right].$$

- The *Bayes premium* (= best experience premium)

$$P^{\text{Bayes}} = \widetilde{\mu(\Theta)} = E\left[\mu(\Theta)|\mathbf{X}\right].$$

  Here, it should be noted that the experience premium is a random variable (it is a function of the observation vector $\mathbf{X}$), the values of which are known at the time at which the risk is to be rated. The Bayes premium is thus a "true" premium, the value of which depends on the claim experience.

2

The Bayes Premium

In this chapter we will study the best experience premium or Bayes premium, which we defined in (1.3),

$$P^{\text{Bayes}} := \widetilde{\mu(\Theta)} = E[\mu(\Theta)|\mathbf{X}].$$

To do this, we will use concepts from statistical decision theory. In particular, we will see in exactly which sense P^{Bayes} is "best" and why it is called the Bayes premium.

2.1 Basic Elements of Statistical Decision Theory

Here we will give an overview of the elements of statistical decision theory which will be necessary for our exposition. For a comprehensive study of statistical decision theory, see for example Lehmann [Leh86]. The raw material for a statistical decision is the observation vector $\mathbf{X} = (X_1, X_2, \ldots, X_n)'$. The distribution function

$$F_\vartheta(\mathbf{x}) = P_\vartheta[\mathbf{X} \leq \mathbf{x}]$$

is completely or partly unknown. (Equivalently: The parameter ϑ is completely or partly unknown.) We are interested in the value of a specific functional $g(\vartheta)$ of the parameter ϑ. We seek a function $T(\mathbf{X})$, which depends only on the observation vector $\mathbf{X}$, which will estimate $g(\vartheta)$ "as well as possible". The function $T(\mathbf{X})$ is called an estimator for $g(\vartheta)$. We will formulate this problem in the following way:

$\vartheta \in \Theta$: The set of parameters, which contains the true value of ϑ,
$T \in D$: The set of functions to which the estimator function must belong.

T is a map from the observation space $\mathbb{R}^n$ into the set of all possible values of the functional g, that is, the set $\{g(\vartheta) : \vartheta \in \Theta\}$.

The idea of "as well as possible" is made precise by the introduction of a *loss function*:

$L(\vartheta, T(\mathbf{x}))$: loss, if ϑ is the "true" parameter and $T(\mathbf{x})$ is the value taken by the estimator when the value $\mathbf{x}$ is observed.

From this we derive the *risk function* of the estimator T

$$R_T(\vartheta) := E_\vartheta\left[L(\vartheta, T)\right] = \int_{\mathbb{R}^n} L(\vartheta, T(\mathbf{x}))\, dF_\vartheta(\mathbf{x}). \tag{2.1}$$

(Only such functions T and L are allowed for which the right-hand side of (2.1) exists.) The goal then is to find an estimator $T \in D$, for which the risk $R_T(\vartheta)$ is as small as possible. In general, it is not possible to do this simultaneously for all values of ϑ. In other words, in general there is no T which minimizes $R_T(\vartheta)$ uniformly over ϑ. In Figure 2.1 we see an example, where depending on the value of ϑ, T_1 or T_2 has the smaller value of the risk function.

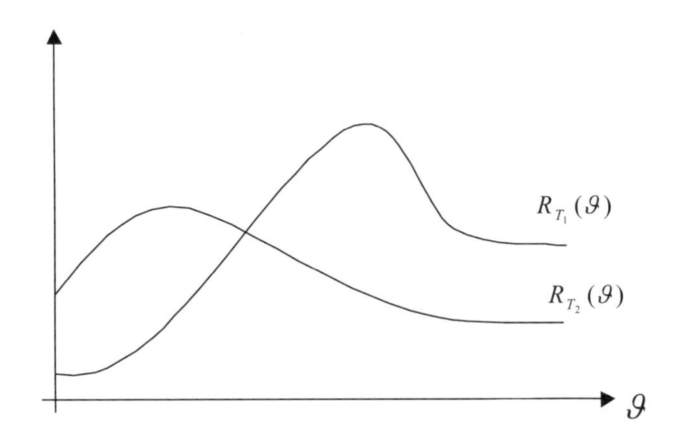

Fig. 2.1. Risk functions R_{T_1} and R_{T_2} for T_1 and T_2

2.2 Bayes Risk and Bayes Estimator

In Bayesian statistics, a smoothed average of the curve $R_T(\vartheta)$ is considered, where the average is weighted by means of a probability distribution $U(\vartheta)$ (called an *a priori* distribution for Θ). In other words, we consider the expected value of $R_T(\Theta)$, by regarding ϑ as the realization of a random variable Θ with probability distribution U.

Definition 2.1. *We define the Bayes risk of the estimator T with respect to the a priori distribution $U(\vartheta)$ as*

$$R(T) := \int_{\Theta} R_T(\vartheta) dU(\vartheta).$$

Assuming that the defined integral makes sense, with this criterion we can always rank estimators by increasing risk. In other words, there is a complete ordering on the set of estimators.

In this book we will use the following notation:

$$\text{If } \widetilde{x} \in D \text{ and } g(\widetilde{x}) \leq g(x) \text{ for all } x \in D,$$
$$\text{we write } \widetilde{x} = \arg\min_{x \in D} g(x).$$

Definition 2.2. *We define the Bayes estimator $\widetilde{T}$ as*

$$\widetilde{T} := \arg\min_{T \in D_1} R(T), \tag{2.2}$$

where D_1 is the set of all mathematically allowable estimators (that is, those estimators with integrable risk functions).

The estimator $\widetilde{T}$ is that estimator (2.2) which minimizes the Bayes risk $R(\cdot)$. Without further ado, we assume that $\min_{T \in D_1} R(T)$ is attained (this will indeed always be the case in the situations that we consider here in this book).

A remark on the notation: We use P to denote the joint distribution of $(\Theta, \mathbf{X})$, F to denote the marginal distribution of $\mathbf{X}$ and $U_{\mathbf{x}}$ to denote the conditional distribution of Θ, given $\mathbf{X} = \mathbf{x}$.

In order to construct the Bayes estimator, we consider the following sequence of equations:

$$\begin{aligned}
R(T) &= \int_{\Theta} R_T(\vartheta) dU(\vartheta) = \int_{\Theta} E_{\vartheta}\left[L(\vartheta, T)\right] dU(\vartheta) \\
&= \int_{\Theta} \int_{\mathbb{R}^n} L(\vartheta, T(\mathbf{x})) dF_{\vartheta}(\mathbf{x}) dU(\vartheta) \\
&= \int_{\mathbb{R}^n} \int_{\Theta} L(\vartheta, T(\mathbf{x})) dU_{\mathbf{x}}(\vartheta) dF(\mathbf{x}).
\end{aligned}$$

From this we deduce the following rule for constructing the Bayes estimator:

Theorem 2.3. *For every possible observation $\mathbf{x}$, $\widetilde{T(\mathbf{x})}$ takes the value which minimizes $\int_{\Theta} L(\vartheta, T(\mathbf{x})) dU_{\mathbf{x}}(\vartheta)$. In other words, for every possible observation $\mathbf{x}$, $\widetilde{T(\mathbf{x})}$ is the Bayes estimator with respect to the distribution $U_{\mathbf{x}}(\vartheta)$.*

Terminology: In Bayesian statistics, $U(\vartheta)$ is called the *a priori* distribution of Θ (before observations have been made), $U_{\mathbf{x}}(\vartheta)$ is called the *a posteriori* distribution of Θ (after observations have been made).

2.3 Bayesian Statistics and the Premium Rating Problem

It is clear from the overview given in Sections 2.1 and 2.2 that we have used Bayesian thinking to formulate the premium rating problem in Chapter 1. In order to express our concepts as introduced in Chapter 1 with Bayesian modelling, it suffices to consider the following correspondences:

$$\text{functional of interest } g(\vartheta) \simeq \text{correct individual premium } \mu(\vartheta),$$

$$\Theta = \begin{array}{l}\text{set of possible} \\ \text{parameter values}\end{array} \simeq \Theta = \begin{array}{l}\text{set of possible} \\ \text{individual risk profiles } \vartheta.\end{array}$$

In later parts of the text the estimand $g(\vartheta)$ might be chosen in a more general way.

As in Chapter 1, we use Θ also as a symbol for a random variable with distribution function $U(\vartheta)$. Hence, in the credibility context, we interpret the a priori distribution $U(\vartheta)$ as the structural distribution on the space Θ.

We now choose a particular form for the loss function, namely

$$L(\vartheta, T(\mathbf{x})) = (\mu(\vartheta) - T(\mathbf{x}))^2 \qquad \text{(quadratic loss function)}.$$

We thus have the following ordering on the space of all estimators for $\mu(\Theta)$:

Definition 2.4. *An estimator $\widehat{\mu(\Theta)}$ is at least as good as another estimator $\widehat{\mu(\Theta)}^*$ if*

$$E\left[\left(\widehat{\mu(\Theta)} - \mu(\Theta)\right)^2\right] \leq E\left[\left(\widehat{\mu(\Theta)}^* - \mu(\Theta)\right)^2\right].$$

$E\left[\left(\widehat{\mu(\Theta)} - \mu(\Theta)\right)^2\right]$ *is called the quadratic loss of the estimator $\widehat{\mu(\Theta)}$.*

The following theorem holds:

Theorem 2.5. *The Bayes estimator with respect to the quadratic loss function is given by*

$$\widetilde{\mu(\Theta)} = E\left[\mu(\Theta)|\,\mathbf{X}\right]. \tag{2.3}$$

Remark:

- This theorem makes precise what we meant in Definition 1.5, when we said that $P^{\text{Bayes}} = \widetilde{\mu(\Theta)}$ is the best possible experience premium.

Proof of Theorem 2.5: Let $\widehat{\mu(\Theta)}$ be an estimator of $\mu(\Theta)$ and $\widetilde{\mu(\Theta)}$ be its a posteriori expectation $E\left[\mu(\Theta)|\,\mathbf{X}\right]$. Then

$$E\left[\left(\widehat{\mu(\Theta)}-\mu(\Theta)\right)^2\right]=E\left[E\left[\left(\widehat{\mu(\Theta)}-\widetilde{\mu(\Theta)}+\widetilde{\mu(\Theta)}-\mu(\Theta)\right)^2\middle|\mathbf{X}\right]\right]$$
$$=E\left[\left(\widehat{\mu(\Theta)}-\widetilde{\mu(\Theta)}\right)^2\right]+E\left[\left(\widetilde{\mu(\Theta)}-\mu(\Theta)\right)^2\right].$$

We thus have that, of all estimators for $\mu(\Theta)$, the estimator $\widetilde{\mu(\Theta)}=E\left[\mu(\Theta)|\,\mathbf{X}\right]$ has the smallest quadratic loss. This proves Theorem 2.5. □

Why do we use the quadratic loss function?

We have seen that the quadratic loss function implies that the best premium is the a posteriori expectation of the individual premium $\mu(\Theta)$. This estimator has another very important property. We write P for the joint distribution of $(\Theta, \mathbf{X})$ so that

$$dP(\vartheta,\mathbf{x}):=dF_\vartheta(\mathbf{x})dU(\vartheta).$$

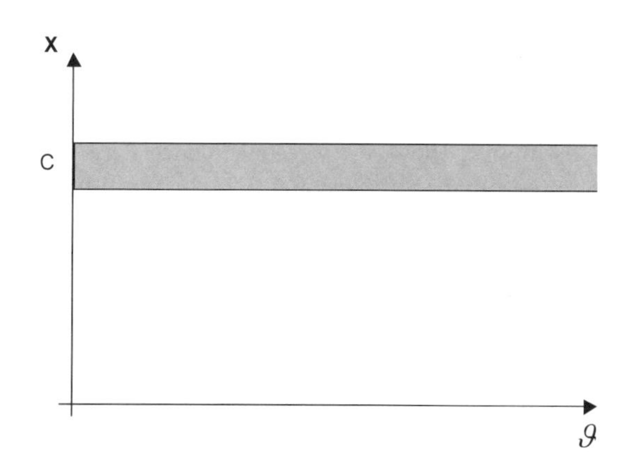

Fig. 2.2. Areas of balance between premiums and claims

For every measurable set $C\in\mathcal{B}^n:=\sigma(X_1,\ldots,X_n)$ (see Figure 2.2) we have, by definition of the conditional expectation of a random variable (see Appendix A) that

$$A:=\int_\Theta\int_C\widetilde{\mu(\vartheta)}\,dP(\vartheta,\mathbf{x})=\int_\Theta\int_C\mu(\vartheta)\,dP(\vartheta,\mathbf{x})=:B. \tag{2.4}$$

We can interpret these equations in the following way: if we set our received premium to be equal to the Bayes premium, then A is the amount taken in from the subcollective with observations $\mathbf{X}\in C$. On the other hand, B is the expected aggregate claim amount paid out to the subcollective with observations $\mathbf{X}\in C$. This means that, over *every* (measurable) subcollective defined in terms of past claims experience, (in expectation) the amount taken in is equal to the amount paid out. Naturally, equation (2.4) is also true for $C = \mathbb{R}^n$, that is, with respect to the whole collective.

Remarks:

- According to the Radon–Nikodym Theorem $\widetilde{\mu(\Theta)}$ is the *only* $\mathcal{B}^n$-measurable function with the property (2.4).
- The quadratic loss function also leads to the collective premium (resp. the individual premium), namely in the case where no loss experience is available (resp. in the case where Θ is known).
- The balance argument (2.4) is a strong reason for using the quadratic loss function. One might advocate other reasons. Economically, the use of a symmetric loss function charging gains and losses with equal penalties expresses a situation of equal bargaining power of the insurer on the one hand and the insured on the other. The resulting premium can therefore be considered as fair for both parts.
- One might argue that, from the insurance company's unilateral point of view, it would be more dangerous to use too low a premium than too high a premium, and that, hence, the loss function should not be symmetrical around the estimand. However, we concentrate here on the pure risk premium. As pointed out in Subsection 1.2.2, the premium to be charged to the policy holder will also contain a risk loading. The riskiness of the underlying business has to be taken into account by this risk loading.

Theorem 2.6.

i) The quadratic loss of the Bayes premium is

$$E\left[\left(\widetilde{\mu(\Theta)} - \mu(\Theta)\right)^2\right] = E\left[\operatorname{Var}\left[\mu(\Theta)\middle|\,\mathbf{X}\right]\right]. \tag{2.5}$$

ii) The quadratic loss of the collective premium is

$$\begin{aligned} E\left[(\mu_0 - \mu(\Theta))^2\right] &= \operatorname{Var}\left[\mu(\Theta)\right] \\ &= \underbrace{E\left[\operatorname{Var}(\mu(\Theta)\,|\mathbf{X})\right]}_{\textit{first variance component}} + \underbrace{\operatorname{Var}(E\left[\mu(\Theta)\,|\mathbf{X}\right])}_{\textit{second variance component}}. \end{aligned} \tag{2.6}$$

Proof: The proof of (2.5) follows directly from the definition of the conditional variance. (2.6) is the well-known decomposition of the variance (see Appendix A). □

Remark:

- Note that the quadratic loss of the Bayes premium is equal to the first variance component of the quadratic loss of the collective premium.

2.4 The Bayes Premium in Three Special Cases

In order to determine the Bayes premium, we must first specify

i) the structural function $U(\vartheta)$ and
ii) the family of conditional distributions $\mathcal{F} := \{F_\vartheta(\mathbf{x}) : \vartheta \in \Theta\}$.

2.4.1 The Poisson–Gamma Case

Motivation: F. Bichsel's Problem

At the end of the 1960s, a Bonus–Malus system was introduced in Switzerland for third-party liability motor insurance. The mathematical arguments underpinning this system are described in the paper of F. Bichsel [Bic64].

The Bonus–Malus system was created in response to the following set of circumstances. In the 1960s, insurers requested approval for the increase of premium rates, claiming that the current level was insufficient to cover their risks. The supervision authority was prepared to give approval only if the rates took into account individual claims experience. It was no longer acceptable that "good" risks, who had never made a claim, should continue to pay premiums which were at the same level as those drivers who had made numerous claims.

At the time, the premium level was based on the horsepower of the car. It was clear, however, that there were huge differences between risks having the same value of this criterion. F. Bichsel was the first non-life actuary in Switzerland, and he was given the task of constructing a risk rating system which was better adjusted to the individual risk profiles. He conjectured that differences between individual risk profiles were best summarized by differences in individual numbers of claims, while claim size, because of the very high variability in the amounts involved, was probably of little predictive value.

Mathematical Modelling

Let N_j be the number of claims made by a particular driver in year j. Its corresponding aggregate claim amount is X_j. The model used by Bichsel was based on the following implicit assumption:

Model Assumptions 2.7 (implicit assumption of Bichsel)
Given the individual risk profile ϑ of the driver, the following holds for the aggregate claim amount X_j:

$E[X_j|\Theta=\vartheta] = CE[N_j|\Theta=\vartheta] \qquad (j=1,2,\ldots),$
where C is a constant depending only on the horsepower of the car and where $E[N_j|\Theta=\vartheta]$ depends only on the driver of the car.

Remarks:

- Assumptions 2.7 hold for any compound model with claim size distribution depending on the horsepower of the car only and claim frequency *not* depending on the horsepower of the car.
- The assumption that the claim frequency does not depend on the horsepower of the car is a simplification not fully reflecting reality.

Because of Model Assumptions 2.7 it suffices to model the *number* of claims. The model used by Bichsel for the claim number is based on the following:

Model Assumptions 2.8 (Poisson–Gamma)

PG1: Conditionally, given $\Theta=\vartheta$, the N_j's $(j=1,2,\ldots,n)$ are independent and Poisson distributed with Poisson parameter ϑ, i.e.

$$P(N_j=k|\,\Theta=\vartheta) = e^{-\vartheta}\frac{\vartheta^k}{k!}.$$

PG2: Θ has a Gamma distribution with shape parameter γ and scale parameter β, i.e. the structural function has density

$$u(\vartheta) = \frac{\beta^\gamma}{\Gamma(\gamma)}\vartheta^{\gamma-1}e^{-\beta\vartheta}, \qquad \vartheta\geq 0.$$

Remarks:

- The first two moments of the Gamma-distributed random variable Θ are

$$E[\Theta] = \frac{\gamma}{\beta}, \qquad \mathrm{Var}[\Theta] = \frac{\gamma}{\beta^2}.$$

- In this book a dot in the index means summation over the corresponding index, for instance

$$N_\bullet = \sum\nolimits_{j=1}^{n} N_j.$$

Proposition 2.9. *Under Model Assumptions 2.8 we have for the claim frequency (denoted by F)*

$$F^{\mathrm{ind}} = E[N_{n+1}|\Theta] = \Theta,$$

$$F^{\mathrm{coll}} = E[\Theta] = \frac{\gamma}{\beta}, \tag{2.7}$$

$$F^{\mathrm{Bayes}} = \frac{\gamma+N_\bullet}{\beta+n} = \alpha\overline{N}+(1-\alpha)\frac{\gamma}{\beta}, \tag{2.8}$$

$$\text{where} \quad \alpha=\frac{n}{n+\beta}, \qquad \overline{N}=\frac{1}{n}\sum_{j=1}^{n}N_j.$$

The quadratic loss of F^{Bayes} is

$$E\left[(F^{\text{Bayes}} - \Theta)^2\right] = (1-\alpha)\, E\left[(F^{\text{coll}} - \Theta)^2\right] \tag{2.9}$$
$$= \alpha\, E\left[(\overline{N} - \Theta)^2\right]. \tag{2.10}$$

Remarks:

- The quantities P^{ind}, P^{coll} and P^{Bayes} can be obtained by multiplying F^{ind}, F^{coll} and F^{Bayes} by the constant C.
- The quadratic loss of F^{Bayes} (resp. P^{Bayes}) is equal to the minimum Bayes risk. Note that the quadratic loss of P^{Bayes} is

$$E\left[(P^{\text{Bayes}} - P^{\text{ind}})^2\right] = C^2 E\left[(F^{\text{Bayes}} - F^{\text{ind}})^2\right].$$

- The Bayes premium CF^{Bayes} is a linear function of the observations (claim numbers). As we shall see later, this is an example of a *credibility premium.*
- F^{Bayes} is an average of

$$\overline{N} = \text{ observed individual claim frequency and}$$
$$\frac{\gamma}{\beta} = E[\Theta] = \text{ a priori expected claim frequency}$$
$$(= \text{expected claim frequency over the whole collective}).$$

- $\alpha = n/(n+\beta)$ is called the *credibility weight.* The greater the number of observation years, n, the larger will be this weight. Similarly, the larger $\beta = E[\Theta]/\text{Var}[\Theta]$ is, the smaller this weight will be. This makes intuitive sense: the more information we have on an individual, the greater the weight we are ready to attach to his individual claim experience, while the more homogeneous our collective is, the more we are ready to use the collective claim experience for the rating of the individual risk.
- (2.9) and (2.10) mean that

$$\text{quadratic loss of } F^{\text{Bayes}} = (1-\alpha)\cdot \text{quadratic loss of } F^{\text{coll}},$$
$$\text{quadratic loss of } F^{\text{Bayes}} = \alpha \cdot \text{ quadratic loss of } \overline{N}.$$

The credibility weight α can thus be interpreted as the factor by which the quadratic loss is reduced if we use the Bayes premium rather than the collective premium or as the factor by which the quadratic loss has to be multiplied if we use F^{Bayes} rather than the observed individual claim frequency $\overline{N}$. Note that F^{coll} is the estimator based only on the a priori knowledge from the collective and neglecting the individual claims experience, whereas $\overline{N}$ is the estimator based only on the individual claims experience and neglecting the a priori knowledge. F^{Bayes} takes both sources of information into account, and we see that the quadratic loss of F^{Bayes} is smaller than the quadratic losses of both F^{coll} and $\overline{N}$.

- The quadratic loss of F^{coll} is

$$E\left[\left(F^{\text{coll}}-\Theta\right)^2\right]=E\left[\left(\frac{\gamma}{\beta}-\Theta\right)^2\right]=\text{Var}\left[\Theta\right]=\frac{\gamma}{\beta^2}. \tag{2.11}$$

The quadratic loss of $\overline{N}$ is

$$E\left\{E\left[\left(\overline{N}-\Theta\right)^2\middle|\Theta\right]\right\}=E\left\{\text{Var}\left(\overline{N}\middle|\Theta\right)\right\}=\frac{E\left[\Theta\right]}{n}=\frac{1}{n}\frac{\gamma}{\beta}. \tag{2.12}$$

Since the quadratic loss of F^{Bayes} is smaller than the quadratic loss for $\overline{N}$ we also see that

$$E\left[\left(F^{\text{Bayes}}-\Theta\right)^2\right]\longrightarrow 0 \qquad \text{for } n\longrightarrow\infty.$$

Proof of Proposition 2.9:
F^{ind} and F^{coll} follow directly from the model assumptions. To obtain F^{Bayes}, we derive first the a posteriori density of Θ given $\mathbf{N}$:

$$u\left(\vartheta\middle|\mathbf{N}\right)=\frac{\frac{\beta^\gamma}{\Gamma(\gamma)}\vartheta^{\gamma-1}e^{-\beta\vartheta}\prod_{j=1}^n e^{-\vartheta}\frac{\vartheta^{N_j}}{N_j!}}{\int\frac{\beta^\gamma}{\Gamma(\gamma)}\vartheta^{\gamma-1}e^{-\beta\vartheta}\prod_{j=1}^n e^{-\vartheta}\frac{\vartheta^{N_j}}{N_j!}d\vartheta}\propto\vartheta^{\gamma+\sum_{j=1}^n N_j-1}e^{-(\beta+n)\vartheta},$$

where $\propto$ is the proportionality operator, i.e. the right-hand side is equal to the left-hand side up to a multiplicative constant not depending on ϑ. We see from the right-hand side of the above equation that $u\left(\vartheta\middle|\mathbf{N}\right)$ is again the density of a Gamma distribution with the updated parameters

$$\begin{aligned}\gamma'&=\gamma+N_\bullet,\\ \beta'&=\beta+n.\end{aligned}$$

From this we arrive directly at the formula (2.8).

From (2.5) and the fact that the a posteriori distribution of Θ given $\mathbf{N}$ is again a Gamma distribution with updated parameters γ' and β', we get

$$\begin{aligned}E\left[\left(F^{\text{Bayes}}-\Theta\right)^2\right]&=E\left[\text{Var}\left[\Theta\middle|\mathbf{N}\right]\right]\\ &=E\left[\frac{\gamma+N_\bullet}{(\beta+n)^2}\right]\\ &=\frac{\gamma}{(n+\beta)^2}\left(1+\frac{n}{\beta}\right)\\ &=\frac{\gamma}{\beta}\frac{1}{n+\beta}\\ &=\frac{\gamma}{\beta^2}\left(1-\alpha\right)\\ &=\frac{1}{n}\frac{\gamma}{\beta}\alpha.\end{aligned}$$

The last equation is obtained by inserting the expression for α into the second to last equation. Since

$$E\left[\left(F^{\text{coll}} - \Theta\right)^2\right] = \frac{\gamma}{\beta^2},$$
$$E\left[\left(F^{\text{coll}} - \overline{N}\right)^2\right] = \left(\frac{1}{n}\frac{\gamma}{\beta}\right),$$

(2.9) and (2.10) are shown. This ends the proof of Proposition 2.9. □

Estimating the Structural Parameters

The Bayes estimator $\widetilde{\Theta}$ depends on the structural function $U(\vartheta)$: in the example here, this means that it depends on the unknown parameters γ and β. These parameters need to be estimated and we now consider the problem of how this should best be done. Not surprisingly, we use data from the collective:

The following table shows the data, generated by Swiss automobile drivers from the year 1961, which were available to F. Bichsel:

k	0	1	2	3	4	5	6	Total
# of policies with k claims	103 704	14 075	1 766	255	45	6	2	119 853

Table: *data of Swiss automobile drivers available to F. Bichsel*

The table can be read in the following way. There are 119 853 "repetitions" of the two-urn experiment, so that for each of the 119 853 drivers, an individual risk profile is drawn from the first urn and then the corresponding number of claims is selected from the second urn, the composition of which depends on the first selection. Intellectually then we have a set of tuples $\{(\Theta_1, N^{(1)}), \ldots, (\Theta_i, N^{(i)}), \ldots, (\Theta_I, N^{(I)})\}$ with $I = 119\ 853$. Thus i is the index for the individual driver and $N^{(i)}$ denotes the number of claims for the driver i in the year 1961.

There are a number of different ways that we can use the above data to estimate the structural parameters γ and β. Bichsel used the so-called *method of moments.* Assuming that the pairs of random variables $\{(\Theta_i, N^{(i)}) : i = 1, 2, \ldots, I\}$ are independent and identically distributed, the following empirical moments are unbiased estimators:

$$\widehat{\mu}_N = \overline{N}_{1961} = \frac{1}{I}\sum_{i=1}^{I} N^{(i)} = 0.155 \quad \text{and}$$

$$\widehat{\sigma}_N^2 = \frac{1}{I-1}\sum_{i=1}^{I}\left(N^{(i)} - \overline{N}_{1961}\right)^2 = 0.179.$$

On the other hand, if we calculate the two first moments of the random variable N, based on the model assumptions, we get

$$\begin{aligned}\mu_N &= E[N] = E[E[N|\Theta]] = E[\Theta] = \frac{\gamma}{\beta},\\ \sigma_N^2 &= \mathrm{Var}[N] = E[\mathrm{Var}[N|\Theta]] + \mathrm{Var}[E[N|\Theta]]\\ &= E[\Theta] + \mathrm{Var}[\Theta] = \frac{\gamma}{\beta}\left(1+\frac{1}{\beta}\right).\end{aligned}$$

The method of moments uses as estimators for the unknown parameters those values which would lead to the identities

$$\widehat{\mu}_N = \mu_N \quad \text{and} \quad \widehat{\sigma}_N^2 = \sigma_N^2,$$

so that we have in this example

$$\begin{aligned}\frac{\widehat{\gamma}}{\widehat{\beta}} &= 0.155,\\ \frac{\widehat{\gamma}}{\widehat{\beta}}\left(1+\frac{1}{\widehat{\beta}}\right) &= 0.155\left(1+\frac{1}{\widehat{\beta}}\right) = 0.179,\\ \widehat{\gamma} = 1.001 \quad &\text{and} \quad \widehat{\beta} = 6.458.\end{aligned}$$

Empirical Bayes Estimator

If we replace the unknown structural parameters γ and β in the formula for the Bayes estimator by the values $\widehat{\gamma}$ and $\widehat{\beta}$ estimated from the collective we get the *empirical Bayes estimator*

$$\widehat{F^{\mathrm{Bayes}}}^{emp} = \frac{n}{n+\widehat{\beta}}\,\overline{N} + \frac{\widehat{\beta}}{n+\widehat{\beta}}\,\frac{\widehat{\gamma}}{\widehat{\beta}}, \qquad \text{where } \overline{N} = \frac{1}{n}\sum_{j=1}^{n} N_j.$$

Remark:
It is usual in the calculation of the Bayes premium that we have to proceed in two steps:

- Derivation of P^{Bayes} under the assumption that the structural parameters are known.
- Estimation of the structural parameters using data from the collective.

Distribution of the Risk in the Collective

How well does the model fit the data of the table on page 25? To check this, we have to first calculate the unconditional distribution of N. In the above example, we can explicitly derive this distribution:

$$
\begin{aligned}
P(N=k) &= \int P(N=k|\Theta=\vartheta)\, u(\vartheta) d\vartheta \\
&= \int e^{-\vartheta} \frac{\vartheta^k}{k!} \frac{\beta^\gamma}{\Gamma(\gamma)} \vartheta^{\gamma-1} e^{-\beta\vartheta} d\vartheta \\
&= \frac{\beta^\gamma}{\Gamma(\gamma)} \frac{1}{k!} \underbrace{\int e^{-(\beta+1)\vartheta} \vartheta^{\gamma+k-1} d\vartheta}_{=\frac{\Gamma(\gamma+k)}{(\beta+1)^{\gamma+k}}} \\
&= \frac{\Gamma(\gamma+k)}{\Gamma(\gamma)k!} \left(\frac{\beta}{\beta+1}\right)^\gamma \left(\frac{1}{\beta+1}\right)^k \\
&= \binom{\gamma+k-1}{k} p^\gamma (1-p)^k, \quad \text{with } p = \frac{\beta}{\beta+1},
\end{aligned}
$$

which is a negative binomial distribution.

We can now compare the observed data with the values we would expect if the data satisfied the model assumption of the two-urn Poisson–Gamma model (fitted negative binomial distribution). We also compare the observed data with the model assumption of a homogeneous portfolio, where all policies are assumed to have the same Poisson parameter (fitted Poisson distribution).

k	observed	Poisson ($\lambda = 0.155$)	Negative binomial ($\gamma = 1.001, \beta = 6.458$)
0	103 704	102 629	103 757
1	14 075	15 922	13 934
2	1 766	1 234	1 871
3	255	64	251
4	45	3	34
5	6	0	5
6	2	0	1
Total	119 853	119 853	119 853

Table: *Number of policies with k claims: observed, expected (fitted) Poisson and expected (fitted) negative binomial*

The above table shows that the negative binomial distribution gives a much better fit.

Application

As explained above, this model and the above data served F. Bichsel as the basis for the construction of the Bonus–Malus system in third-party liability motor insurance in Switzerland. His goal was to adjust the individual premium according to the estimated expected claim frequency of the individual driver. The idea was that, if a given risk was expected to make twice as many claims as the average driver, then such a person should pay a premium twice as high as that of an "average" risk (malus of 100%). If, on the other hand, a driver was expected to make only half as many claims as the average, he should receive a bonus of 50%.

Based on the mathematical model and data presented above, the formula for the Bonus–Malus factor (the factor by which the collective premium should be multiplied) is:

$$\textit{Bonus–Malus factor} = \frac{\widehat{F^{\text{Bayes}}}}{\lambda_0},$$

where $\lambda_0 = 0.155$ = average observed claim frequency. Tabulated below are the values of the Bonus–Malus factor derived for this example. In 1964, this table formed the basis of the Bonus–Malus system in Switzerland.

$n \backslash k$	0	1	2	3
1	87%	173%	260%	346%
2	76%	153%	229%	305%
3	68%	136%	205%	273%
4	62%	123%	185%	247%
5	56%	113%	169%	225%
6	52%	104%	155%	207%

Table: *Bonus–Malus factor, where n = number of observation years, k = number of observed claims*

Remarks:

- The Bonus–Malus system introduced in 1964 used other factors. The modifications were due to the desire for a recursive premium on the one hand and solidarity considerations on the other.
- In Bichsel's work, it is implicitly assumed that the a priori claim number distribution does not depend on the horsepower class, which is of course a simplifying assumption not fully reflecting reality.
- Nowadays the tariffs are more refined and based on many rating factors. However, experience shows that the individual claim experience still remains one of the most significant variables for predicting future claims. But, in calculating the Bayes premium the differentiation of the tariff based on the other explanatory variables has to be taken into account (see [Gis89]).

A Model Where the a Priori Expected Claim Frequencies Differ Between Years and Risks

In the model used by Bichsel it was assumed that all drivers are a priori equal risks, i.e. that they all have the same a priori expected claim frequency. However, in many situations in practice this is not the case. For instance, if the risks considered are fleets of cars, then the a priori expected claim frequency will depend on the composition of the types of car of the fleet and possibly on other explanatory variates like region or mileage. Or if we want to make individual estimates of mortality or disability rates in group life or group accident insurance, then the a priori expected rates will depend on things like age and sex structure of the considered group. There might also be changes or trends in the claim frequencies over the years. The Poisson–Gamma model can, however, also be used for such situations after a slight modification and extension. Applications in group life insurance of this model can be found e.g. in [Kef29] and [Nor89]. It is worthwhile to mention that the Poisson–Gamma model was known in the insurance field as early as in 1929 [Kef29].

Before we do the modification of the Poisson–Gamma model to the case of known differences of the a priori claim frequencies, it is worthwhile to reconsider Model Assumptions 2.8 and to write them in a slightly different way.

The following assumptions are equivalent to Model Assumptions 2.8:

PG1: *Conditionally, given* $\Theta = \vartheta$, *the* N_j*'s* $(j = 1, 2, \ldots, n)$ *are independent and Poisson distributed with Poisson parameter* $\vartheta \cdot \lambda_0$, *where* $\lambda_0 = \frac{\gamma}{\beta} =$ *a priori expected claim frequency.*

PG2: Θ *has a Gamma distribution with* $E[\Theta] = 1$ and *shape parameter* γ.

Remarks:

- From $E[\Theta] = 1$ it follows that the scale parameter of the Gamma distribution must be equal to γ. Note that $\gamma = \lambda_0 \beta$ and that
$$\operatorname{Var}[\Theta] = \frac{1}{\gamma}.$$
- Note also that
$$F^{\text{ind}} = E[N_j | \Theta] = \lambda_0 \Theta.$$
In the problem and terminology of Bichsel, Θ denotes directly the Bonus–Malus factor.
- A good intuitive measure for the heterogeneity of the portfolio is the coefficient of variation $\operatorname{CoVa}(F^{\text{ind}})$. Note that in this model $\operatorname{CoVa}(F^{\text{ind}}) = \sqrt{\operatorname{Var}[\Theta]}$ and that the parameter γ has now a direct interpretation, since $\gamma = \left(\operatorname{CoVa}(F^{\text{ind}})\right)^{-2}$.
- For estimating Θ it is natural to consider in addition or instead of the "absolute" claim numbers N_j the "relative" claim frequencies
$$\widetilde{F}_j = \frac{N_j}{\lambda_0}.$$
Note that
$$\widetilde{F}^{\text{ind}} = E\left[\widetilde{F}_j \middle| \Theta\right] = \Theta.$$

- By dividing F^{Bayes} from (2.8) by λ_0, we directly obtain

$$\widetilde{F}^{\text{Bayes}} = \widetilde{\Theta} = E\left[\Theta | \mathbf{N}\right] = 1 + \alpha \left(\frac{N_\bullet}{\nu_\bullet} - 1 \right), \tag{2.13}$$

where

$$\begin{aligned} N_\bullet &= \sum\nolimits_{j=1}^{n} N_j = \text{ observed number of claims,} \\ \nu_\bullet &= n \cdot \lambda_0 \;=\; \text{a priori expected number of claims,} \\ \alpha &= \frac{\nu_\bullet}{\nu_\bullet + \lambda_0 \beta} = \frac{\nu_\bullet}{\nu_\bullet + \gamma}. \end{aligned}$$

Note also that

$$\frac{N_\bullet}{\nu_\bullet} = \overline{\widetilde{F}} := \frac{1}{n} \sum_{j=1}^{n} \widetilde{F}_j,$$

and hence

$$\widetilde{\Theta} = 1 + \alpha \left(\overline{\widetilde{F}} - 1 \right).$$

We now consider the situation where the a priori expected claim frequencies vary between years and between risks in the portfolio. We consider a particular risk and we denote by N_j the observed and by λ_j the a priori expected number of claims of that particular risk in year j. Of course λ_j may now depend on exogenous and explanatory variables like the composition of the particular fleet or the sex and age structure of the particular risk group in question. We make the following assumptions:

Model Assumptions 2.10 (Poisson–Gamma model II)

PG1′: The claim numbers N_j, $j = 1, 2, \ldots$, are conditionally, given Θ, independent and Poisson distributed with Poisson parameter $\Theta \lambda_j$, where λ_j is the a priori expected claim number in year j.

PG2′: Θ has a Gamma distribution with $E[\Theta] = 1$ and shape parameter γ.

Proposition 2.11. *Under Model Assumptions 2.10 we have*

$$\widetilde{F}^{Bayes} = \widetilde{\Theta} = E\left[\Theta | \mathbf{N}\right] = 1 + \alpha \left(\frac{N_\bullet}{\nu_\bullet} - 1 \right), \tag{2.14}$$

where

$$\begin{aligned} N_\bullet &= \sum\nolimits_{j=1}^{n} N_j \;=\; \textit{observed number of claims,} \\ \nu_\bullet &= \sum\nolimits_{j=1}^{n} \lambda_j \;=\; \textit{a priori expected number of claims,} \\ \alpha &= \frac{\nu_\bullet}{\nu_\bullet + \gamma}. \end{aligned}$$

Remarks:

- Note that formula (2.14) is identical to formula (2.13). The interpretation has, however, become wider: $\nu_\bullet$ may now vary between risks and depend on explanatory variables as for instance on the composition of the types of car for fleets of cars or sex and age structure in group life insurance.
- Formula (2.14) is a very convenient and intuitive one. It says that $\widetilde{F}^{\text{Bayes}}$ is equal to one plus a correction term and the correction term is equal to the credibility weight α times the deviation of the ratio "observed number of claims divided by expected number of claims" from one. The constant γ in the denominator of α is the coefficient of variation of F^{ind} to the power minus 2.

Proof of Proposition 2.11:
The proof is analogous to the proof of Proposition 2.9.

2.4.2 The Binomial–Beta Case

Motivation
In group life insurance or group accident insurance we are, among other things, interested in the number of disability cases or equivalently, the disability frequency for a particular group. For simplicity we assume that each member of the group has the same probability of disablement and that individual disabilities occur independently. Let us also assume that disabled members leave the group.

Define the following random variables for each year $j = 1, 2, \ldots$:

$$
\begin{aligned}
N_j &= \text{number of new disabilities occurring in the group in the year } j, \\
V_j &= \text{number of (not disabled) members of the group at the beginning of year } j, \\
X_j &:= \tfrac{N_j}{V_j} = \text{observed disablement frequency in year } j.
\end{aligned}
$$

At time n all random variables with indices $\leq n$ are known and we are interested in

$$
X_{n+1} = \frac{N_{n+1}}{V_{n+1}}.
$$

Observe that on the right-hand side of the last equation only N_{n+1} is not yet known, hence we need to model only N_{n+1}.

Model Assumptions 2.12 (Binomial–Beta)

BB1: Conditionally, given $\Theta = \vartheta$, N_j $(j = 1, 2, \ldots)$ are independent and binomial distributed, i.e.

$$P[N_j = k | \Theta = \vartheta] = \binom{V_j}{k} \vartheta^k (1-\vartheta)^{V_j - k}.$$

BB2: Θ has a Beta(a,b) distribution with $a, b > 0$, equivalently, the structural function has density

$$u(\vartheta) = \frac{1}{B(a,b)} \vartheta^{a-1} (1-\vartheta)^{b-1}, \qquad 0 \le \vartheta \le 1,$$

$$\textit{where} \qquad B(a,b) = \frac{\Gamma(a)\,\Gamma(b)}{\Gamma(a+b)}.$$

Remarks:

- The first two moments of the Beta-distributed random variable Θ are

$$E[\Theta] = \frac{a}{a+b}, \qquad \mathrm{Var}[\Theta] = \frac{ab}{(1+a+b)(a+b)^2}.$$

- Note that Θ is the "true" underlying disablement probability we want to determine.
- The family of the Beta distributions for the structure function is quite large. The density functions for different values of the structural parameters a and b can be seen from Figure 2.3.

Proposition 2.13. *Under Model Assumptions 2.12 we have for the frequency*

$$F^{\text{ind}} = E[X_{n+1} | \Theta] = \Theta,$$

$$F^{\text{coll}} = E[\Theta] = \frac{a}{a+b}, \tag{2.15}$$

$$F^{\text{Bayes}} = \frac{a + N_\bullet}{a+b+V_\bullet} = \alpha \overline{N} + (1-\alpha) \frac{a}{a+b}, \tag{2.16}$$

$$\textit{where } \overline{N} = \frac{N_\bullet}{V_\bullet}, \qquad \alpha = \frac{V_\bullet}{a+b+V_\bullet}.$$

The quadratic loss of F^{Bayes} is

$$E\left[(F^{\text{Bayes}} - \Theta)^2\right] = (1-\alpha)\, E\left[(F^{\text{coll}} - \Theta)^2\right] \tag{2.17}$$

$$= \alpha\, E\left[(\overline{N} - \Theta)^2\right]. \tag{2.18}$$

Proof:

F^{ind} and F^{coll} follow directly from the model assumptions. To obtain F^{Bayes}, we derive the posterior density of Θ given $N_1, N_2, \ldots, N_n$

$$\begin{aligned} u(\vartheta | \mathbf{N}) &\propto \prod_{j=1}^{n} \vartheta^{N_j} (1-\vartheta)^{V_j - N_j} \vartheta^{a-1} (1-\vartheta)^{b-1} \\ &= \vartheta^{a+N_\bullet - 1} (1-\vartheta)^{b+V_\bullet - N_\bullet - 1}. \end{aligned}$$

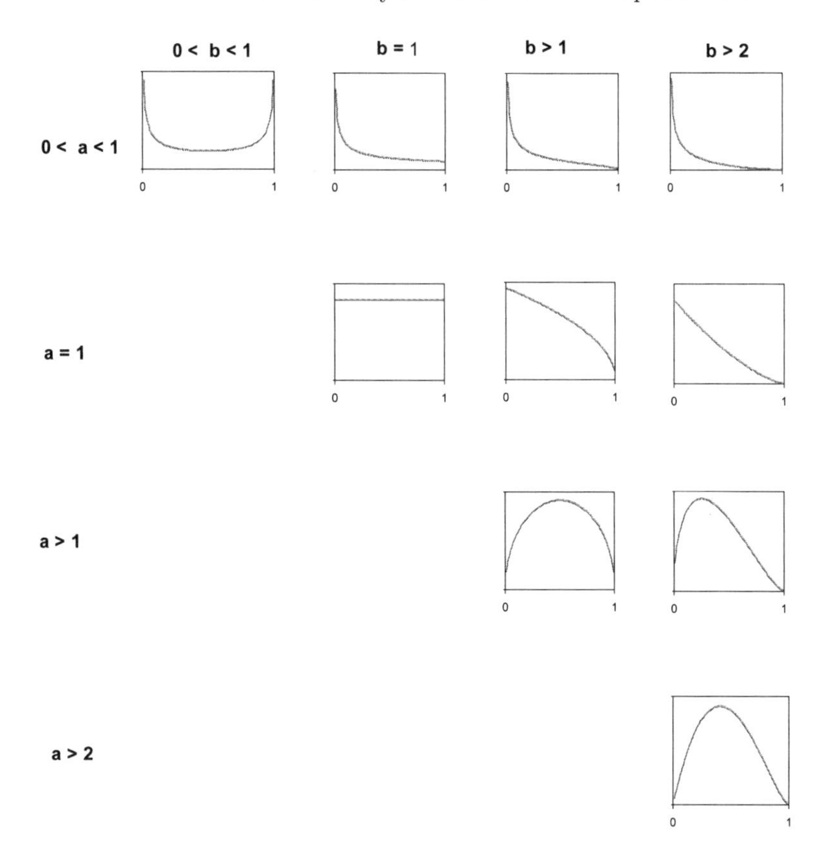

Fig. 2.3. Beta densities for different parameter values a and b

Hence the posterior distribution is again Beta, but with *updated parameter values*

$$\begin{aligned} a' &= a + N_\bullet, \\ b' &= b + V_\bullet - N_\bullet. \end{aligned}$$

Hence

$$F^{\text{Bayes}} = \widetilde{\Theta} := E\left[\Theta|\,\mathbf{N}\right] = \frac{a'}{a'+b'},$$

which is the same as (2.16).

For the quadratic loss of F^{Bayes} we obtain

$$E\left[\left(\widetilde{\Theta}-\Theta\right)^2\right]=E\left[E\left[\left(\widetilde{\Theta}-\Theta\right)^2|\Theta\right]\right]$$

$$=\alpha^2\,E\left[\operatorname{Var}\left[\overline{N}\,|\Theta\right]\right]+(1-\alpha)^2\operatorname{Var}\left[\Theta\right].$$

$$E\left[\operatorname{Var}\left[\overline{N}\,|\Theta\right]\right]=\frac{1}{V_\bullet}\,E\left[\Theta\,(1-\Theta)\right]$$

$$=\frac{1}{V_\bullet}\left(E\left[\Theta\right]-\operatorname{Var}\left[\Theta\right]-E^2\left[\Theta\right]\right)$$

$$=\frac{1}{V_\bullet}\left(\frac{ab}{(a+b)^2}-\operatorname{Var}\left[\Theta\right]\right)$$

$$=\frac{1}{V_\bullet}\,(a+b)\operatorname{Var}\left[\Theta\right].$$

Hence

$$E\left[\left(\widetilde{\Theta}-\Theta\right)^2\right]=\frac{V_\bullet}{(a+b+V_\bullet)^2}\,(a+b)\operatorname{Var}\left[\Theta\right]+\frac{a+b}{(a+b+V_\bullet)^2}\operatorname{Var}\left[\Theta\right],$$

$$=\frac{a+b}{a+b+V_\bullet}\operatorname{Var}\left[\Theta\right],$$

$$=(1-\alpha)\operatorname{Var}\left[\Theta\right],$$

which is the same as (2.17). From the above equations and the expression for α we also get

$$\frac{a+b}{a+b+V_\bullet}\operatorname{Var}\left[\Theta\right]=\alpha\,E\left[\operatorname{Var}\left[\overline{N}\,|\Theta\right]\right].$$

Noting that

$$E\left[\operatorname{Var}\left[\overline{N}\,|\Theta\right]\right]=E\left[\left(\overline{N}-\Theta\right)^2\right],$$

we see that (2.18) is also fulfilled, which ends the proof of Proposition 2.13.□

2.4.3 The Normal–Normal Case

We will present this model here without any practical motivation.

Let us again consider an individual risk and let $\mathbf{X}=(X_1,X_2,\ldots,X_n)'$ be the observation vector, where X_j is, for example, the aggregate claim amount in the jth year.

Model Assumptions 2.14 (normal–normal)

- *Conditionally, given $\Theta = \vartheta$, the X_j's $(j = 1, 2, \ldots, n)$ are independent and normally distributed, that is,*

$$X_j \sim \mathcal{N}\left(\vartheta, \sigma^2\right).$$

- *$\Theta \sim \mathcal{N}\left(\mu, \tau^2\right)$, that is the structural function has density*

$$u\left(\vartheta\right) = \frac{1}{\sqrt{2\pi}\tau}\, e^{-\frac{1}{2}\left(\frac{\vartheta-\mu}{\tau}\right)^2}.$$

Remark:

- Insurance data are mostly not normally distributed. However, because of the Central Limit Theorem, Model Assumptions 2.14 are sometimes appropriate for a portfolio of large risk groups.

Proposition 2.15. *Under Model Assumptions 2.14 we have*

$$\begin{aligned} P^{\text{ind}} &= E\left[X_{n+1} \left|\Theta\right.\right] = \Theta, \\ P^{\text{coll}} &= E\left[\Theta\right] = \mu, && (2.19) \\ P^{\text{Bayes}} &= \frac{\tau^2\,\mu + \sigma^2 X_\bullet}{\tau^2 + n\,\sigma^2} = \alpha\overline{X} + (1-\alpha)\,E\left[\Theta\right], && (2.20) \\ &\qquad \textit{where } \alpha = \frac{n}{n + \frac{\sigma^2}{\tau^2}}, \qquad \overline{X} = \frac{1}{n} X_\bullet. \end{aligned}$$

The quadratic loss of P^{Bayes} is

$$\begin{aligned} E\left[\left(P^{\text{Bayes}} - \Theta\right)^2\right] &= (1-\alpha)\,E\left[\left(P^{\text{coll}} - \Theta\right)^2\right] && (2.21) \\ &= \alpha\; E\left[\left(\overline{X} - \Theta\right)^2\right]. && (2.22) \end{aligned}$$

Proof:
P^{ind} and P^{coll} follow directly from the model assumptions. To obtain $P^{\text{Bayes}} = \widetilde{\Theta} := E\left[\Theta\right|\mathbf{X}]$, we derive the a posteriori density of Θ given $\mathbf{X}$

$$u\left(\vartheta \mid \mathbf{X}\right) \propto \frac{1}{\sqrt{2\pi}\tau}\, e^{-\frac{1}{2}\left(\frac{\vartheta-\mu}{\tau}\right)^2} \prod_{j=1}^{n} \left(\frac{1}{\sqrt{2\pi}\sigma}\, e^{-\frac{1}{2}\left(\frac{X_j-\vartheta}{\sigma}\right)^2}\right). \tag{2.23}$$

For the exponents on the right-hand side of (2.23) we get (β_1 and β_2 denote the terms, which are independent of ϑ):

$$\begin{aligned} &-\frac{1}{2}\left\{\left(\tau^{-2} + n\sigma^{-2}\right)\vartheta^2 - 2\left(\tau^{-2}\mu + n\sigma^{-2}\overline{X}\right)\vartheta + \beta_1\right\} \\ &= -\frac{1}{2}\left(\tau^{-2} + n\sigma^{-2}\right)\left\{\left(\vartheta - \frac{\tau^{-2}\,\mu + n\,\sigma^{-2}\,\overline{X}}{\tau^{-2} + n\,\sigma^{-2}}\right)^2 + \beta_2\right\} \\ &= -\frac{1}{2}\left(\tau^{-2} + n\sigma^{-2}\right)\left\{\left(\vartheta - \frac{\sigma^2\,\mu + n\,\tau^2\,\overline{X}}{\sigma^2 + n\,\tau^2}\right)^2 + \beta_2\right\}. \end{aligned}$$

From this, it is obvious that the a posteriori distribution of Θ given $\mathbf{X}$ is again a normal distribution, but with updated parameters

$$\mu' = \frac{\sigma^2 \mu + n \tau^2 \overline{X}}{\sigma^2 + n \tau^2},$$
$$(\tau')^2 = \frac{\tau^2 \sigma^2}{\sigma^2 + n \tau^2}.$$

From the identity

$$\frac{\sigma^2 \mu + n \tau^2 \overline{X}}{\sigma^2 + n \tau^2} = \frac{n}{n + \frac{\sigma^2}{\tau^2}} \overline{X} + \left(1 - \frac{n}{n + \frac{\sigma^2}{\tau^2}}\right) \mu$$

we get the formula for P^{Bayes}.

For the quadratic loss of P^{Bayes} we get

$$\begin{aligned} E\left[\left(\widetilde{\Theta} - \Theta\right)^2\right] &= E\left[\mathrm{Var}\left[\Theta \,|\mathbf{X}\right]\right] \\ &= (\tau')^2 \\ &= (1 - \alpha)\, \tau^2, \end{aligned}$$

which is the same as (2.21). From

$$(1 - \alpha)\, \tau^2 = \frac{\sigma^2}{n + \frac{\sigma^2}{\tau^2}} = \alpha \frac{\sigma^2}{n} \quad \text{and}$$

$$E\left[(\overline{X} - \Theta)^2\right] = E\left\{E\left[(\overline{X} - \Theta)^2 \middle| \Theta\right]\right\} = \frac{\sigma^2}{n}$$

follows that (2.22) is also fulfilled, which ends the proof of Proposition 2.15. □

Figures 2.4 and 2.5 illustrate the conversion of the a priori distribution into the a posteriori distribution. Notice the shift in the expected value and the fact that the variance decreases with increasing n.

2.4.4 Common Features of the Three Special Cases Considered

i) In all three cases, the Bayes premium is a linear function of the observations and is therefore a credibility premium.
ii) In all three cases, P^{Bayes} can be expressed as a weighted mean, that is,

$$P^{\mathrm{Bayes}} = \alpha \overline{X} + (1 - \alpha)\, P^{\mathrm{coll}}. \tag{2.24}$$

iii) In all three special cases we see that the weight α is given by

$$\alpha = \frac{n}{n + \kappa}, \qquad \text{where } \kappa \text{ is an appropriate constant.} \tag{2.25}$$

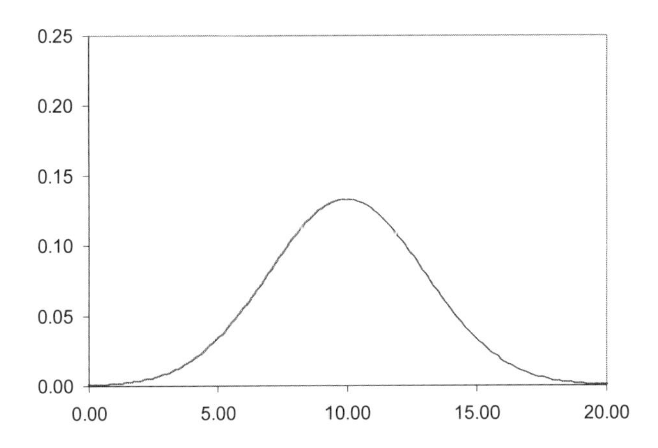

Fig. 2.4. A priori distribution of Θ: $\Theta \sim$ normal $\left(\mu = 10, \tau^2 = 3^2\right)$

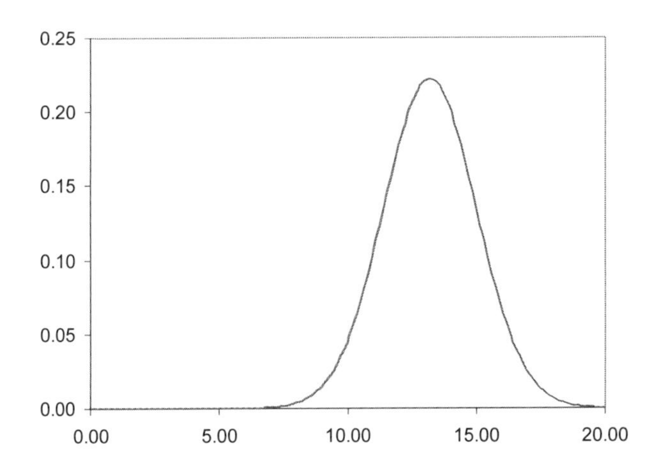

Fig. 2.5. A posteriori distribution of Θ, given 5 observations with $\overline{X} = 15$ and $\sigma^2 = 5^2$

iv) In all three cases the quadratic loss of the Bayes premium is given by

$$\begin{aligned} E\left[\left(P^{\text{Bayes}} - \Theta\right)^2\right] &= (1-\alpha)\, E\left[\left(P^{\text{coll}} - \Theta\right)^2\right] \\ &= \alpha\, E\left[\left(\overline{X} - \Theta\right)^2\right]. \end{aligned}$$

v) In all three cases, we find that the a posteriori distribution of Θ belongs to the same family as the a priori distribution. We will see in the following section that this did not happen purely by chance.

2.5 Conjugate Classes of Distributions

The above three cases are contained in the general framework that we are going to discuss in this section. Let the observation vector $\mathbf{X}$ have independent components, which – up to some weighting – have the same conditional distribution F_ϑ given $\Theta = \vartheta$. We look at the family of possible marginal distributions $\mathcal{F} = \{F_\vartheta : \vartheta \in \Theta\}$. In insurance practice, the specification of the family $\mathcal{F}$ is in itself a problem. Sometimes, there are indications about which families might be appropriate. For example, for modelling the random variable of the number of claims, the Poisson distribution is often a reasonable choice. However, in many situations, we have no idea about the distribution of the observations and specifying $\mathcal{F}$ is very problematic. How we deal with such situations will be discussed in Chapter 3.

The choice of the family $\mathcal{F}$ is not easy, but the choice of the structural function $U(\vartheta)$, or more precisely, the choice of a family $\mathcal{U} = \{U_\gamma(\vartheta) \,|\gamma \in \Gamma\}$ of structural functions indexed with the *hyperparameter* $\gamma \in \Gamma$, to which $U(\vartheta)$ belongs, is even more difficult.

The following conditions should be satisfied so that the whole framework provides an instrument which is suitable for use in practice:

i) The family $\mathcal{U}$ must be big enough so that it contains sufficiently many distributions which could, in reality, describe the collective.
ii) The family $\mathcal{U}$ should be as small as possible to reflect our understanding of the collective.
iii) Ideally, the families $\mathcal{F}$ and $\mathcal{U}$ should be chosen so that the model is mathematically tractable and in particular, so that the Bayes premium can be written in an analytic form.

In order to guarantee such behaviour, a useful concept is that of conjugate families of distributions.

Definition 2.16. *The family $\mathcal{U}$ is conjugate to the family $\mathcal{F}$ if for all $\gamma \in \Gamma$ and for all realizations $\mathbf{x}$ of the observation vector $\mathbf{X}$ there exists a $\gamma' \in \Gamma$ such that*

$$U_\gamma\left(\vartheta\right|\mathbf{X} = \mathbf{x}) = U_{\gamma'}(\vartheta) \qquad \textit{for all } \vartheta \in \Theta, \tag{2.26}$$

i.e. the a posteriori distribution of Θ given $\mathbf{X}$ is again in $\mathcal{U}$ for every a priori distribution from $\mathcal{U}$.

Remarks:

- The biggest possible family $\mathcal{U}$ (that which contains all distribution functions) is conjugate to any given family $\mathcal{F}$. However, in order that our resulting model be useful, it is important that $\mathcal{U}$ should be as small as possible, while still containing the range of "realistic" distributions for the collective.
- In Section 2.4 we saw three examples of conjugate classes of distributions (Poisson–gamma, binomial–beta, normal–normal).

2.5.1 The Exponential Class and Their Associated Conjugate Families

Definition 2.17. *A distribution is said to be of the exponential type, if it can be expressed as*

$$dF(x) = \exp\left[\frac{x\vartheta - b(\vartheta)}{\sigma^2/w} + c(x, \sigma^2/w)\right] d\nu(x), \quad x \in A \subset \mathbb{R}. \tag{2.27}$$

In (2.27) ν denotes either the Lebesgue measure or counting measure, $b(\cdot)$ is some real-valued twice-differentiable function of ϑ, and w and σ^2 are some real-valued constants.

Definition 2.18. *The class of distributions of the exponential type as defined in (2.27) is referred to as the one (real-valued) parameter exponential class*

$$\mathcal{F}_{exp} = \{F_\vartheta : \vartheta \in \Theta\}. \tag{2.28}$$

Remarks:

- The one parametric exponential class $\mathcal{F}_{exp}$ covers a large class of families of distributions. It includes, among others, the families of the Poisson, Bernoulli, gamma, normal and inverse-Gaussian distributions. It plays a central role in the framework of general linear models (GLM), which itself belongs to the standard repertoire of tools used in the calculation of premiums depending on several rating factors.
- Each of such families within $\mathcal{F}_{exp}$ is characterized by the specific form of $b(.)$ and $c(.,.)$. We will denote such a specified family by $\mathcal{F}_{exp}^{b,c}$.
- The above parameterization of the density is the standard notation used in the GLM literature (see, for example, McCullagh and Nelder [MN89] or the description of the GENMOD procedure [SAS93]), and is often referred to as the natural parametrization. The term σ^2/w in (2.27) could be replaced by a more general function $a(\sigma^2)$. In the literature, the exponential class is often defined in this slightly more general way. However, this more general form is of a rather theoretical nature, since in practical applications and also in the GENMOD procedure, it is almost always true that $a(\sigma^2) = \sigma^2/w$.
- The parameter ϑ is referred to as the canonical parameter. In our context, ϑ is to be interpreted as the risk profile taking values in Θ. Observe that ϑ is here one-dimensional and real. The parameter σ^2 is called the dispersion parameter, assumed to be fixed. Lastly, the quantity w denotes a prior known weight (w = weight) associated with the observation. Whereas the dispersion parameter σ^2 is constant over observations, the weight w may vary among the components of the observation vector.
- Another interesting parametrization can be found in [Ger95].

Let us now take a specific family $\mathcal{F}_{exp}^{b,c} \in \mathcal{F}_{exp}$ characterized by the specific form of $b(.)$ and $c(.,.)$.

Theorem 2.19. *We assume that for a given ϑ, the components of the vector $\mathbf{X} = (X_1, \dots, X_n)$ are independent with distribution $F_\vartheta \in \mathcal{F}_{exp}^{b,c}$, each with the same dispersion parameter σ^2 and with weights w_j, $j = 1, 2, \dots, n$.*
We consider the family

$$\mathcal{U}_{exp}^b = \left\{ u_\gamma(\vartheta) : \gamma = \left(x_0, \tau^2\right) \subset \mathbb{R} \times \mathbb{R}^+ \right\}, \tag{2.29}$$

where

$$u_\gamma(\vartheta) = \exp\left[\frac{x_0\vartheta - b(\vartheta)}{\tau^2} + d(x_0, \tau^2)\right], \ \vartheta \in \Theta, \tag{2.30}$$

are densities (with respect to the Lebesgue measure).

Then it holds that $\mathcal{U}_{exp}^b$ is conjugate to $\mathcal{F}_{exp}^{b,c}$.

Remarks:

- x_0 and τ^2 are hyperparameters.
- Note that $\exp\left[d(x_0, \tau^2)\right]$ in (2.30) is simply a normalizing factor.
- Note that the family conjugate to $\mathcal{F}_{exp}^{b,c}$ depends only on the function $b\left(\vartheta\right)$ and not on $c\left(x, \sigma^2/w\right)$.

Proof of Theorem 2.19:
For the a posteriori density of Θ given $\mathbf{X}$ we get

$$\begin{aligned} u\left(\vartheta \middle| \mathbf{X} = \mathbf{x}\right) &\propto \prod_{j=1}^{n} \exp\left[\frac{x_j\vartheta - b(\vartheta)}{\sigma^2/w_j}\right] \cdot \exp\left[\frac{x_0\vartheta - b(\vartheta)}{\tau^2}\right] \\ &= \exp\left[\frac{\left[\frac{\sigma^2}{\tau^2} + w_\bullet\right]^{-1}\left[\frac{\sigma^2}{\tau^2}x_0 + w_\bullet\overline{x}\right]\vartheta - b(\vartheta)}{\sigma^2/\left(\frac{\sigma^2}{\tau^2} + w_\bullet\right)}\right], \\ \text{where } \overline{x} &= \sum\nolimits_{j=1}^{n} \frac{w_j}{w_\bullet}x_j \quad \text{and} \quad w_\bullet = \sum\nolimits_{j=1}^{n} w_j. \end{aligned}$$

We see immediately that with the a priori distribution (with the hyperparameters x_0, τ^2) the a posteriori distribution, given $\mathbf{X} = \mathbf{x}$, is again in $\mathcal{U}_{exp}^b$, with updated hyperparameters

$$x_0' = \left(\frac{\sigma^2}{\tau^2}x_0 + w_\bullet\overline{x}\right)\left(\frac{\sigma^2}{\tau^2} + w_\bullet\right)^{-1} \quad \text{and} \quad \tau'^2 = \sigma^2\left(\frac{\sigma^2}{\tau^2} + w_\bullet\right)^{-1}. \tag{2.31}$$

This proves Theorem 2.19. □

We now want to determine P^{ind}, P^{coll} and P^{Bayes}.

Theorem 2.20. *For the family $\mathcal{F}_{exp}^{b,c}$ and its conjugate family $\mathcal{U}_{exp}^{b}$ we have*

i)

$$P^{\text{ind}}(\vartheta) = b'(\vartheta) \quad \textit{and} \quad \operatorname{Var}[X_j | \Theta = \vartheta, w_j] = b''(\vartheta)\sigma^2/w_j. \tag{2.32}$$

If the region Θ is such that $\exp[x_0\vartheta - b(\vartheta)]$ disappears on the boundary of Θ for each possible value x_0, then we have

ii)

$$P^{\text{coll}} = x_0, \tag{2.33}$$

iii)

$$P^{\text{Bayes}} = \alpha \overline{X} + (1-\alpha) P^{\text{coll}}, \tag{2.34}$$

where

$$\overline{X} = \sum\nolimits_j \frac{w_j}{w_\bullet} X_j,$$
$$\alpha = \frac{w_\bullet}{w_\bullet + \frac{\sigma^2}{\tau^2}}.$$

Remarks:

- Note that P^{Bayes} is a weighted average of the individual claims experience and the collective premium. It is a linear function of the observations, and therefore a credibility premium. The case where P^{Bayes} is of a credibility type is often referred to as *exact credibility* in the literature.
- The credibility weight α has the same form as in (2.25) and $w_\bullet$ now plays the role of the number of observation years.

Proof of Theorem 2.20:
For the proof of (2.32) we consider the moment-generating function of X given $\Theta = \vartheta$ and given the weight w.

$$\begin{aligned}
m_X(r) &= E\left[e^{rX} \middle| \Theta = \vartheta\right] \\
&= \int e^{rx} \exp\left[\frac{x\vartheta - b(\vartheta)}{\sigma^2/w} + c\left(x, \sigma^2/w\right)\right] d\nu(x) \\
&= \int \exp\left[\frac{x\left(\vartheta + r\sigma^2/w\right) - b\left(\vartheta + r\sigma^2/w\right)}{\sigma^2/w} + c\left(x, \sigma^2/w\right)\right] d\nu(x) \\
&\qquad \times \exp\left[\frac{b\left(\vartheta + r\sigma^2/\omega\right) - b(\vartheta)}{\sigma^2/w}\right] \\
&= \exp\left[\frac{b\left(\vartheta + r\sigma^2/w\right) - b(\vartheta)}{\sigma^2/w}\right],
\end{aligned}$$

where the last equality follows because the integral term is equal to 1 (integration of a probability density function). The cumulant-generating function is therefore given by

$$k_X(r) := \ln m_X(r) = \frac{b\left(\vartheta + r\sigma^2/w\right) - b\left(\vartheta\right)}{\sigma^2/w}.$$

From this we get

$$\begin{aligned} P^{\text{ind}}\left(\vartheta\right) := \quad & E\left[X|\Theta = \vartheta\right] = k_X'(0) = b'\left(\vartheta\right), \\ \operatorname{Var}\left[X|\,\Theta = \vartheta\right] &= k_X''(0) = b''(\vartheta)\,\sigma^2/w. \end{aligned}$$

We have thus proved (2.32).

For the proof of (2.33) we see that

$$\begin{aligned} P^{\text{coll}} &= \int_\Theta \mu\left(\vartheta\right) \exp\left[\frac{x_0\vartheta - b\left(\vartheta\right)}{\tau^2}\right] d\vartheta \cdot \exp\left\{d\left(x_0, \tau^2\right)\right\} \\ &= \int_\Theta b'\left(\vartheta\right) \exp\left[\frac{x_0\vartheta - b\left(\vartheta\right)}{\tau^2}\right] d\vartheta \cdot \exp\left\{d\left(x_0, \tau^2\right)\right\} \\ &= x_0 - \int_\Theta \left(x_0 - b'\left(\vartheta\right)\right) \exp\left[\frac{x_0\vartheta - b\left(\vartheta\right)}{\tau^2}\right] d\vartheta \cdot \exp\left\{d\left(x_0, \tau^2\right)\right\} \\ &= x_0 - \tau^2 \exp\left[x_0\vartheta - b\left(\vartheta\right)\right]\big|_{\partial\Theta}, \end{aligned} \tag{2.35}$$

where $\partial\Theta$ denotes the boundary of Θ. The choice of the parameter region Θ turns out to be crucial. Under the technical assumption made in Theorem 2.20, that Θ has been chosen in such a way, that the boundary term disappears for every x_0, we get

$$P^{\text{coll}} = x_0,$$

so that we have proved (2.33).

From the proof of Theorem 2.19, where we showed that the a posteriori distribution of Θ given $\mathbf{X} = \mathbf{x}$ also belongs to $\mathcal{U}_{exp}^b$ with hyperparameters

$$\begin{aligned} x_0' &= \left(\frac{\sigma^2}{\tau^2}x_0 + w_\bullet \overline{x}\right)\left(\frac{\sigma^2}{\tau^2} + w_\bullet^{-1}\right)^{-1}, \\ \tau'^2 &= \sigma^2\left(\frac{\sigma^2}{\tau^2} + w_\bullet\right)^{-1}, \end{aligned}$$

we get without further ado

$$P^{\text{Bayes}} = \widetilde{\mu\left(\Theta\right)} = E\left[\mu\left(\Theta\right)|\mathbf{X}\right] = \alpha\overline{X} + (1-\alpha)\,x_0,$$

$$\text{where } \overline{X} = \sum_{j=1}^n \frac{w_j}{w_\bullet} X_j, \qquad w_\bullet = \sum_{j=1}^n w_j, \qquad \alpha = \frac{w_\bullet}{w_\bullet + \sigma^2/\tau^2}.$$

This proves (2.34) and thus ends the proof of Theorem 2.20. □

Under the following special assumptions, we get the classical examples (including those already seen in Section 2.4):

a) **Poisson–Gamma:** We have frequency observations $X_j = \frac{N_j}{w_j}$, where, conditional on ϑ, N_j is Poisson distributed with parameter $\lambda_j = w_j \vartheta$. The density of X_j is then given by

$$f_\vartheta(x) = e^{-w_j \vartheta} \frac{(w_j \vartheta)^{w_j x}}{(w_j x)!} \qquad \text{for } x = \frac{k}{w_j}, \; k \in \mathbb{N}_0.$$

This can be written in the form (2.27) as follows:

$$\begin{aligned} \widetilde{\vartheta} &= \log \vartheta, \quad b\left(\widetilde{\vartheta}\right) = \exp\left(\widetilde{\vartheta}\right), \quad \widetilde{\sigma}^2 = 1, \quad \widetilde{w} = w_j, \\ c\left(x, \widetilde{\sigma}^2 / \widetilde{w}\right) &= -\log\left((\widetilde{w} x)!\right) + \widetilde{w} x \log \widetilde{w}. \end{aligned}$$

To find the conjugate family of prior distributions $\mathcal{U}^b_{exp}$ we insert $\widetilde{\vartheta}$ and $b\left(\widetilde{\vartheta}\right)$ into (2.30) and we get

$$u_\gamma(\widetilde{\vartheta}) \propto \exp\left[\frac{x_0 \widetilde{\vartheta} - \exp\left(\widetilde{\vartheta}\right)}{\tau^2}\right].$$

This density, expressed in terms of the original variable ϑ rather than $\widetilde{\vartheta}$, becomes

$$\begin{aligned} u_\gamma(\vartheta) &\propto \frac{1}{\vartheta} \cdot \exp\left[\log\left(\vartheta^{x_0/\tau^2}\right) - \frac{\vartheta}{\tau^2}\right] \qquad (2.36) \\ &= \vartheta^{\frac{x_0}{\tau^2} - 1} e^{-\frac{1}{\tau^2} \vartheta}. \end{aligned}$$

Note that by the change of variables from $\widetilde{\vartheta}$ to ϑ we have to take into account the first derivative $d\widetilde{\vartheta}/d\vartheta$ (term $1/\vartheta$ on the right-hand side of (2.36)).
Hence

$$\mathcal{U}^b_{exp} = \left\{u(\vartheta) : u(\vartheta) \propto \vartheta^{\frac{x_0}{\tau^2} - 1} e^{-\frac{1}{\tau^2} \vartheta}; \; x_0, \tau^2 > 0\right\},$$

which is the family of the Gamma distributions.
For P^{coll} we get

$$P^{\text{coll}} = E[\Theta] = \frac{x_0}{\tau^2} \tau^2 = x_0,$$

hence (2.33) is fulfilled.

b) **Binomial–Beta:** We have frequency observations $X_j = \frac{N_j}{w_j}$, where conditional on ϑ, N_j has a Binomial distribution with $n = w_j$ and $p = \vartheta$. The density of X_j is then given by

$$f_\vartheta(x) = \binom{w_j}{w_j x} \vartheta^{w_j x} (1-\vartheta)^{w_j - w_j x} \quad \text{for} \quad x = k/w_j,\ k = 0, \ldots, w_j, w_j \in \mathbb{N}.$$

This can be written in the form of (2.27) as follows:

$$\begin{aligned} \widetilde{\vartheta} &= \log\left(\tfrac{\vartheta}{1-\vartheta}\right), \quad b\left(\widetilde{\vartheta}\right) = \log\left(1 + e^{\widetilde{\vartheta}}\right), \quad \widetilde{\sigma}^2 = 1, \\ \widetilde{w} &= w_j, \quad c\left(x, \widetilde{\sigma}^2/\widetilde{w}\right) = \log\left(\tbinom{\widetilde{w}}{\widetilde{w}x}\right). \end{aligned}$$

From (2.30) we get

$$\mathcal{U}^b_{exp} = \left\{u(\vartheta) : u(\vartheta) \propto \vartheta^{\frac{x_0}{\tau^2}-1} (1-\vartheta)^{\frac{1-x_0}{\tau^2}-1}; \quad 0 < x_0 < 1, \tau^2 > 0\right\},$$

which is the family of beta distributions.
For P^{coll}, we get

$$P^{\text{coll}} = E[\Theta] = \frac{\frac{x_0}{\tau^2}}{\frac{x_0}{\tau^2} + \frac{1-x_0}{\tau^2}} = x_0.$$

c) **Gamma–Gamma:** We have observations X_j, that are, conditional on ϑ, Gamma distributed with shape parameter $w_j\gamma$ and scale parameter $w_j\gamma\vartheta$, where w_j is the weight associated with X_j. In particular, this is the case if the observations X_j are the average of w_j independent claim sizes, each of them Gamma distributed with shape parameter γ and scale parameter $\vartheta\gamma$. The density of X_j is then given by

$$f_\vartheta(x) = \frac{(w_j\gamma\vartheta)^{w_j\gamma}}{\Gamma(w_j\gamma)} x^{w_j\gamma-1} e^{-w_j\gamma\vartheta x}.$$

This can be written in the form (2.30) as follows:

$$\begin{aligned} &\widetilde{\vartheta} = -\vartheta, \quad b\left(\widetilde{\vartheta}\right) = -\log\left(-\widetilde{\vartheta}\right), \quad \widetilde{\sigma}^2 = \gamma^{-1}, \quad \widetilde{w} = w_j, \\ &c\left(x, \widetilde{\sigma}^2/\widetilde{w}\right) = (\widetilde{w}\gamma - 1)\log(x) - \log(\Gamma(\widetilde{w}\gamma)) + \widetilde{w}\gamma\log(\widetilde{w}\gamma). \end{aligned}$$

Note that

$$\mu(\vartheta) := E_\vartheta(X_j) = \vartheta^{-1}.$$

From (2.30) we get

$$\mathcal{U}^b_{exp} = \left\{u(\vartheta) : u(\vartheta) \propto \vartheta^{\frac{1}{\tau^2}} e^{-\frac{x_0}{\tau^2}\vartheta}; \quad \tau^2 \in (0,1), x_0 > 0\right\},$$

which is the family of gamma distributions.
One can easily check that

$$P^{\text{coll}} = E[\mu(\Theta)] = x_0.$$

d) **Normal–Normal:** We have observations X_j that are, conditional on ϑ, normally distributed with expected value ϑ and variance σ^2/w_j. The density of X_j is then given by

$$f_{\vartheta,\gamma,w_j}(x) = (2\pi\sigma^2/w_j)^{-1/2} \exp\left\{-\frac{(x-\vartheta)^2}{2\sigma^2/w_j}\right\}.$$

This can be written in the form of (2.27) as follows:

$$\begin{aligned}
&\widetilde{\vartheta} = \vartheta, \quad b\left(\widetilde{\vartheta}\right) = \widetilde{\vartheta}^2/2, \quad \widetilde{\sigma}^2 = \sigma^2, \quad \widetilde{w} = w_j, \\
&c\left(x, \widetilde{\sigma}^2/\widetilde{w}\right) = -\tfrac{1}{2}\left(\tfrac{x^2}{\widetilde{\sigma}^2/\widetilde{w}} + \log\left(2\pi\widetilde{\sigma}^2/\widetilde{w}\right)\right).
\end{aligned}$$

From (2.30) we get that

$$\mathcal{U}^b_{exp} = \left\{u(\vartheta) : u(\vartheta) \propto \exp\left(-\frac{(\vartheta - x_0)^2}{2\tau^2}\right); \quad x_0, \tau^2 \subset \mathbb{R} \times \mathbb{R}^+\right\},$$

i.e. the family of conjugate distributions is the family of normal distributions. From this we can immediately see that

$$P^{\text{coll}} = E[\Theta] = x_0.$$

e) **Geometric–Beta:** We have observations X_j, that are, conditional on ϑ, distributed with a geometric distribution with parameter ϑ (i.e. without weights, or equivalently, all weights w_j are equal to 1). The density of X_j is then given by

$$f_\vartheta(x) = (1-\vartheta)^x\, \vartheta, \quad x \in \mathbb{N}.$$

This can be written in the form of (2.27) as follows:

$$\begin{aligned}
&\widetilde{\vartheta} = \log(1-\vartheta), \quad b\left(\widetilde{\vartheta}\right) = -\log\left(1 - e^{\widetilde{\vartheta}}\right), \\
&\widetilde{\sigma}^2 = 1, \quad \widetilde{w} = 1, \quad c\left(x, \widetilde{\sigma}^2/\widetilde{w}\right) = 0.
\end{aligned}$$

Note that

$$\mu(\vartheta) = E_\vartheta(X_j) = \frac{1-\vartheta}{\vartheta}.$$

From (2.30) we get that

$$\mathcal{U}^b_{exp} = \left\{u(\vartheta) : u(\vartheta) \propto (1-\vartheta)^{\frac{x_0}{\tau^2}-1}\, \vartheta^{\frac{1}{\tau^2}}; \quad x_0 > \tau^2, 0 < \tau^2 < 1\right\},$$

which is the family of the Beta distributions.
Again, one can easily verify, that

$$P^{\text{coll}} = E[\mu(\Theta)] = x_0.$$

2.5.2 Construction of Conjugate Classes

The following theorem is often helpful when we are looking for a conjugate family $\mathcal{U}$ to the family $\mathcal{F} = \{F_\vartheta : \vartheta \in \Theta\}$.

Theorem 2.21.
Assumptions: $\mathcal{F}$ *and* $\mathcal{U}$ *satisfy the following conditions:*

i) The likelihood functions $l(\vartheta) = f_\vartheta(\mathbf{x})$, $\mathbf{x}$ *fixed, are proportional to an element of* $\mathcal{U}$, *i.e. for every possible observation* $\mathbf{x} \in A$, *there exists a* $u_\mathbf{x} \in \mathcal{U}$, *such that* $u_\mathbf{x}(\vartheta) = f_\vartheta(\mathbf{x}) \left(\int f_\vartheta(\mathbf{x}) d\vartheta\right)^{-1}$.
ii) $\mathcal{U}$ *is closed under the product operation, i.e. for every pair* $u, v \in \mathcal{U}$ *we have that* $u(\cdot)v(\cdot) \left(\int u(\vartheta)v(\vartheta)d\vartheta\right)^{-1} \in \mathcal{U}$.

Then it holds that $\mathcal{U}$ *is conjugate to* $\mathcal{F}$.

Proof: Given $\mathbf{X} = \mathbf{x}$, we find the following a posteriori distribution:

$$
\begin{aligned}
u(\vartheta \mid \mathbf{x}) &= \frac{f_\vartheta(\mathbf{x})u(\vartheta)}{\int f_\vartheta(\mathbf{x})u(\vartheta)d\vartheta} \\
&= \frac{\int f_\vartheta(\mathbf{x})d\vartheta}{\int f_\vartheta(\mathbf{x})u(\vartheta)d\vartheta} \, u_\mathbf{x}(\vartheta)\, u(\vartheta) \in \mathcal{U}.
\end{aligned}
$$

This proves Theorem 2.21. □

Theorem 2.21 indicates how we should construct a family $\mathcal{U}$, which is conjugate to the family $\mathcal{F}$. Define

$$
\mathcal{U}' = \left\{ u_\mathbf{x} : \; u_\mathbf{x}(\vartheta) = f_\vartheta(\mathbf{x}) \left(\int f_\vartheta(\mathbf{x}) d\vartheta \right)^{-1}, \; \mathbf{x} \in A \right\}.
$$

If $\mathcal{U}'$ is closed under the product operation then $\mathcal{U}'$ is conjugate to $\mathcal{F}$. If not, it can often be extended in a natural way.

Example:
Let $\mathcal{F}$ be the family of binomial distributions

$$
\mathcal{F} = \{f_\vartheta(\mathbf{x}) : \vartheta \in [0,1]\} \text{ with}
$$

$$
f_\vartheta(\mathbf{x}) = \binom{n}{x_\bullet} \vartheta^{x_\bullet} (1-\vartheta)^{n-x_\bullet}, \qquad x_\bullet = \sum_{j=1}^n x_j, \qquad x_j \in \{0,1\}.
$$

Then we see that $\mathcal{U}'$ consists of the Beta distributions with $(a,b) \in \{(1, n+1), (2,n), (3, n-1), \ldots, (n+1, 1)\}$.

$\mathcal{U}'$ is not closed under the product operation. A natural extension is

$$
\mathcal{U} = \{\text{Beta}(a,b) : a, b > 0\}.
$$

It is easy to check that $\mathcal{U}$ is closed under the product operation and that $\mathcal{U}$ is therefore, by Theorem 2.21, conjugate to $\mathcal{F}$.

2.6 Another Type of Example: the Pareto–Gamma Case

The model and the results of this section are taken from [Ryt90].

Motivation

A frequent assumption in the practice of reinsurance is that the claim amounts of those claims which exceed a given limit x_0 are Pareto distributed. Typically, based on information from numerous portfolios, the reinsurer has an "a priori" idea about the level of the "Pareto parameter" ϑ. He also collects information from the primary insurer about all claims exceeding a particular limit c_0, and he can also use this information to estimate ϑ. The question then arises as to how he can best combine both sources of information in order to estimate ϑ as accurately as possible.

Let $\mathbf{X}' = (X_1, \dots, X_n)$ be the vector of observations of all the claim sizes, of those claims belonging to a given contract, whose size exceeds x_0. We assume that the X_j $(j = 1, \dots, n)$ are independent and Pareto distributed,

$$X_j \sim \text{Pareto}\ (x_0, \vartheta), \tag{2.37}$$

with density and distribution function

$$f_\vartheta(x) = \frac{\vartheta}{x_0}\left(\frac{x}{x_0}\right)^{-(\vartheta+1)} \quad \text{and} \quad F_\vartheta\,(x) = 1 - \left(\frac{x}{x_0}\right)^{-\vartheta} \quad \text{for } x \geq x_0, \tag{2.38}$$

and with moments

$$\mu(\vartheta) = x_0 \frac{\vartheta}{\vartheta - 1}, \quad \text{if } \vartheta > 1,$$

$$\sigma^2(\vartheta) = x_0^2 \frac{\vartheta}{(\vartheta-1)^2(\vartheta-2)}, \quad \text{if } \vartheta > 2.$$

In order to incorporate the a priori knowledge, we further assume that the Pareto parameter ϑ is itself the realization of a random variable Θ with distribution function $U\,(\vartheta)$. In order to specify an appropriate class $\mathcal{U}$ of a priori distributions, to which $U(\vartheta)$ belongs, we use the technique from Theorem 2.21.

The likelihood function leads to the following family:

$$\mathcal{U}' = \left\{ u_{\mathbf{x}} : \ u_{\mathbf{x}}(\vartheta) \propto \vartheta^n \exp\left(- \left(\sum_{j=1}^n \ln\left(\frac{x_j}{x_0}\right)\right) \vartheta \right) \right\}.$$

The elements of $\mathcal{U}'$ are Gamma distributions. The natural extension of $\mathcal{U}'$ is therefore the family of the Gamma distributions, i.e.

$$\mathcal{U} = \left\{ \text{distributions with densities } u(\vartheta) = \frac{\beta^\gamma}{\Gamma(\gamma)} \vartheta^{\gamma-1} \exp\{-\beta\vartheta\} \right\}. \tag{2.39}$$

Since the family $\mathcal{U}$ is closed with respect to the product operation, we have that $\mathcal{U}$ is conjugate to

$$\mathcal{F} = \{\text{Pareto}(x_0, \vartheta) : \ \vartheta, \ x_0 > 0\}.$$

We are looking for the best estimator of the Pareto parameter Θ.

From the form of the likelihood function (2.38) and with the form of the a priori distribution (2.39) we get for the a posteriori density of Θ given $\mathbf{x}$

$$u_{\mathbf{x}}(\vartheta) \propto \vartheta^{\gamma+n-1} \exp\left\{-\left(\beta + \sum_{j=1}^{n} \ln\left(\frac{x_j}{x_0}\right)\right)\vartheta\right\}.$$

We see that $u_{\mathbf{x}}(\vartheta)$ is again the density of a Gamma distribution with updated parameters

$$\gamma' = \gamma + n \qquad \text{and} \qquad \beta' = \beta + \sum\nolimits_{j=1}^{n} \ln\left(\frac{x_j}{x_0}\right).$$

The Bayes estimator for Θ is therefore given by

$$\widetilde{\Theta} = E[\Theta|\mathbf{X}] = \frac{\gamma+n}{\beta + \sum_{j=1}^{n} \ln\left(\frac{X_j}{x_0}\right)}. \tag{2.40}$$

Formula (2.40) can be written in another, more easily interpretable form. In order to do this, we first consider the maximum-likelihood estimator for the Pareto parameter ϑ, i.e. the estimator that one might use if we only had data and no a priori information. This is given by

$$\widehat{\vartheta}_{MLE} = \frac{n}{\sum_{j=1}^{n} \ln\left(\frac{X_j}{x_0}\right)}.$$

(2.40) can now be written as

$$\widetilde{\Theta} = \alpha \cdot \widehat{\vartheta}_{MLE} + (1-\alpha) \cdot \frac{\gamma}{\beta},$$

where

$$\alpha = \left(\sum_{j=1}^{n} \ln\left(\frac{X_j}{x_0}\right)\right)\left(\beta + \sum_{j=1}^{n} \ln\left(\frac{X_j}{x_0}\right)\right)^{-1}.$$

The Bayes estimator for Θ is therefore a weighted average of the maximum-likelihood estimator and the a priori expected value $E[\Theta] = \gamma/\beta$, where, however, in contrast to the earlier examples, the weight α depends on the observations and is not a credibility weight in the strict sense of Theorem 2.20.

2.7 Summary

Here, we give a short summary of the results of this chapter.

Bayes model:

- $\mathcal{F} = \{F_\vartheta : \vartheta \in \Theta\}$: family of distributions indexed with the parameter $\vartheta \in \Theta$.
- $U(\vartheta)$: structural function, the a priori distribution of Θ.
- $P^{\text{Bayes}} = \widetilde{\mu(\Theta)} = E[\mu(\Theta)|\mathbf{X}]$: Bayes estimator (with respect to the quadratic loss function).

In a number of well-structured models, it is possible to explicitly calculate the Bayes estimator, and in some of these cases (one-dimensional exponential families and conjugate families, see Section 2.5) this estimator has a linear form, i.e. it is a credibility estimator. We refer to such cases as *exact credibility*.

If the structural function U is not known exactly, and we only know that U belongs to a certain family $\mathcal{U}$ of distributions, it may be possible to estimate the unknown parameters of the distribution from the data of the collective (empirical Bayes).

Various open questions:

- In the cases studied in this chapter, we could write down, in closed form, formulae for the Bayes premium. Typically, however, this is not the case. Usually the Bayes estimator can be calculated only using relatively complicated numerical procedures.
- How restrictive are the assumptions that we have made about the distribution functions (e.g. Poisson–Gamma)? Do we really need to know the whole distribution function or would it be enough to know only the first and second moments?

Requirements and constraints in practice:

- *Simplicity:* The formula for the premium should be as simple and as intuitive as possible. This is in general not the case for the exact Bayes estimator.
- *Structure:* In order to calculate the Bayes estimator we would need to specify the family of conditional distributions $\mathcal{F}$ and the structural function U, respectively the family $\mathcal{U}$ of structural functions, from which U comes. In practice such a specification is either quite artificial (in the case where $\mathcal{U}$ and $\mathcal{F}$ are small) or not helpful (in the case where $\mathcal{U}$ and $\mathcal{F}$ are large). In other words: in practice we usually know neither $\mathcal{F}$ nor U, respectively $\mathcal{U}$. In general, we must content ourselves with being able to make reasonable estimates of the first few moments of the distributions.

In practice then, the exact Bayes formula is often not a realistic method. One way out of this dilemma is the *credibility technique*, which will be introduced in the next chapter.

2.8 Exercises

Exercise 2.1

We consider an individual risk which can produce a total claim amount in one year of either 10 000 or 0 (claim free year).

This individual risk is taken from a collective, which contains three types of risks: good (65%), medium (30%), bad (5%). The conditional probabilities are given in the following table:

	Conditional probabilities		
Total claim amount	Good	Medium	Bad
0	97%	95%	90%
10 000	3%	5%	10%

a) Calculate P^{ind} and P^{coll}.
b) There is one year's claim experience.
Calculate P^{Bayes} for all possible observations (claim amount 0 or 10 000).
c) Calculate P^{Bayes} for all possible observations in the case where we have two years' claim experience, and under the usual assumption, the claims in the two years are conditionally independent, given the risk parameter Θ.

Exercise 2.2

A company wants to introduce a no-claims discount in household insurance: the insured gets a discount of 15% after three consecutive claim-free years.

The premium is proportional to the sum insured and it is assumed that the following assumptions hold true:

- The claim number N_{ij} of the insured i in year j is Poisson distributed with Poisson parameter Θ_i. This Poisson parameter does not depend on the sum insured, and it is assumed that the claim numbers N_{ij} satisfy the assumptions of the Poisson–Gamma model (Model Assumptions 2.8).
- Claim number and claim sizes are independent.
- The expected value of the claim size is proportional to the sum insured.

The following statistical data are available:

k	0	1	2	3	4	≥ 5	Total
# of policies with k claims in one year	143 991	11 588	1 005	85	6	2	156 677

a) Fit a negative binomial distribution to the observations of the above table.
b) Calculate the coefficient of variation $\text{CoVa}(\Theta_i)$, which is a good measure of the heterogeneity of the portfolio.

c) Calculate the impact of the introduction of such a no-claims discount on the premiums (premium reduction in % due to the no claims discount). For simplicity reasons we assume that all policies have an observation period of $\geqq 3$ years.
d) Is the discount rate of 15% for three claim free years reasonable? Compare it with the value obtained for the corresponding empirical Bayes estimator.

Exercise 2.3

In group life insurance, for an age segment of interest (e.g. ages between 25 and 65) and for larger collectives, we would like to use "individual mortality tables". In order to model this, we assume that the age profile in the individual mortality table is equal to that in the "basis mortality table" multiplied by a factor Θ_i (so that e.g. $q_x^{ind} = \Theta_i q_x$). We further assume that Θ_i has a Gamma distribution with expected value 1 and a coefficient of variation of 20%.

We have the following observations from the past:
observed number of years at risk: 6 500
expected number of deaths based on the basis mortality table: 25
observed number of deaths: 15

Find the associated Bayes estimator Θ_i^{Bayes} of Θ_i.

Exercise 2.4

We consider a fleet of cars. On the basis of suitable statistics, we know the expected "a priori" claim frequencies. On the other hand, factors such as driver education, deadline expectations and car equipment have an influence on the claim frequency and lead to large differences between the risk quality of different fleets. We assume that the assumptions of the Poisson–Gamma model II (Model Assumptions 2.10) are fulfilled and we estimate that the differences between the risk quality of different fleets is about ± 20% (i.e. $\mathrm{CoVa}(\Theta) = 20\%$).

a) Determine the formula for the Bayes premium based on the total claim numbers in one year of an entire fleet with the following vehicles: 30 lorries, 30 delivery vans, 10 passenger cars. The a priori claim frequencies for the different vehicle types are

Vehicle type	"a priori" frequency in ‰
Lorry	300
Delivery van	100
Passenger car	75

b) How big is the "experience correction" in %, if
   - 16 claims are observed in one year,
   - 20 claims are observed in two years.
c) Make a graphical display of the a priori distribution of Θ and of the a posteriori distribution of Θ given 20 observed claims in two years.

Exercise 2.5

In workers' compensation (collective accident insurance), the probability of disability is of interest. For a particular portfolio of collective policies, we know that, on average, the probability of disability is $p_0 = 10$ ‰. The probabilities however differ between types of business. We use the Beta–Binomial model, and we assume that the coefficient of variation of Θ_i is 50%.

a) Determine the parameters of the a priori distribution and make a graph of the a priori density $u(\vartheta)$.
b) Find the Bayes estimator Θ_i^{Bayes} for business type i where we have observed 15 disability cases out of 1 000 observed insured.
c) A company would like to divide all business types into three groups and apply to them the following tariff rules:
   - a surcharge of 20%, if $E[\Theta_i | N_i, w_i] \geq 1.3\ p_0$,
   - a rebate of 20%, if $E[\Theta_i | N_i, w_i] \leq 0.7\ p_0$,

   where N_i is the observed number of disability cases, w_i the observed number of years at risk and p_0 the a priori probability of disability.

   What is the corresponding decision rule expressed in the observed number of disability cases
   i) for a business with 500 observed years at risk,
   ii) for a business with 2 000 observed years at risk?

Exercise 2.6

An actuary in a reinsurance company has to quote an excess of loss reinsurance cover with retention 2 Millions CHF and no upper limit for a particular ceding company in the motor liability line. For this purpose he assumes that the claims above 1 Million CHF are Pareto distributed, and he wants to estimate the Pareto parameter Θ.

During the last three years ten claims exceeding 1 Million CHF (statistical limit) have been observed with the following (inflation adjusted) claim sizes in Millions CHF: $\{1.1, 1.2, 1.2, 1.4, 1.5, 1.8, 2.0, 2.4, 3.0, 4.2\}$. We estimate the expected number of claims above this statistical limit of 1 Million as 3.5 for the next year to be quoted.

In addition, from experience with lots of such treaties, the following a priori information is available: $E[\Theta] = 2.5$, $\sigma(\Theta) = 0.5$.

a) Calculate
$\widehat{\vartheta}_{MLE}$ (maximum-likelihood estimator) and Θ^{Bayes} (Bayes estimator of Θ based on the Pareto–Gamma model (see Section 2.6)).
b) Calculate the pure risk premium for the excess of loss cover with retention 2 Millions based on the Pareto distribution with Θ^{Bayes}, $\widehat{\vartheta}_{MLE}$ and Θ^{coll}.

Exercise 2.7

In the accident line of business, we would like to determine the expected average claim amount for various risk classes. The data can be well described by a Lognormal distribution. In the model we assume that the coefficient of variation is the same for all risk classes and has the value 4, but that the expected values differ from one class to another. We therefore assume, conditionally on Θ_i, that the claim amounts have a lognormal distribution with parameters Θ_i and σ.

a) Find the natural conjugate family.
b) Denote by $Y_i^{(\nu)}, \nu = 1, 2, ..., n$ the observed claim amounts of risk class i. Find the Bayes estimator for Θ_i with the a priori distribution from the natural conjugate family and under the assumptions that $E[\Theta_i] = 7.3$ and that the coefficient of variation of Θ_i is 10%. Compare it with the maximum-likelihood estimator (for σ known, ϑ unknown).
c) For two classes $i = 1, 2$ we have observed the following claim amounts $Y_i^{(\nu)}$:

$\nu =$	1	2	3	4	5
$Y_1^{(\nu)}$	125	187	210	326	470
$Y_2^{(\nu)}$	161	1 100	1 899	6 106	7 469

$\nu =$	6	7	8	9	10
$Y_1^{(\nu)}$	532	1 001	1 095	9 427	140 591

Estimate the expected value of the claim size for each of the two classes by estimating the parameter Θ_i of the lognormal distribution and inserting this estimate into the formula of the first moment of the Lognormal distribution. Estimate Θ_i by three different methods: $\widehat{\vartheta}_i^{\text{MLE}}$ (maximum-likelihood estimate), Θ_i^{coll}, Θ_i^{Bayes}. Make a table of the obtained estimates of the expected value of the claim amount and the observed mean of the claim amounts. Comment on the results.

Exercise 2.8

Consider the family $\mathcal{F} = \{f_\vartheta(k) \,|\, \vartheta \in (0,1)\}$ of negative binomial distributions, where $f_\vartheta(k) = P_\vartheta(X = k) \propto \vartheta^\gamma (1-\vartheta)^k$ and where γ is fixed.

i) Find the family $\mathcal{U}$ conjugate to $\mathcal{F}$.
ii) Assume that $\{X_j : j = 1, 2, \ldots, n\}$ have, conditionally on $\Theta = \vartheta$, a negative binomial distribution with density $f_\vartheta(k)$ and that they are conditionally independent. It is further assumed that Θ has a Beta distribution with parameters (a, b). Derive the Bayes estimator for Θ.

Exercise 2.9

Consider the standard model:

- $X_1, X_2 \ldots$ are conditionally on $\Theta = \vartheta$ independent and identically distributed.
- Θ is a random variable with distribution $U(\vartheta)$.

We define the Bayes predictor X_{n+1}^{Bayes} of X_{n+1} based on the observations $X_1, \ldots, X_n$ as the best predictor of X_{n+1} with respect to expected quadratic loss.

Show that the Bayes predictor of X_{n+1} is identical to the Bayes estimator of $\mu(\Theta) = E[X_{n+1} |\Theta]$.

3

Credibility Estimators

We have seen that the Bayes premium $\widetilde{\mu(\Theta)} = E\left[\mu(\Theta)|\,\mathbf{X}\right]$ is the best possible estimator in the class of all estimator functions. In general, however, this estimator cannot be expressed in a closed analytical form and can only be calculated by numerical procedures. Therefore it does not fulfil the requirement of simplicity. Moreover, to calculate $\widetilde{\mu(\Theta)}$, one has to specify the conditional distributions as well as the a priori distribution, which, in practice, can often neither be inferred from data nor guessed by intuition.

The basic idea underlying credibility is to force the required simplicity of the estimator by *restricting* the class of allowable estimator functions to those which are *linear* in the observations $\mathbf{X} = (X_1, X_2, \ldots, X_n)'$. In other words, we look for the best estimator in the class of all *linear estimator functions.* "Best" is to be understood in the Bayesian sense and the optimality criterion is again quadratic loss. *Credibility estimators* are therefore *linear Bayes estimators.*

In the previous chapters $\mathbf{X}$ stands for the observations of the individual risk and $\mu(\Theta)$ for the individual premium. Also, until now, we have essentially worked under the rather simple Assumption 1.1, which in the language of Bayesian statistics (see Subsection 1.2.4) means that the components of $\mathbf{X}$ are, conditional on $\Theta = \vartheta$, independent and identically distributed.

In this chapter we will first continue within this framework, but later (see Subsection 3.2.1 and further) we shall pass to a more general interpretation and we shall define the credibility estimator in a general set-up. We will also see that the credibility estimators can be understood as orthogonal projections in the Hilbert space of square integrable random variables, and we will prove some general characteristics and properties.

3.1 Credibility Estimators in a Simple Context

3.1.1 The Credibility Premium in a Simple Credibility Model

We consider the following simple credibility model:

Model Assumptions 3.1 (simple credibility model)

i) The random variables X_j ($j = 1, \ldots, n$) are, conditional on $\Theta = \vartheta$, independent with the same distribution function F_ϑ with the conditional moments

$$\mu(\vartheta) = E[X_j | \Theta = \vartheta],$$
$$\sigma^2(\vartheta) = \mathrm{Var}[X_j | \Theta = \vartheta].$$

ii) Θ is a random variable with distribution $U(\vartheta)$.

In this model we have

$$P^{\mathrm{ind}} = \mu(\Theta) = E[X_{n+1} | \Theta],$$
$$P^{\mathrm{coll}} = \mu_0 = \int_\Theta \mu(\vartheta)\, dU(\vartheta).$$

Our aim is again to find an estimator for the individual premium $\mu(\Theta)$, but now we concentrate on estimators, which are linear in the observations. We will denote the best estimator within this class by P^{cred} or $\widehat{\widehat{\mu(\Theta)}}$, which we are going to derive now.

By definition, $\widehat{\widehat{\mu(\Theta)}}$ has to be of the form

$$\widehat{\widehat{\mu(\Theta)}} = \widehat{a}_0 + \sum_{j=1}^{n} \widehat{a}_j X_j,$$

where the real coefficients $\widehat{a}_0, \widehat{a}_1, \ldots, \widehat{a}_n$ need to solve

$$E\left[\left(\mu(\Theta) - \widehat{a}_0 - \sum_{j=1}^{n} \widehat{a}_j X_j\right)^2\right] = \min_{a_0, a_1, \ldots, a_n \in \mathbb{R}} E\left[\left(\mu(\Theta) - a_0 - \sum_{j=1}^{n} a_j X_j\right)^2\right].$$

Since the probability distribution of $X_1, \ldots, X_n$ is invariant under permutations of X_j and $\widehat{\widehat{\mu(\Theta)}}$ is uniquely defined it must hold that

$$\widehat{a}_1 = \widehat{a}_2 = \cdots = \widehat{a}_n,$$

i.e. the estimator $\widehat{\widehat{\mu(\Theta)}}$ has the form

$$\widehat{\widehat{\mu(\Theta)}} = \widehat{a} + \widehat{b}\,\overline{X},$$

where

$$\overline{X} = \frac{1}{n}\sum_{j=1}^{n} X_j$$

and where $\widehat{a}$ and $\widehat{b}$ are the solution of the minimizing problem

$$E\left[\left(\mu(\Theta) - \widehat{a} - \widehat{b}\,\overline{X}\right)^2\right] = \min_{a,b\in\mathbb{R}} E\left[\left(\mu(\Theta) - a - b\,\overline{X}\right)^2\right].$$

Taking partial derivatives with respect to a, resp. b, we get

$$\begin{aligned}
&E\left[\mu(\Theta) - a - b\overline{X}\right] = 0, \\
&\operatorname{Cov}\left(\overline{X}, \mu(\Theta)\right) - b\operatorname{Var}\left(\overline{X}\right) = 0.
\end{aligned}$$

From the dependency structure imposed by Model Assumptions 3.1 we have

$$\begin{aligned}
\operatorname{Cov}\left(\overline{X}, \mu(\Theta)\right) &= \operatorname{Var}\left(\mu(\Theta)\right) =: \tau^2, \\
\operatorname{Var}\left(\overline{X}\right) &= \frac{E\left[\sigma^2(\Theta)\right]}{n} + \operatorname{Var}\left(\mu(\Theta)\right) =: \frac{\sigma^2}{n} + \tau^2,
\end{aligned}$$

from which we get

$$\begin{aligned}
b &= \frac{\tau^2}{\tau^2 + \sigma^2/n} = \frac{n}{n + \sigma^2/\tau^2}, \\
a &= (1-b)\,\mu_0.
\end{aligned}$$

We have thus proved the following theorem:

Theorem 3.2. *The credibility estimator under Model Assumptions 3.1 is given by*

$$\widehat{\widehat{\mu(\Theta)}} = \alpha\,\overline{X} + (1-\alpha)\,\mu_0, \tag{3.1}$$

where

$$\mu_0 = E[\mu(\Theta)], \tag{3.2}$$

$$\alpha = \frac{n}{n + \sigma^2/\tau^2}. \tag{3.3}$$

Remarks:

- In Theorem 3.2 we have found a formula for the credibility premium, the structure of which we have already seen for the Bayes premium in various specific cases, where the distribution of the X_j was assumed to belong to an exponential family (see Chapter 2).
- P^{cred} is a weighted mean of P^{coll} and the individual observed average $\overline{X}$.

- The quotient $\kappa = \sigma^2/\tau^2$ is called the *credibility coefficient,* which can also be written as $\kappa = (\sigma/\mu_0)^2 (\tau/\mu_0)^{-2}$. Note that τ/μ_0 is the coefficient of variation of $\mu(\Theta)$, which is a good measure for the heterogeneity of the portfolio, whereas $\sigma/\mu_0 = \sqrt{E[\text{Var}[X_j|\Theta]]}/E[X_j]$ is the expected standard deviation within risk divided by the overall expected value, which is a good measure for the within risk variability.
- The credibility weight α increases as
  - the number of observed years n increases,
  - the heterogeneity of the portfolio (as measured by the coefficient of variation τ/μ_0) increases,
  - the within risk variability (as measured by σ/μ_0) decreases.
- The formula for P^{cred} involves the structural parameters σ^2, τ^2 and μ_0. If there exists a collective composed of similar risks, these parameters can be estimated using the data from this collective (empirical Bayes procedure). In the next chapter we will propose estimators for these parameters in a more general model. The sizes of the structural parameters could also be intuitively "decided" using the a priori knowledge of an experienced actuary or underwriter (pure Bayesian procedure).
- The class of estimators that are linear in the observations is a subclass of the class of all estimators based on the observations. Hence the Bayes estimator is equal to the credibility estimator, if the former is linear. We refer to such cases as *exact credibility.*

3.1.2 A General Intuitive Principle

The credibility premium as a weighted mean of P^{coll} and $\overline{X}$ can also be interpreted as follows:

General Intuitive Principle 3.3

- $P^{\text{coll}} = \mu_0$ is the best estimator based on the a priori knowledge alone. It has the quadratic loss

$$E\left[(\mu_0 - \mu(\Theta))^2\right] = \text{Var}(\mu(\Theta)) = \tau^2.$$

- $\overline{X}$ is the best possible linear and individually unbiased (i.e. conditionally unbiased given Θ) estimator, based only on the observation vector $\mathbf{X}$. It has the quadratic loss

$$E\left[\left(\overline{X} - \mu(\Theta)\right)^2\right] = E\left[\sigma^2(\Theta)/n\right] = \sigma^2/n.$$

- P^{cred} is a weighted mean of these two, where the weights are *proportional to the inverse quadratic loss (precision)* associated with each of the two components, i.e.

$$\widehat{\widehat{\mu(\Theta)}} = \alpha \overline{X} + (1-\alpha)\mu_0,$$

$$\text{where } \alpha = \frac{n/\sigma^2}{n/\sigma^2 + 1/\tau^2} = \frac{\tau^2}{\tau^2 + \sigma^2/n} = \frac{n}{n + \sigma^2/\tau^2}.$$

This is a very intuitive principle. For the estimation of the pure risk premium, two sources of information are available, namely, a) the a priori knowledge and b) the individual observations. First, one looks and sees what one can learn from each of these two sources on its own. The a priori knowledge contains information about the collective and the best estimator that we can derive from this information is the a priori expectation μ_0. On the other hand, the observations contain information about the individual risk and the individual risk profile Θ. It is reasonable here to consider linear estimators that are individually, i.e. conditionally, unbiased and to choose from these the one with the greatest precision, i.e. the smallest variance. Finally, we have to weight the estimators derived from each of these information sources. It is intuitively reasonable to weight them according to their precision, respectively the inverse value of the quadratic loss.

This simple principle applies also in more general probability frameworks and will be a big help in finding the credibility estimator in the more complicated models.

3.1.3 The Quadratic Loss of the Credibility Premium

Theorem 3.4. *The quadratic loss of the credibility estimator* $\widehat{\widehat{\mu(\Theta)}}$ *found in Theorem 3.2 is given by*

$$E\left[\left(\widehat{\widehat{\mu(\Theta)}} - \mu(\Theta)\right)^2\right] = (1-\alpha)\,\tau^2 \tag{3.4}$$

$$= \alpha \frac{\sigma^2}{n}. \tag{3.5}$$

Remarks:

- Note that τ^2 is equal to the quadratic loss of P^{coll} (best estimator based only on the a priori knowledge and neglecting the individual claims experience) and that σ^2/n is equal to the quadratic loss of $\overline{X}$ (best estimator based only on the individual claims experience and neglecting the a priori knowledge). We can see from Theorem 3.4 that the quadratic loss of the credibility estimator is smaller than the quadratic loss of both P^{coll} and $\overline{X}$.
- We have already encountered these shrinkage formulae in Chapter 2, Propositions 2.9, 2.13, 2.15.

Proof of Theorem 3.4:
By straightforward calculations we find

$$\begin{aligned}
E\left[\left(\mu(\Theta)-\widehat{\widehat{\mu(\Theta)}}\right)^2\right] &= E\left[\left(\alpha\left(\mu(\Theta)-\overline{X}\right)+(1-\alpha)\left(\mu(\Theta)-\mu_0\right)\right)^2\right] \\
&= \alpha^2\frac{\sigma^2}{n}+(1-\alpha)^2\tau^2 \\
&= \left(\frac{n}{n+\sigma^2/\tau^2}\right)^2\frac{\sigma^2}{n}+\left(\frac{\sigma^2/\tau^2}{n+\sigma^2/\tau^2}\right)^2\tau^2 \\
&= \frac{\sigma^2}{n+\sigma^2/\tau^2}=\alpha\,\frac{\sigma^2}{n} \\
&= \frac{\sigma^2/\tau^2}{n+\sigma^2/\tau^2}\,\tau^2=(1-\alpha)\,\tau^2.
\end{aligned}$$

□

The following corollary gives an overview of the quadratic loss of the different estimators for $\mu(\Theta)$:

Corollary 3.5. *With Model Assumptions 3.1 we have:*

	Premium	*Quadratic loss*
a)	$P^{\text{coll}}=\mu_0,$	$\operatorname{Var}\left(\mu(\Theta)\right),$
b)	$P^{\text{cred}}=\widehat{\widehat{\mu(\Theta)}},$	$(1-\alpha)\operatorname{Var}\left(\mu(\Theta)\right),$
c)	$P^{\text{Bayes}}=\widetilde{\mu(\Theta)},$	$E\left[\operatorname{Var}\left(\mu(\Theta)\mid\mathbf{X}\right)\right].$

For the quadratic loss we have that *a)* $\geq$ *b)* $\geq$ *c)*. The improvement from *a)* to *b)* is, as a rule, considerable. The closer that the Bayes premium is to a function which is linear in the observations, the smaller is the improvement from *b)* to *c)*. The advantage of the credibility premium is its simplicity and the fact that, contrary to P^{Bayes}, we don't have to specify the family of kernel distributions and the prior distribution, a specification which would imply an additional model risk. The somewhat greater loss associated with *b)* over *c)* has therefore to be looked at in relative terms and is often acceptable.

3.1.4 The Simple Bühlmann Model and the Homogeneous Credibility Estimator

So far we have only considered one particular risk and we have derived the credibility estimator based only on the observations of this particular risk. In practice, however, one usually has observations of a whole portfolio of similar risks numbered $i = 1, 2, \ldots, I$.

We denote by $\mathbf{X}_i = (X_{i1}, X_{i2}, \ldots, X_{in})'$ the observation vector of risk i and by Θ_i its risk profile. We now assume that for each risk in the portfolio, Θ_i and $\mathbf{X}_i$ fulfil Model Assumptions 3.1 of the simple credibility model, i.e. Θ_i and $\mathbf{X}_i$ are the outcomes of the two-urn model described in Subsection 1.2.4 applied independently to all risks i. We then arrive at the simple Bühlmann model published in the seminal paper "Experience Rating and Credibility" [Büh67].

Model Assumptions 3.6 (simple Bühlmann model)

B1: The random variables X_{ij} ($j = 1, \ldots, n$) are, conditional on $\Theta_i = \vartheta$, independent with the same distribution function F_ϑ and the conditional moments

$$\mu(\vartheta) = E\left[X_{ij} \middle| \Theta_i = \vartheta\right],$$
$$\sigma^2(\vartheta) = \operatorname{Var}\left[X_{ij} \middle| \Theta_i = \vartheta\right].$$

B2: The pairs $(\Theta_1, \mathbf{X}_1), \ldots, (\Theta_I, \mathbf{X}_I)$ are independent and identically distributed.

Remark:

- By Assumption B2, a heterogeneous portfolio is modelled. The risk profiles $\Theta_1, \Theta_2, \ldots, \Theta_I$ are independent random variables drawn from the same urn with structural distribution $U(\vartheta)$. Hence the risks in the portfolio have different risk profiles (heterogeneity of the portfolio). On the other hand, the risks in the portfolio have something in common: a priori they are equal, i.e. a priori they cannot be recognized as being different.

We now want to find the credibility estimator in the simple Bühlmann model. The first question is: the credibility estimator of what? Of course, we want to estimate for each risk i its individual premium $\mu(\Theta_i)$. Hence there is not just one credibility estimator, but rather we want to find the credibility estimators $\widehat{\widehat{\mu(\Theta_i)}}$ of $\mu(\Theta_i)$ for $i = 1, 2, \ldots, I$. By definition, these credibility estimators $\widehat{\widehat{\mu(\Theta_i)}}$ have to be linear functions of the observations. But here the second question arises: linear in what? Should the credibility estimator of $\mu(\Theta_i)$ just be a linear function of the observations of risk i, or should we allow for linear functions of all observations in the portfolio? Hence there are always two items to be specified with credibility estimators: the quantity that we want to estimate and the statistics that the credibility estimator should be based on.

Generally, the credibility estimator is defined as the best estimator which is a linear function of all observations in the portfolio, i.e. the credibility estimator $\widehat{\widehat{\mu(\Theta_i)}}$ of $\mu(\Theta_i)$ is by definition the best estimator in the class

$$\left\{ \widehat{\mu\left(\Theta_i\right)} : \widehat{\mu\left(\Theta_i\right)} = a + \sum_{k=1}^{I} \sum_{j=1}^{n} b_{kj} X_{kj}; \quad a, b_{kj} \in \mathbb{R} \right\}.$$

We now want to derive the formula for $\widehat{\widehat{\mu\left(\Theta_i\right)}}$. By the same invariance and permutation argument as in the derivation of Theorem 3.2 we find that $\widehat{\widehat{\mu(\Theta_i)}}$ must be of the form

$$\widehat{\widehat{\mu(\Theta_i)}} = \widehat{a}_0^{(i)} + \sum_{k=1}^{I} \widehat{b}_k^{(i)} \overline{X}_k$$

$$\text{where } \overline{X}_k = \frac{1}{n} \sum_{j=1}^{n} X_{kj}.$$

To find the coefficients $\widehat{a}_0^{(i)}$ and $\widehat{b}_k^{(i)}$ we have to minimize

$$E\left[\left(\mu(\Theta_i) - a_0^{(i)} - \sum_{k=1}^{I} b_k^{(i)} \overline{X}_k\right)^2\right].$$

Taking partial derivatives with respect to $b_k^{(i)}$ and $a_0^{(i)}$ we find

$$\operatorname{Cov}\left(\mu\left(\Theta_i\right), \overline{X}_k\right) = b_k^{(i)} \operatorname{Var}\left[\overline{X}_k\right].$$

Since the left-hand side is equal to zero for $i \neq k$, it follows that $\widehat{b}_k^{(i)} = 0$ for $k \neq i$. Hence in this model the credibility estimator of $\mu\left(\Theta_i\right)$ depends only on $\overline{X}_i$, the observed mean of risk i, and not on the observations of the other risks in the collective. Therefore it coincides with the credibility estimator found in Theorem 3.2, i.e.

$$\widehat{\widehat{\mu(\Theta_i)}} = \alpha \overline{X}_i + (1 - \alpha) \mu_0, \tag{3.6}$$

where α is given by (3.3). But note the difference to Theorem 3.2: we have now proved that $\widehat{\widehat{\mu(\Theta_i)}}$ given by formula (3.6) is not only the credibility estimator based on $\mathbf{X}_i$, but also the credibility estimator based on all observations of the portfolio, i.e. based on the observation vector $\mathbf{X} = (\mathbf{X}_1', \mathbf{X}_2', \ldots, \mathbf{X}_I')'$.

Given a whole portfolio of risks, we can also consider another type of credibility estimator, which is referred to in the literature as the *homogeneous credibility estimator*. We define the *homogeneous credibility estimator* of $\mu\left(\Theta_i\right)$ as the best estimator in the class of collectively unbiased estimators

$$\left\{ \widehat{\mu\left(\Theta_i\right)} : \widehat{\mu\left(\Theta_i\right)} = \sum_{k=1}^{I} \sum_{j=1}^{n} b_{kj} X_{kj}, \quad E\left[\widehat{\mu\left(\Theta_i\right)}\right] = E\left[\mu\left(\Theta_i\right)\right], b_{kj} \in \mathbb{R} \right\}.$$

We denote this estimator by $\widehat{\widehat{\mu(\Theta_i)}}^{\text{hom}}$. Contrary to $\widehat{\widehat{\mu(\Theta_i)}}$, the homogeneous estimator $\widehat{\widehat{\mu(\Theta_i)}}^{\text{hom}}$ does not contain a constant term, but it is required that $\widehat{\widehat{\mu(\Theta_i)}}^{\text{hom}}$ be unbiased over the collective, a condition which is automatically fulfilled for the inhomogeneous credibility estimator.

To find $\widehat{\widehat{\mu(\Theta_i)}}^{\text{hom}}$ we have to solve the minimization problem

$$E\left[\left(\mu\left(\Theta_i\right)-\sum_{k=1}^{I}\sum_{j=1}^{n}b_{kj}X_{kj}\right)^2\right]=min! \tag{3.7}$$

under the side constraint

$$\sum_{k=1}^{I}\sum_{l=1}^{n}b_{kl}=1. \tag{3.8}$$

The side constraint follows from the unbiasedness requirement

$$E\left[\widehat{\widehat{\mu(\Theta_i)}}^{\text{hom}}\right]=E\left[\mu\left(\Theta_i\right)\right]$$

and the fact that

$$E\left[\mu\left(\Theta_i\right)\right]=E\left[X_{kj}\right]=\mu_0 \quad \text{for all } i,j,k.$$

Instead of formally solving (3.7), we take a closer look at $\widehat{\widehat{\mu(\Theta_i)}}$ given by (3.6). In order to transfer this (inhomogeneous) credibility estimator to a homogeneous one, we have to replace the constant term $(1-\alpha)\,\mu_0$ in (3.6) by a suitable linear function of all observations. It is very natural to replace the overall expected value μ_0 by the observed collective mean

$$\overline{X}=\frac{1}{I\,n}\sum_{i=1}^{I}\sum_{j=1}^{n}X_{ij}$$

in the portfolio. One can easily check that the resulting estimator is indeed the solution of the minimizing problem (3.7) and hence the homogeneous credibility estimator. Another formal proof for this statement can also be found in the next chapter: take all weights equal to one in Theorem 4.4. The following theorem summarizes the results found in this subsection:

Theorem 3.7. *The (inhomogeneous) credibility estimator* $\widehat{\widehat{\mu(\Theta_i)}}$ *and the homogeneous credibility estimator* $\widehat{\widehat{\mu(\Theta_i)}}^{\text{hom}}$ *in the simple Bühlmann model (Model Assumptions 3.1) are given by*

$$\widehat{\widehat{\mu(\Theta_i)}} = \alpha\,\overline{X}_i + (1-\alpha)\,\mu_0, \tag{3.9}$$

$$\widehat{\widehat{\mu(\Theta_i)}}^{\text{hom}} = \alpha\,\overline{X}_i + (1-\alpha)\,\overline{X}, \tag{3.10}$$

where

$$\alpha = \frac{n}{n+\frac{\sigma^2}{\tau^2}},$$

$$\overline{X}_i = \frac{1}{n}\sum_{j=1}^{n} X_{ij},$$

$$\overline{X} = \frac{1}{I\,n}\sum_{i=1}^{I}\sum_{j=1}^{n} X_{ij}.$$

Remark:

- Note that the homogeneous credibility estimator (3.10) contains a built-in estimator for the overall mean μ_0.

3.2 Credibility Estimators in a General Set-Up

We have already seen in the simple Bühlmann model (Subsection 3.1.4) that there are always two items to be specified with credibility estimators:

i) The credibility estimator of what? What quantity do we want to estimate?
ii) What statistics should the credibility estimator be based on? Credibility estimators are linear estimators, but linear in what?

Usually the quantity to be estimated will be the individual premium in some future period of a particular risk, which we denote by $\mu(\Theta)$. For instance, if we consider the particular risk i in the simple Bühlmann model, then this quantity is $\mu(\Theta) = \mu(\Theta_i) = E[X_{i,n+1}\,|\Theta_i]$.

The data available to the insurer are usually some claims data of a portfolio of similar risks. Therefore, the underlying statistics for the credibility estimator will usually be the vector $\mathbf{X}$ containing the observations from a whole portfolio of risks (cf. simple Bühlmann model, Subsection 3.1.4).

In the simple Bühlmann model, the quantity to be estimated and the observation vector $\mathbf{X}$ were well defined, and their probability structure was specified by Model Assumptions 3.6.

In the *general set-up*, the task remains the same, namely to find the credibility estimator of $\mu(\Theta)$ based on some observation vector $\mathbf{X}$. However, we do not define $\mu(\Theta)$ and $\mathbf{X}$ exactly, nor do we specify their probability structure. Hence the *only mathematical structure in the general set-up* is that we want to estimate some unknown real-valued random variable $\mu(\Theta)$ based on some known random vector $\mathbf{X} = (X_1, \ldots, X_n)'$.

Remarks:

- Other interpretations of $\mu(\Theta)$ (quantity to be estimated) than the one given above will occur occasionally (see e.g. Section 4.14).
- We will also encounter examples where the elements of the vector $\mathbf{X}$ are not the original claim observations but rather transformations of the original data or where the elements of $\mathbf{X}$ are maximum-likelihood estimators.
- To make precise exactly what the quantity to be estimated and the vector $\mathbf{X}$ are, as well as to specify their probability structure, will be the content of the specific credibility models treated in subsequent chapters.
- The mathematical structure of the general set-up also allows us to replace $\mu(\Theta)$ by any other square integrable random variable Y (most general set-up). Indeed, all results in this and the following two subsections are still valid if we replace $\mu(\Theta)$ by Y.

Credibility estimators are always the best estimators in an a priori given class. Given the vector $\mathbf{X}$ we define below two such classes $L(\mathbf{X}, \mathbf{1})$ and $L_e(\mathbf{X})$ and their corresponding credibility estimators.

Definition 3.8. *The credibility estimator of $\mu(\Theta)$ based on $\mathbf{X}$ is the best possible estimator in the class*

$$L(\mathbf{X}, 1) := \left\{\widehat{\mu(\Theta)} : \widehat{\mu(\Theta)} = a_0 + \sum_j a_j X_j, \quad a_0, a_1, \dots \in \mathbb{R}\right\}.$$

In the following we will use P^{cred} or $\widehat{\widehat{\mu(\Theta)}}$ to denote this estimator. We therefore have

$$P^{\text{cred}} = \widehat{\widehat{\mu(\Theta)}} = \underset{\widehat{\mu(\Theta)} \in L(\mathbf{X},1)}{\arg\min} \; E\left[\left(\widehat{\mu(\Theta)} - \mu(\Theta)\right)^2\right]. \tag{3.11}$$

In order to make a clear distinction from the homogeneous credibility estimator (defined below), this estimator is often referred to in the literature as the *inhomogeneous* credibility estimator.

Definition 3.9. *The homogeneous credibility estimator of $\mu(\Theta)$ based on $\mathbf{X}$ is the best possible estimator in the class of collectively unbiased estimators*

$$L_e(\mathbf{X}) := \left\{\widehat{\mu(\Theta)} : \widehat{\mu(\Theta)} = \sum a_j X_j, \; a_1, a_2, \dots \in \mathbb{R}, \; E\left[(\widehat{\mu(\Theta)}\right] = E[\mu(\Theta)]\right\}.$$

In the following we will use $P^{\text{cred}_{\text{hom}}}$ or $\widehat{\widehat{\mu(\Theta)}}^{\text{hom}}$ to denote this estimator. Note that, in contrast to the inhomogeneous estimator, it has no constant term a_0, and that the definition requires that it be unbiased over the collective. However, it is *not* required that the estimator be unbiased for any individual value of ϑ. Formally we write

$$P^{\text{cred}_{\text{hom}}} = \widehat{\widehat{\mu(\Theta)}}^{\text{hom}} = \underset{\widehat{\mu(\Theta)} \in L_e(\mathbf{X})}{\arg\min} \; E\left[\left(\widehat{\mu(\Theta)} - \mu(\Theta)\right)^2\right]. \tag{3.12}$$

Whereas Definition 3.8 poses no problems, one has to be careful with Definition 3.9 for the following reasons:

- The class $L_e(\mathbf{X})$ in Definition 3.9 is the class of all collectively unbiased estimators for $\mu(\Theta)$. More explicitly this means that the unbiasedness condition should be read as follows:

$$E_P\left[\widehat{\mu(\Theta)}\right] = E_P\left[\mu(\Theta)\right]$$

  for all model consistent probability measures

$$dP(\vartheta, \mathbf{x}) = dF_\vartheta(\mathbf{x})\, dU(\vartheta).$$

  It follows that the estimator $\widehat{\widehat{\mu(\Theta)}}^{\text{hom}}$ may *not* depend on P. In particular, if

$$\widehat{\widehat{\mu(\Theta)}}^{\text{hom}} = \sum \widehat{a}_j X_j,$$

  then the coefficients $\widehat{a}_j$ are not allowed to depend on

$$\mu_0 = E_P\left[\mu(\Theta)\right].$$

- It may of course happen that no collectively unbiased estimator exists. Then $L_e(\mathbf{X})$ is empty, and no homogeneous credibility estimator as defined in Definition 3.9 exists.
- In later chapters we will also consider situations where $\mu(\Theta)$ is a linear function of other variables, i.e.

$$\mu(\Theta) = \sum_{l=1}^{p} y_l\, \beta_l(\Theta).$$

  Then (see Exercise 3.1)

$$\widehat{\widehat{\mu(\Theta)}}^{\text{hom}} = \sum_{l=1}^{p} y_l \widehat{\widehat{\beta_l(\Theta)}}^{\text{hom}},$$

  i.e. we can find the homogeneous estimator of $\mu(\Theta)$ from the homogeneous estimators of $\beta_l(\Theta)$ $l = 1, 2, \ldots, p$.
- The homogeneous credibility estimator only makes sense if the observation vector $\mathbf{X}$ incorporates collateral data, i.e. data of a portfolio of similar risks (and not only data of one particular risk).
- Forcing the constant term to be zero, together with the collectively unbiasedness condition, automatically implies a built-in estimator of μ_0, which is the main reason for considering the homogeneous credibility estimator.

- The above Definition 3.9 of the homogeneous credibility estimator is different to the one usually found in the literature (see e.g. [DKG96], [DV76b], [DV81a], [Hac75], [GH87]). Often it is only required that

$$E\left[\widehat{(\mu(\Theta)}\right] = E[\mu(\Theta)],$$

but not that this equation should be fulfilled whatever the value of $E[\mu(\Theta)]$ or whatever the probability measure $dP(\vartheta, \mathbf{x})$ might be. But then the coefficients $\widehat{a}_j$ of the homogeneous estimator defined in such a way may depend on the overall mean μ_0. However, a homogeneous estimator depending on μ_0 is not of much use. If μ_0 is known, we would better use the inhomogeneous estimator. For a more thorough discussion on homogeneous credibility estimators we refer to Sundt [Sun98].

3.2.1 Credibility Estimators as Orthogonal Projections in the $\mathcal{L}^2$ Hilbert Space

The estimators $\widehat{\widehat{\mu(\Theta)}}$ and $\widehat{\widehat{\mu(\Theta)}}^{\text{hom}}$, defined in (3.11) and (3.12) as a solution to the least squares problem, are most elegantly understood as projections in the Hilbert space of all square integrable functions $\mathcal{L}^2$. One of the first to have this idea in the context of credibility was F. De Vylder [DV76a]. A good introduction into Hilbert spaces can be found in Kolmogorov and Fomin [KF70].

However, in the following we will not assume a special knowledge of the theory of Hilbert spaces, so that the derivations and proofs can also be understood by the reader who does not have an extensive mathematical background. Of course, the least squares problems could always be solved by purely computational techniques. As we shall see, this involves typically the inversion of a large matrix. As soon as one goes beyond very simple model structures this matrix inversion is not only cumbersome, but entails the loss of intuitive insight for the resulting credibility formulae. The advantage of applying Hilbert space theory to credibility is that we can then immediately apply our intuitive understanding of the properties of linear vector spaces to the credibility problem. In this way, certain features of credibility theory come much closer to our intuition.

A concise summary of those properties of Hilbert space and in particular of $\mathcal{L}^2$, which are necessary for the understanding of credibility theory, is given in Appendix B. We quickly review here those points that we will need in this section.

We define

$$\mathcal{L}^2 := \left\{ X : \ X = \text{random variable with } E[X^2] = \int X^2 dP < \infty \right\}.$$

Two random variables X and X' with $E\left[(X - X')^2\right] = 0$ will be considered to be identical: in other words, we do not differentiate between random variables differing from each other only on a set of measure 0. If X and Y belong to $\mathcal{L}^2$ then the *inner product* associated with the Hilbert space is given by

$$< X, Y >:= E[XY]$$

and the corresponding *norm* is

$$\|X\| :=< X, X >^{1/2}= E[X^2]^{1/2}.$$

We consider the random variables

$$\begin{aligned} &\mu(\Theta) \in \mathcal{L}^2 && \text{the individual premium to be estimated,} \\ &\mathbf{X}' = (X_1, X_2, \ldots, X_n) && \text{the observation vector with elements } X_j \in \mathcal{L}^2, \end{aligned}$$

as elements of the Hilbert space $\mathcal{L}^2$ with the corresponding moments

$$\begin{aligned} \mu_0 &: = E\left[\mu(\Theta)\right], \\ \boldsymbol{\mu}_{\mathbf{X}}' &: = (\mu_{X_1}, \ldots, \mu_{X_n}) = (E\left[X_1\right], \ldots, E\left[X_n\right]) \end{aligned}$$

and the covariance matrix of $\mathbf{X}$

$$\Sigma_{\mathbf{X}} := \operatorname{Cov}(\mathbf{X}, \mathbf{X}')$$

$$= \begin{pmatrix} \operatorname{Var}[X_1] & \operatorname{Cov}(X_1, X_2) \ldots & \operatorname{Cov}(X_1, X_n) \\ \operatorname{Cov}(X_2, X_1) & \operatorname{Var}[X_2] & \operatorname{Cov}(X_2, X_n) \\ \vdots & \ddots & \vdots \\ \vdots & & \vdots \\ \operatorname{Cov}(X_n, X_1) & \ldots \quad \ldots & \operatorname{Var}[X_n] \end{pmatrix}.$$

Definition 3.10.

i) Two elements X and $Y \in \mathcal{L}^2$ are **orthogonal** *($X \perp Y$), if the inner product $<X, Y>$ equals 0.*

ii) For a closed subspace (or a closed affine subspace) $M \subset \mathcal{L}^2$ we define the orthogonal projection of $Y \in \mathcal{L}^2$ on M as follows: $Y^* \in M$ is the ***orthogonal projection of** Y **on** M ($Y^* = \operatorname{Pro}(Y|\, M)$) if $Y - Y^* \perp M$, i.e. $Y - Y^* \perp Z_1 - Z_2$ for all $Z_1, Z_2 \in M$.*

One can prove that the random variable Y^* always exists and is unique.

Theorem 3.11. *The following statements are equivalent:*

$$i)\quad Y^* = \operatorname{Pro}(Y|\, M). \tag{3.13}$$

$$ii)\quad Y^* \in M \text{ and } <Y - Y^*, Z - Y^*> = 0 \quad \text{for all } Z \in M. \tag{3.14}$$

$$iii)\quad Y^* \in M \text{ and } \|Y - Y^*\| \leq \|Y - Z\| \quad \text{for all } Z \in M. \tag{3.15}$$

Remarks:

- We call (3.14) the orthogonality condition.
- If M is a subspace, then the condition (3.14) can also be written as

$$<Y - Y^*, Z> = 0 \text{ for all } Z \in M. \qquad (3.16)$$

- Remember the definition of the following spaces (Definitions 3.8 and 3.9):

$$L(\mathbf{X}, 1) := \left\{\widehat{\mu(\Theta)} : \widehat{\mu(\Theta)} = a_0 + \sum a_j X_j, \quad a_0, a_1, \dots \in \mathbb{R}\right\},$$

$$L_e(\mathbf{X}) := \left\{\widehat{\mu(\Theta)} : \widehat{\mu(\Theta)} = \sum a_j X_j, \ a_1, a_2, \dots \in \mathbb{R}, \ E\left[\widehat{\mu(\Theta)}\right] = E[\mu(\Theta)]\right\}.$$

We further introduce

$$G(\mathbf{X}) := \{Z : \ Z = g(\mathbf{X}), g = \text{real-valued function and } g(\mathbf{X}) \text{ square integrable}\}.$$

$L(\mathbf{X}, 1)$ and $G(\mathbf{X})$ are closed subspaces, $L_e(\mathbf{X})$ is a closed affine subspace of $\mathcal{L}^2$.

Using this notation and these results, we can now reformulate the definition of the credibility estimator as well as that of the Bayes premium. The following alternative definitions of the credibility estimators and the Bayes premium are direct consequences of Theorem 3.2.1 and Definitions 3.8 and 3.9.

Definition 3.12 (credibility estimators as orthogonal projections).

i) The (inhomogeneous) credibility estimator of $\mu(\Theta)$ based on $\mathbf{X}$ is defined as

$$\widehat{\widehat{\mu(\Theta)}} = \operatorname{Pro}(\mu(\Theta)| \, L(\mathbf{X}, 1)). \qquad (3.17)$$

ii) The homogeneous credibility estimator of $\mu(\Theta)$ based on $\mathbf{X}$ is defined as

$$\widehat{\widehat{\mu(\Theta)}}^{\text{hom}} = \operatorname{Pro}(\mu(\Theta)| \, L_e(\mathbf{X})). \qquad (3.18)$$

iii) The Bayes premium (of $\mu(\Theta)$ based on $\mathbf{X}$) is defined as

$$\widetilde{\mu(\Theta)} = \operatorname{Pro}(\mu(\Theta)| \, G(\mathbf{X})). \qquad (3.19)$$

We thus interpret both the credibility estimator and the Bayes premium as *orthogonal projections* of the (unknown) individual premium $\mu(\Theta)$ on appropriately defined subspaces (respectively affine subspaces) of $\mathcal{L}^2$.

A property of the projection operator which is generally valid, and one which is easy to understand in the case of linear vector spaces, is given here without proof. This result will be used frequently in the rest of this section and in later chapters.

Theorem 3.13 (iterativity of projections). *Let M and M' be closed subspaces (or closed affine subspaces) of $\mathcal{L}^2$ with $M \subset M'$, then we have*

$$\operatorname{Pro}(Y\,|M) = \operatorname{Pro}(\operatorname{Pro}(Y\,|M')\,|M) \tag{3.20}$$

and

$$\begin{aligned}\|Y - \operatorname{Pro}(Y\,|M)\|^2 &= \|Y - \operatorname{Pro}(Y \mid M')\|^2 \\ &\quad + \|\operatorname{Pro}(Y\,|M') - \operatorname{Pro}(Y\,|M)\|^2 .\end{aligned} \tag{3.21}$$

Note, that geometrically (3.21) corresponds to the Theorem of Pythagoras.

To finish this section we state a result about the relationship between the credibility estimators and the Bayes premium.

Theorem 3.14. *It holds that:*

i) The credibility premium $\widehat{\widehat{\mu(\Theta)}}$ is the best linear approximation to the Bayes premium $\widetilde{\mu(\Theta)}$.

ii)

$$\underbrace{E\left[(\widehat{\widehat{\mu(\Theta)}} - \mu(\Theta))^2\right]}_{''total\ error''} = \underbrace{E\left[(\widehat{\widehat{\mu(\Theta)}} - \widetilde{\mu(\Theta)})^2\right]}_{''approximation\ error''} + \underbrace{E\left[(\widetilde{\mu(\Theta)} - \mu(\Theta))^2\right]}_{''Bayes\ risk''}.$$

iii) The homogeneous credibility premium $\widehat{\widehat{\mu(\Theta)}}^{\text{hom}}$ is the best homogeneous linear approximation to the credibility premium $\widehat{\widehat{\mu(\Theta)}}$ as well as to the Bayes premium $\widetilde{\mu(\Theta)}$.

Proof: $G(\mathbf{X})$ and $L(\mathbf{X},1)$ are subspaces of $\mathcal{L}^2$ and $L(\mathbf{X},1) \subset G(\mathbf{X})$. From Definition 3.12 and Theorem 3.13 it follows directly that

$$\begin{aligned}\widehat{\widehat{\mu(\Theta)}} &= \operatorname{Pro}(\mu(\Theta)|\, L(\mathbf{X},\ 1)) \\ &= \operatorname{Pro}(\underbrace{\operatorname{Pro}\left(\mu(\Theta)|\, G(\mathbf{X})\right)}_{=\widetilde{\mu(\Theta)}}|L(\mathbf{X},\ 1)),\end{aligned}$$

so that i) is proved. From Theorem 3.13 it then also follows that

$$\left\|\widehat{\widehat{\mu(\Theta)}} - \mu(\Theta)\right\|^2 = \left\|\mu(\Theta) - \widetilde{\mu(\Theta)}\right\|^2 + \left\|\widetilde{\mu(\Theta)} - \widehat{\widehat{\mu(\Theta)}}\right\|^2 .$$

iii) follows analogously to i) and from the fact that $L_e(\mathbf{X})$ is a subspace of $L(\mathbf{X},\ 1)$. This proves Theorem 3.14. □

3.3 Orthogonality Conditions and Normal Equations

Theorem 3.15 (orthogonality conditions for the credibility estimator). *$\widehat{\mu(\Theta)}$ is the (inhomogeneous) credibility estimator for $\mu(\Theta)$ based on $\mathbf{X}$, if and only if the following orthogonality conditions are satisfied:*

$$i) \; < \mu(\Theta) - \widehat{\mu(\Theta)}, 1> \; = 0, \tag{3.22}$$

$$ii) \; < \mu(\Theta) - \widehat{\mu(\Theta)}, X_j> \; = 0 \quad \textit{for } j = 1, 2, \ldots, n. \tag{3.23}$$

Theorem 3.16 (orthogonality conditions for the homogeneous credibility estimator). *$\widehat{\mu(\Theta)}^{\mathrm{hom}}$ is the homogeneous credibility estimator of $\mu(\Theta)$ based on $\mathbf{X}$, if and only if the following orthogonality conditions are satisfied:*

$$i) <\mu(\Theta) - \widehat{\mu(\Theta)}^{\mathrm{hom}}, 1> \; = 0, \tag{3.24}$$

$$ii) <\mu(\Theta) - \widehat{\mu(\Theta)}^{\mathrm{hom}}, \widehat{\mu(\Theta)} - \widehat{\mu(\Theta)}^{\mathrm{hom}} > \; = 0 \quad \textit{for all } \widehat{\mu(\Theta)} \in L_e(\mathbf{X}). \tag{3.25}$$

Remark:

- The credibility estimator and the homogeneous credibility estimator are both unbiased in the collective (this follows from (3.22) and (3.24)). For the homogeneous credibility estimator, however, the collective unbiasedness, i.e. the condition (3.24), is forced by the definition, while this follows automatically for the (inhomogeneous) credibility estimator.

Proof of Theorem 3.15:
Since $\widehat{\mu(\Theta)} = \mathrm{Pro}(\mu(\Theta) | \, L(\mathbf{X}, 1))$ we can apply Theorem 3.2, formula (3.14). Observe that it is sufficient to check the validity of (3.14) for $1, X_1, \ldots, X_n$ which form a basis of $L(\mathbf{X}, 1)$. □

Proof of Theorem 3.16:
Analogously as for Theorem 3.15. □

The above orthogonality conditions can also be written in a somewhat different form, which are called the *normal equations* in the literature and which play a central role in abstract least squares theory.

Corollary 3.17 (normal equations). *$\widehat{\mu(\Theta)}$ is the (inhomogeneous) credibility estimator of $\mu(\Theta)$ based on $\mathbf{X}$, if and only if the following normal equations are satisfied:*

$$i)\ E\left[\widehat{\widehat{\mu(\Theta)}}-\mu(\Theta)\right]=0, \tag{3.26}$$

$$ii)\ \operatorname{Cov}\left[\mu(\Theta),X_j\right]=\operatorname{Cov}\left[\widehat{\widehat{\mu(\Theta)}},X_j\right] \quad \textit{for } j=1,2,\ldots,n. \tag{3.27}$$

Remarks:

- Let $\widehat{a}_0$ and $\widehat{\mathbf{a}}=(\widehat{a}_1,\ldots,\widehat{a}_n)'$ be the coefficients of the credibility estimator $\widehat{\widehat{\mu(\Theta)}}$, that is
$$\widehat{\widehat{\mu(\Theta)}}=\widehat{a}_0+\sum_{j=1}^{n}\widehat{a}_jX_j.$$
Then the normal equations (3.26) and (3.27) can be written as
$$i)\quad \widehat{a}_0=\mu_0-\sum_{j=1}^{n}\widehat{a}_j\,\mu_{X_j}, \tag{3.28}$$
$$ii)\quad \sum_{j=1}^{n}\widehat{a}_j\operatorname{Cov}(X_j,X_k)=\operatorname{Cov}(\mu(\Theta),X_k), \quad k=1,\ldots,n. \tag{3.29}$$
(3.28) and (3.29) explain why we use the term "normal equations" for Corollary 3.17.
- As $\widehat{\widehat{\mu(\Theta)}}$ is by definition a linear combination of $X_1,\ldots,X_n$, it follows from (3.27), that
$$\operatorname{Var}\left[\widehat{\widehat{\mu(\Theta)}}\right]=\operatorname{Cov}\left(\mu(\Theta),\widehat{\widehat{\mu(\Theta)}}\right).$$
- The normal equations imply that the credibility estimator depends only on the first two moments of the joint distribution of $\mu(\Theta)$ and $\mathbf{X}$. We do not need to know the full distributions $\{F_\vartheta:\vartheta\in\Theta\}$ or the a priori distribution $\mathcal{U}(\vartheta)$. This is an enormous advantage in practice. For credibility one only needs to know the first- and second-order moments (or estimates of them).
- The normal equations hold very generally. We have not assumed any special structure, not even a Bayes structure with an a priori distribution and conditional distributions. The result is also true when instead of $\mu(\Theta)$, some other random variable from $\mathcal{L}^2$ is to be predicted, and also, when the X_j's are any random variables from $\mathcal{L}^2$.
- From the normal equations follows that the "residual" $\widehat{\widehat{\mu(\Theta)}}-\mu(\Theta)$ is uncorrelated with the components of the observation vector $\mathbf{X}$.

Proof of Corollary 3.17: From Theorem 3.15 it follows that

$$<\widehat{\widehat{\mu(\Theta)}}-\mu(\Theta),1>=0\Longleftrightarrow E\left[\widehat{\widehat{\mu(\Theta)}}-\mu(\Theta)\right]=0.$$

For the proof of (3.27) we have

$$
\begin{aligned}
&<\widehat{\widehat{\mu(\Theta)}} - \mu(\Theta), X_j> = 0 \\
&\qquad \Longleftrightarrow \quad <\widehat{\widehat{\mu(\Theta)}} - \mu(\Theta), X_j - \mu_{X_j}> = 0 \\
&\qquad \Longleftrightarrow \quad <(\widehat{\widehat{\mu(\Theta)}} - \mu_0) - (\mu(\Theta) - \mu_0), X_j - \mu_{X_j}> = 0 \\
&\qquad \Longleftrightarrow \quad \mathrm{Cov}(\widehat{\widehat{\mu(\Theta)}}, X_j) = \mathrm{Cov}(\mu(\Theta), X_j).
\end{aligned}
$$

This ends the proof of Corollary 3.17. □

Corollary 3.18. *If the covariance matrix $\Sigma_{\mathbf{X}}$ is non-singular, then it follows that*

$$
\begin{aligned}
&\widehat{\widehat{\mu(\Theta)}} = \mu_0 + \mathbf{c}^{'} \, \Sigma_{\mathbf{X}}^{-1} \, (\mathbf{X} - \boldsymbol{\mu}_{\mathbf{X}}), \\
&\textit{where } \; \mathbf{c}^{'} = (\mathrm{Cov}\,(\mu(\Theta), \mathbf{X}_1)\,, \ldots, \mathrm{Cov}(\mu(\Theta), \; \mathbf{X}_n))\,.
\end{aligned}
$$

Proof: Let $\widehat{\widehat{\mu(\Theta)}} = \widehat{a}_0 + \widehat{\mathbf{a}}' \, \mathbf{X}$ be the credibility estimator. Then condition ii) of the normal equations written in matrix notation is

$$\widehat{\mathbf{a}}^{'} \, \Sigma_{\mathbf{X}} = \mathbf{c}^{'}.$$

For $\Sigma_{\mathbf{X}}$ non-singular we get $\widehat{\mathbf{a}}^{'} = \mathbf{c}^{'} \Sigma_{\mathbf{X}}^{-1}$. This expression substituted in condition i) of the normal equations gives $\widehat{a}_0 = \mu_0 - \mathbf{c}^{'} \Sigma_{\mathbf{X}}^{-1} \boldsymbol{\mu}_{\mathbf{X}}$. This proves Corollary 3.18. □

Corollary 3.19 (normal equations for the homogeneous credibility estimator).
$\widehat{\widehat{\mu(\Theta)}}^{\mathrm{hom}}$ *is the homogeneous credibility estimator of $\mu(\Theta)$ based on $\mathbf{X}$, if and only if the following equations are satisfied:*

$$
\begin{aligned}
&i) \;\; E\left[\widehat{\widehat{\mu(\Theta)}}^{\mathrm{hom}} - \mu(\Theta)\right] = 0, \\
&ii) \;\; \mathrm{Cov}\left(\widehat{\widehat{\mu(\Theta)}}^{\mathrm{hom}} - \mu(\Theta), \widehat{\widehat{\mu(\Theta)}}^{\mathrm{hom}}\right) = \mathrm{Cov}\left(\widehat{\widehat{\mu(\Theta)}}^{\mathrm{hom}} - \mu(\Theta), \widehat{\mu(\Theta)}\right) \\
&\qquad\qquad\qquad\qquad\qquad\qquad\qquad\qquad \textit{for all } \widehat{\mu(\Theta)} \in L_e(\mathbf{X})\,.
\end{aligned}
$$

We state this corollary without proof, which is similar to the reasoning for Corollary 3.17.

Summary Discussion:
This section has provided us with a wide range of tools for finding the credibility formulae. Corollary 3.17 gives the computational approach. As stated earlier, usually we shall not follow this route. Instead, our derivation of credibility formulae follows the more intuitive way:

i) Try to identify a credibility formula by intuitive reasoning, which will often be based on the General Intuitive Principle 3.3.
ii) Check the correctness of such formulae by verifying that they fulfil the orthogonality conditions or the normal equations (Theorem 3.15 or Corollary 3.17).

In the following chapters we shall discuss specific model assumptions and the corresponding credibility formulae.

3.4 Exercises

Exercise 3.1 (linearity property)

Let

$$\mu(\Theta) = \sum_{k=1}^{p} a_k \mu_k(\Theta) \tag{3.30}$$

and let $\widehat{\widehat{\mu_k(\Theta)}}$ $\left(\text{resp. } \widehat{\widehat{\mu_k(\Theta)}}^{\text{hom}} \right)$ be the credibility estimators (resp. the homogeneous credibility estimators) of $\mu_k(\Theta)$ based on $\mathbf{X} = (X_1, X_2, ..., X_n)'$. Show that the following linearity properties are satisfied:

a) linearity property for the credibility estimator:

$$\widehat{\widehat{\mu(\Theta)}} = \sum_{k=1}^{p} a_k \, \widehat{\widehat{\mu_k(\Theta)}}. \tag{3.31}$$

b) linearity property for the homogeneous credibility estimator:

$$\widehat{\widehat{\mu(\Theta)}}^{\text{hom}} = \sum_{k=1}^{p} a_k \, \widehat{\widehat{\mu_k(\Theta)}}^{\text{hom}}.$$

Remark:
We will encounter situations with (3.30) in the multidimensional and in the regression case (Chapters 7 and 8), where the quantities to be estimated are linear functions of other quantities.

Exercise 3.2 (independence property)

Let $\mu(\Theta)$ be the quantity to be estimated and $\mathbf{X} = (\mathbf{X}_1', \mathbf{X}_2')'$, where $\mathbf{X}_2$ is independent of $\mathbf{X}_1$ and $\mu(\Theta)$. Show that the credibility estimator $\widehat{\widehat{\mu(\Theta)}}$ depends on $\mathbf{X}$ only through $\mathbf{X}_1$.

Exercise 3.3

Consider the simple credibility model (Model Assumptions 3.1) and denote by $\widehat{\widehat{X}}_{n+1}$ the best linear predictor of X_{n+1} with respect to quadratic loss. Show that

a)
$$\widehat{\widehat{X}}_{n+1} = \widehat{\widehat{\mu(\Theta)}},$$

b)
$$E\left[\left(\widehat{\widehat{X}}_{n+1} - X_{n+1}\right)^2\right] = \sigma^2\left(1 + \frac{\alpha}{n}\right).$$

Exercise 3.4

Consider the situation of Exercise 2.1. Calculate the credibility premium P^{cred} for all possible observations:

a) if there is only one year's claim experience;
b) if there are two years' claim experience.
c) Compare with the results with P^{Bayes} obtained in Exercise 2.1 and comment on the findings.

Exercise 3.5

In many lines of business, a small number of "large claims" are responsible for more than half of the total claims load. In this context an actuary wants to estimate for a particular risk the probability Θ that a (randomly drawn) claim is a large claim.

From portfolio information he knows that about 2% of the claims are large claims, but that this proportion varies quite a lot between the risks. Based on this portfolio information he makes the following a priori assumptions:

$$E[\Theta] = 2\%, \qquad \sqrt{\operatorname{Var}(\Theta)} = 1\%.$$

For a particular risk, there were six large claims observed out of a total number of 120 observed claims. Determine the credibility estimator $\widehat{\widehat{\Theta}}$ of Θ for that particular risk based on this information.

4

The Bühlmann–Straub Model

This model, which was developed by Bühlmann and Straub in 1970 (see [BS70]) in connection with determining claims ratios in reinsurance, has over time found a multitude of applications in insurance practice: in life and non-life insurance, in primary insurance as well as in reinsurance. It is still by far the most used and the most important credibility model for insurance practice.

4.1 Motivation

We consider the situation where an actuary has to rate a risk in motor third-party liability insurance. On the basis of certain risk characteristics, the risks have been grouped into various risk classes (tariff positions) and now a risk premium must be calculated for each of these classes.

Statistical information from this branch is available. Typically, for the ith risk class we have the following information:

S_{ij}	aggregate claim amount in year j,
V_{ij}	number of years at risk in year j (contracts which were in force for the whole year count as one, the others pro rata temporis),
$X_{ij} = S_{ij}/V_{ij}$	average claim costs per year at risk in year j (claims ratio),
N_{ij}	number of claims in year j,
$F_{ij} = N_{ij}/V_{ij}$	claim frequency in year j,
$Y_{ij} = S_{ij}/N_{ij}$	average claim size in year j.

The actuary's task consists of calculating for the ith risk class the "true" individual pure risk premium for a future period. This may be done directly on the basis of the observations X_{ij}, or indirectly, by analysis and calculation of the two components "claim frequency" and "average claim size". To fix the idea, we work in the following with the average claim costs X_{ij}.

Assume as in the simple Bühlmann model (Model Assumptions 3.6) that each risk class i is then characterized by its specific risk parameter Θ_i. Observe, however, that the assumption $\operatorname{Var}[X_{ij}|\Theta_i] = \sigma^2(\Theta_i)$ is no longer reasonable. The variance should depend on the volume measure V_{ij}. Let us look at the special case where each risk has been in force for a full year, so that the volume measure, i.e. the number of years at risk, is a whole number. Notice that in this case $X_{ij} = \frac{1}{V_{ij}} \sum_{\nu=1}^{Vij} S_{ij}^{(\nu)}$, where $S_{ij}^{(\nu)}$ is the aggregate claim amount of the νth risk in the class i. Since the X_{ij} is an average of V_{ij} risks, which can be considered as, conditionally, independent of each other, we have $\operatorname{Var}[X_{ij}|\Theta_i] = \sigma^2(\Theta_i)/V_{ij}$. It is reasonable to model the conditional variance as inversely proportional to the known volume measure (here, the number of years at risk) also in the case where V_{ij} is the sum of parts of years at risk (because some contracts have been in force only for part of the ith year). If we are interested in the average claim size or in the claim frequency, we have the same type of behaviour, except that in the first case, the volume measure is the observed number of claims n_{ij}.

We should also note that in the above example, the total pure risk premium for the risk class i is the product of the number of years at risk times $E[X_{ij}|\Theta_i]$. In other words, the total risk premium grows linearly with the volume measure.

That the premium grows linearly with the size of the insured unit, as quantified by an appropriate volume measure, is a feature that is common to many other lines of business. For example, the premium in fire insurance depends on the insured value of the building and on the insured value of its contents. In collective accident insurance by which all the employees of a firm are insured against accidents, the size of the firm must be taken into account, in some fashion, in order to set the premium.

Indeed in many lines of business, the premiums are calculated as "volume measure" times "premium rate" where the volume measure is appropriately chosen for the particular line of business under consideration.

Practical examples of volume measures:

- number of years at risk in motor insurance,
- total amount of annual (or monthly) wages in the collective health or collective accident insurance,
- sum insured in fire insurance,
- annual turnover in commercial liability insurance,
- annual premium written (or earned) by the ceding company in excess of loss reinsurance.

For the calculation of the premium rate it is important to consider the claim amounts in relation to the volume measure, that is, to consider the quantities $X_{ij} = S_{ij}/V_{ij}$. These are called claims ratios (or loss ratios) and can be regarded as "standardized" aggregate claim amounts. For each risk i, we have to determine the corresponding individual claims ratio $\mu(\Theta_i) = E[X_{ij}|\Theta_i]$, where here and otherwise in this book, the term risk is used to describe either an individual risk in the physical sense or a risk class.

4.2 Model Assumptions

We are given a portfolio of I risks or "risk categories", and we use the notation

X_{ij} claims ratio (other possible interpretations are claim frequency or average claim size) of the risk i in year j,
w_{ij} associated known weight.

The weights w_{ij} are in general interpreted as volume measures as described in Section 4.1. However, other interpretations are possible. To indicate such wider interpretations we use the symbol w_{ij} instead of V_{ij}.

Model Assumptions 4.1 (Bühlmann–Straub model)
The risk i is characterized by an individual risk profile ϑ_i, which is itself the realization of a random variable Θ_i, and we have that:

BS1: Conditionally, given Θ_i, the $\{X_{ij} : j = 1, 2, \ldots, n\}$ are independent with

$$E\left[X_{ij}\left|\Theta_i\right.\right] = \mu\left(\Theta_i\right), \tag{4.1}$$

$$\operatorname{Var}\left[X_{ij}\left|\Theta_i\right.\right] = \frac{\sigma^2(\Theta_i)}{w_{ij}}. \tag{4.2}$$

BS2: The pairs $(\Theta_1, \mathbf{X}_1)$, $(\Theta_2, \mathbf{X}_2)$, ... are independent, and Θ_1, Θ_2, ... are independent and identically distributed.

Technical Remark:

- The number of observation years n may also vary between risks. This could be formally expressed by setting $w_{ij} = 0$ for non-observed years. Note, however, that the estimator of the structural parameters needs to be changed in this case (see Section 4.11).

Remarks on Model Assumptions 4.1:

- The Bühlmann–Straub model is a two-urn model (see Figure 1.6 on page 12). From the first urn we draw the risk profile Θ_i, which determines the "content" of the second urn. In the second step, a random variable X_{ij} is drawn from the second urn. In this way, a heterogeneous portfolio is modelled. The risks in the portfolio have different risk profiles (heterogeneity in the group). But the risks in the portfolio also have something in common: *a priori, they cannot be recognized as being different (short terminology: they are a priori equal).* This is modelled by the fact that the risk profiles Θ_i are all drawn from the same urn.

- The conditional independence condition in BS1 can be weakened. All results in this chapter also hold if the entries of the vector $\mathbf{X}_i$ are on the average conditionally uncorrelated, i.e. if $E\left[\mathrm{Cov}\left(X_{ik}, X_{il}\left|\Theta_i\right.\right)\right] = 0$ for $k \neq l$.
- Comments on Assumption BS2:
  a) The risks are independent, i.e. random variables that belong to different contracts are independent. This assumption, as a rule, is not a problem in practice.
  b) The risks are *a priori equal:*
  This assumption is not always valid in practice and in concrete applications we need to consider if this is reasonable. For example, in fire insurance, we know by common sense and without looking at data, that for instance a carpenter's workshop is a greater risk than an office building. In Section 4.13 we will see what can be done in such situations, and how known a priori differences can be built into the model.
- Comments on Assumption BS1:
  a) The "true" individual claims ratio $\mu\left(\Theta_i\right)$ is constant over time:
  We have again an assumption about homogeneity over time, although it is weaker than that made in the simple Bühlmann model, in that the variance is allowed to change with varying values of the volume measure. In practice, it is often the case that the assumption of a constant claims ratio over time is not valid, for example when the X_{ij}'s under consideration depend on inflation or when insurance conditions have changed during the observation period. Sometimes, one can eliminate this problem by an *as-if* transformation of the data.
  b) The conditional variance is inversely proportional to the known weight w_{ij}. In many applications in insurance practice we can write the X_{ij} as $X_{ij} = S_{ij}/V_{ij}$, where S_{ij} is the aggregate claim amount and V_{ij} is a given volume measure (see Section 4.1). In general, these volume measures V_{ij} are then used as weights w_{ij}. If, as in the example of Section 4.1, the X_{ij} can be written as $X_{ij} = \frac{1}{V_{ij}} \sum_{\nu=1}^{V_{ij}} S_{ij}^{(\nu)}$, that is, as some observed average of conditionally independent random variables, then (4.2) is an appropriate assumption. In other situations, for example in fire insurance, where mostly the insured sum is taken as the volume measure, it is not so obvious that V_{ij} is also the appropriate weight w_{ij} and one can consider choosing w_{ij} different from V_{ij}, for instance a linear function of V_{ij}.

Further remark:

- Assume that the aggregate claim amount S_{ij} in cell ij is conditionally compound Poisson distributed with parameter $\lambda_{ij}(\Theta_i, w_{ij})$ and claim size Y whose distribution F_Y does not depend on Θ_i. Let $X_{ij} = S_{ij}/w_{ij}$. If $\lambda_{ij}\left(w_{ij,}\Theta_i\right) = w_{ij}\lambda\left(\Theta_i\right)$, then we have

$$E[X_{ij}|\,\Theta_i] = \lambda(\Theta_i)\, E[Y|\,\Theta_i]$$

and

$$\mathrm{Var}[X_{ij}|\,\Theta_i] = \frac{1}{w_{ij}}\lambda(\Theta_i)\, E\left[Y^2\middle|\,\Theta_i\right].$$

Thus, with $\mu(\Theta_i) = \lambda(\Theta_i)\, E[Y|\,\Theta_i]$ and $\sigma^2(\Theta_i) = \lambda(\Theta_i)\, E\left[Y^2\middle|\,\Theta_i\right]$, conditions (4.1) and (4.2) are satisfied.

The following quantities are of interest:

risk i	**collective/portfolio**
$\mu(\Theta_i) := E[X_{ij}\|\,\Theta_i]$, $\sigma^2(\Theta_i) := w_{ij}\mathrm{Var}[X_{ij}\|\,\Theta_i]$.	$\mu_0 := E[\mu(\Theta_i)]$, $\sigma^2 := E\left[\sigma^2(\Theta_i)\right]$, $\tau^2 := \mathrm{Var}[\mu(\Theta_i)]$.
Interpretation: $\mu(\Theta_i)$: individual risk premium, $\sigma^2(\Theta_i)$: variance within individual risk (normalized for weight 1).	Interpretation: μ_0: collective premium, σ^2: average variance within individual risk (normalized for weight 1; first variance component), τ^2: variance between individual risk premiums (second variance component).

4.3 The Credibility Premium in the Bühlmann–Straub Model

We want to estimate for each risk i its individual claims ratio $\mu(\Theta_i)$. We have available to us the data $\mathcal{D} = \{\mathbf{X}_i : i = 1, 2, \ldots, I\}$, where $\mathbf{X}_i = (X_{i1}, X_{i2}, \ldots, X_{in})'$ is the observation vector of the ith risk. We seek the credibility estimator based on the data $\mathcal{D}$ from all the risks in the portfolio.

It is easy to see that the credibility estimator for $\mu(\Theta_i)$ depends only on the observations from the ith risk. Let

$$\widehat{\widehat{\mu(\Theta_i)}} = a_{i_0} + \sum\nolimits_j a_{ij} X_{ij} \tag{4.3}$$

be the credibility estimator based on $\mathbf{X}_i$. Because of the independence of the risks, it follows that for $k \neq i$ and all l

$$\mathrm{Cov}\left(\widehat{\widehat{\mu(\Theta_i)}}, X_{kl}\right) = \mathrm{Cov}(\mu(\Theta_i), X_{kl}) = 0, \tag{4.4}$$

i.e. the normal equations (Corollary 3.17) are satisfied and $\widehat{\mu(\Theta_i)}$ is therefore the credibility estimator based on all data.

That the credibility premium of the risk i depends only on the individual claim experience, and not on the claim experience of the other risks, is intuitively obvious. If we know the a priori expected value μ_0 (collective premium), then the other risks cannot supply any extra information, because they are independent of the risk being rated. In practice, however, μ_0 is usually unknown. The other risks then contain information for the estimation of this a priori expected value. We will learn more of this later when we discuss the homogeneous credibility estimator.

Here and in the following we will use the notation that a point in the subscript denotes summation over the corresponding subscript, so that for example $w_{i\bullet} = \sum_j w_{ij}$. According to (4.1) and (4.2) the observations X_{ij} are individually (i.e. conditionally, given Θ_i) unbiased, with a conditional variance that is inversely proportional to the weights w_{ij}. It is easy to check that the best linear estimator, which is individually unbiased and which has the smallest conditional variance, is given by the weighted average

$$X_i = \sum_j \frac{w_{ij}}{w_{i\bullet}} X_{ij}.$$

For X_i we have

$$\begin{aligned} E[X_i|\Theta_i] &= \mu(\Theta_i), \\ \operatorname{Var}[X_i|\Theta_i] &= \sum_j \left(\frac{w_{ij}}{w_{i\bullet}}\right)^2 \operatorname{Var}[X_{ij}|\Theta_i] = \frac{\sigma^2(\Theta_i)}{w_{i\bullet}}. \end{aligned}$$

As a further useful property we record that X_i is also the homogeneous credibility estimator based on the observation vector $\mathbf{X}_i$, because

$$X_i = \operatorname{Pro}\left(\mu(\Theta_i)\,|L_e(\mathbf{X}_i)\right). \tag{4.5}$$

We now derive the credibility estimator based on X_i and then show that this is also the credibility estimator based on all data. This then also implies that the compressed data $\{X_i : i = 1, 2, \ldots, I\}$ is a *linear sufficient statistic*, i.e. the credibility estimator depends only on the observations via X_i. Because of the normal equations (Corollary 3.17) and the fact that $E[X_i] = \mu_0$, the credibility estimator must be of the form (4.6) and satisfy (4.7),

$$\widehat{\mu(\Theta_i)} = \alpha_i X_i + (1 - \alpha_i)\,\mu_0, \tag{4.6}$$

$$\operatorname{Cov}\left(\widehat{\mu(\Theta_i)}, X_i\right) = \alpha_i \operatorname{Cov}(X_i, X_i) = \operatorname{Cov}(\mu(\Theta_i), X_i). \tag{4.7}$$

From the fact that

$$\begin{aligned}\operatorname{Var}[X_i] &= E\left[\operatorname{Var}\left[X_i \mid \Theta_i\right]\right] + \operatorname{Var}\left[E\left[X_i \mid \Theta_i\right]\right] \\ &= \frac{\sigma^2}{w_{i\bullet}} + \tau^2,\end{aligned}$$

$$\begin{aligned}\operatorname{Cov}(\mu(\Theta_i), X_i) &= E\left[\operatorname{Cov}(\mu(\Theta_i), X_i \mid \Theta_i)\right] + \operatorname{Cov}\left(\mu(\Theta_i), E\left[X_i \mid \Theta_i\right]\right) \\ &= 0 + \operatorname{Var}\left[\mu(\Theta_i)\right] \\ &= \tau^2,\end{aligned}$$

it follows that

$$\alpha_i = \frac{\tau^2}{\frac{\sigma^2}{w_{i\bullet}} + \tau^2} = \frac{w_{i\bullet}}{w_{i\bullet} + \frac{\sigma^2}{\tau^2}}. \tag{4.8}$$

Now we want to show that $\widehat{\widehat{\mu(\Theta_i)}}$ is also the credibility estimator for the risk i based on all data, which, based on the properties given above for X_i, we would expect. We have already shown on page 81 that the credibility estimator of risk i depends only on the data of risk i (this follows also from the independence property shown in Exercise 3.2 on page 75). Hence it remains to show that

$$\operatorname{Cov}\left(\widehat{\widehat{\mu(\Theta_i)}}, X_{ij}\right) = \operatorname{Cov}(\mu(\Theta_i), X_{ij}) = \tau^2 \quad \text{for } j = 1, 2, \ldots, n. \tag{4.9}$$

It holds that

$$\begin{aligned}\operatorname{Cov}\left(\widehat{\widehat{\mu(\Theta_i)}}, X_{ij}\right) &= \alpha_i \left(\sum\nolimits_k \frac{w_{ik}}{w_{i\bullet}} \operatorname{Cov}(X_{ik}, X_{ij})\right) \\ &= \alpha_i \left(\sum\nolimits_k \frac{w_{ik}}{w_{i\bullet}} \left(\frac{\sigma^2}{w_{ik}} \delta_{kj} + \tau^2\right)\right) \\ &= \alpha_i \left(\sum\nolimits_k \frac{w_{ik}}{w_{i\bullet}} \tau^2 + \frac{\sigma^2}{w_{i\bullet}}\right) \\ &= \tau^2, \\ \text{where } \delta_{kj} &= \begin{cases} 1 \text{ for } k = j \\ 0 \text{ otherwise.} \end{cases}\end{aligned}$$

In the last step we have used the explicit form of α_i given by (4.8).

Hence (4.9) holds true, and we have thus proved the following result:

Theorem 4.2 (credibility estimator). *The (inhomogeneous) credibility estimator in the Bühlmann–Straub model (Model Assumptions 4.1) is given by*

$$\widehat{\widehat{\mu(\Theta_i)}} = \alpha_i X_i + (1 - \alpha_i)\, \mu_0 \;=\; \mu_0 + \alpha_i \left(X_i - \mu_0\right), \tag{4.10}$$

$$\text{where} \quad X_i = \sum_j \frac{w_{ij}}{w_{i\bullet}} X_{ij},$$
$$w_{i\bullet} = \sum_j w_{ij},$$
$$\alpha_i = \frac{w_{i\bullet}}{w_{i\bullet} + \frac{\sigma^2}{\tau^2}} = \frac{w_{i\bullet}}{w_{i\bullet} + \kappa}.$$

Remarks:

- $\kappa = \sigma^2/\tau^2$ is called the credibility coefficient.
- Note that the credibility estimator has the same form as the Bayes estimator in Theorem 2.20. This means that the credibility estimator coincides with the Bayes estimator for exponential families with priors within the associated conjugate family.

4.4 Discussion and Interpretation of the Credibility Estimator

We can give a very nice and intuitive interpretation for the form of the credibility estimator given in Theorem 4.2.

- μ_0 is the best estimator based only on the a priori knowledge and has the quadratic loss

$$E\left[(\mu_0 - \mu(\Theta_i))^2\right] = \operatorname{Var}(\mu(\Theta_i)) = \tau^2.$$

- X_i is the best linear and individually unbiased estimator based only on the observation vector $\mathbf{X}_i$ and has the quadratic loss

$$E\left[(X_i - \mu(\Theta_i))^2\right] = E\left\{E\left[\left(\sum_j \frac{w_{ij}}{w_{i\bullet}} (X_{ij} - \mu(\Theta_i))\right)^2 \middle| \Theta_i\right]\right\}$$
$$= E\left[\sigma^2(\Theta_i)/w_{i\bullet}\right] = \sigma^2/w_{i\bullet}\,.$$

- The credibility estimator is a weighted mean of the two estimators where the weight assigned to each summand is proportional to its inverse quadratic loss (precision), i.e.

$$\widehat{\mu(\Theta_i)} = \alpha_i X_i + (1 - \alpha_i)\mu_0,$$

where

$$\alpha_i = \frac{\left(\sigma^2/w_{i\bullet}\right)^{-1}}{\left(\tau^2\right)^{-1} + \left(\sigma^2/w_{i\bullet}\right)^{-1}}.$$

The General Intuitive Principle 3.3, which we already stated for the simple Bühlmann Model (Model Assumptions 3.6), is therefore also valid here.

Remarks:

- The credibility coefficient can also be written as

$$\kappa = \frac{(\sigma/\mu_0)^2}{(\tau/\mu_0)^2}.$$

Note that:
- τ/μ_0 is the coefficient of variation (standard deviation divided by the expected value) of $\mu(\Theta_i)$, which is a good measure for the heterogeneity of the portfolio, or in other words, a good measure for the between risk variability.
- σ is the average standard deviation within risk normalized for weight 1, and μ_0 is the expected value averaged over the whole portfolio. Hence σ/μ_0 can be interpreted as some kind of an average within risk coefficient of variation and is a good measure of the within risk variability.

- From the formula for the credibility weight α_i we see that (Figure 4.1):
  - the greater the weight $w_{i\bullet}$, summed over the years, the greater is α_i (notice also that $w_{i\bullet}$ assumes the role of the number of observation years n in the simple Bühlmann model),
  - the smaller the credibility coefficient $\kappa = \sigma^2/\tau^2$, the greater is α_i.

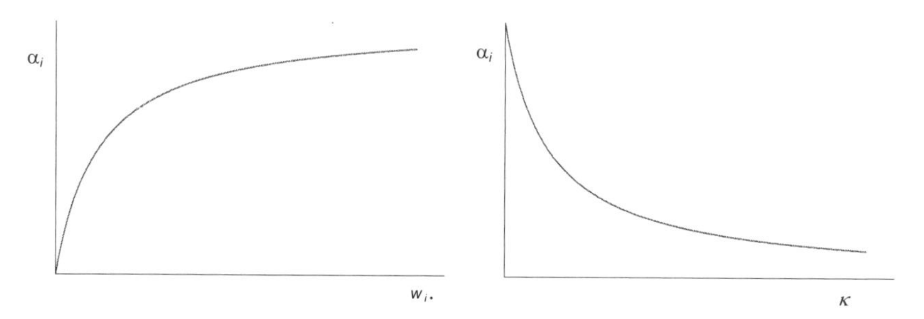

Fig. 4.1. Credibility weights as a function of the parameters

- $E\left[\widehat{\widehat{\mu(\Theta_i)}}\right] = \mu_0$, i.e. on average, over the collective, we have correctly rated the risk (unbiased in the collective).
- For the credibility estimator, we need, besides the observed claims ratios X_{ij} and their corresponding weights w_{ij}, the so-called structural parameters μ_0, σ^2 and τ^2. These can be determined based either on a priori knowledge, for example from the opinions of experts (pure Bayesian procedure), or they can be determined from observations of a collective of similar risks (empirical Bayes). Estimators for the structural parameters when we are in the latter situation are given below.

- In the Bühlmann–Straub model, if we were to replace the variance condition (4.2) by the condition $\operatorname{Var}[X_{ij}|\Theta_i] = \sigma_i^2/w_{ij}$, thus modelling the conditional variances normalized for weight 1 as constants σ_i^2, this would result in the term σ_i^2 appearing in the credibility weight instead of σ^2. In this way, the risk-specific variability of the claims ratios X_{ij} would be taken into account, whereas in the Bühlmann–Straub credibility estimator only the "average" of this variability over the collective is used. The risk-specific approach means that, for risks with more stable claims ratios, more weight is given to the individual experience and vice versa. At first glance, this appears to be intuitively reasonable. This would, without doubt, also be the case, if the structural parameters σ_i^2 were known. As a rule, however, they are not known, and in such a modified model, instead of the *single* unknown structural parameter σ^2 we would have a considerably larger number I of unknown structural parameters σ_i^2, all of which would have to be estimated. The increase of the estimation error resulting from the necessity of estimating so many structural parameters will typically not be offset by the improved optimality of the credibility estimator.

4.5 Quadratic Loss of the Credibility Estimator

Theorem 4.3. *Given Model Assumptions 4.1 we have that the quadratic loss of the credibility estimator (4.10) is given by*

$$E\left[\left(\widehat{\widehat{\mu(\Theta_i)}} - \mu(\Theta_i)\right)^2\right] = (1-\alpha_i)\tau^2 = \alpha_i \frac{\sigma^2}{w_{i\bullet}}. \tag{4.11}$$

Proof: The proof is left to the reader as an exercise. It is analogous to the proof in the Bühlmann model. □

Remarks:

- Note that τ^2 is the quadratic loss of the collective premium μ_0. By the use of the credibility estimator instead, this quadratic loss is reduced by the factor $1-\alpha_i$.
- Note also that $\sigma^2/w_{i\bullet}$ is the quadratic loss of X_i, which is the best linear estimator, when looking only at the data. When using the credibility estimator instead, this quadratic loss is reduced by the factor α_i.

4.6 The Homogeneous Credibility Estimator in the Bühlmann–Straub Model

Since the homogeneous credibility estimator has no constant term, the overall expected value μ_0 appearing in the inhomogeneous credibility estimator (4.10) must be replaced by a collectively unbiased estimator $\widehat{\mu}_0$, which is a linear function of the observations. Intuitively, we might first think of using the weighted average of the observed claims ratios,

$$\overline{X} := \frac{\sum_{i,j} w_{ij} X_{ij}}{w_{\bullet\bullet}},$$

where $w_{\bullet\bullet} = \sum_{ij} w_{ij}$. But here, as we shall see, intuition fails.

From Theorem 3.14 iii) (iterativity of projections) we have

$$\widehat{\widehat{\mu(\Theta_i)}}^{\text{hom}} = \operatorname{Pro}\left(\widehat{\widehat{\mu(\Theta_i)}} \middle| L_e(\mathcal{D}) \right),$$
$$\text{where } \mathcal{D} = \{X_{ij},\ i = 1, 2, \ldots, I,\ j = 1, 2, \ldots, n\}.$$

From the normed linearity property in affine subspaces (see Theorem B.11 of Appendix B), we have

$$\widehat{\widehat{\mu(\Theta_i)}}^{\text{hom}} = \alpha_i \operatorname{Pro}(X_i | L_e(\mathcal{D})) + (1 - \alpha_i) \operatorname{Pro}(\mu_0 | L_e(\mathcal{D})).$$

The individual mean X_i is in $L_e(\mathcal{D})$, since $E[X_i] = E\{E[X_i|\Theta_i]\} = E[\mu(\Theta_i)] = \mu_0$. Hence

$$\widehat{\widehat{\mu(\Theta_i)}}^{\text{hom}} = \alpha_i X_i + (1 - \alpha_i) \widehat{\widehat{\mu_0}},$$
$$\text{where } \widehat{\widehat{\mu_0}} = \operatorname{Pro}(\mu_0 | L_e(\mathcal{D})).$$

In Section 4.4 we have seen that X_i is the best linear, individually unbiased estimator of $\mu(\Theta_i)$ based on the individual observation vector $\mathbf{X}_i$. One would presume then, that X_i contains all the information in $\mathbf{X}_i$ relating to $\mu(\Theta_i)$, that one can derive from a linear estimator. It seems reasonable then to suppose that $\widehat{\widehat{\mu_0}}$ depends only on the data $\{X_i : i = 1, 2, \ldots, I\}$. We proceed therefore by first determining

$$\widehat{\mu_0} = \operatorname{Pro}(\mu_0 | L_e(X_1, \ldots, X_I)) \tag{4.12}$$

and then showing that

$$\widehat{\mu_0} = \widehat{\widehat{\mu_0}} = \operatorname{Pro}(\mu_0 | L_e(\mathcal{D})). \tag{4.13}$$

Let $\widehat{\mu_0} = \sum_i a_i X_i = \operatorname{Pro}(\mu_0 | L_e(X_1, \ldots, X_I))$. Because of the unbiasedness in the collective, we must have

$$\sum_i a_i = 1. \tag{4.14}$$

Because of (4.12) it must also be true that

$$\mu_0 - \widehat{\mu_0} \perp X_k - X_r \quad \text{for } k, r = 1, 2, \ldots, I, \tag{4.15}$$

which is equivalent to

$$\operatorname{Cov}(\widehat{\mu_0}, X_k) = \operatorname{Cov}(\widehat{\mu_0}, X_r).$$

From this we get directly that

$$a_i \operatorname{Var}[X_i] = const. \tag{4.16}$$

The weights a_i are therefore inversely proportional to $\operatorname{Var}[X_i]$. Notice also that

$$\operatorname{Var}[X_i] = E\left[(X_i - \mu_0)^2\right].$$

The variance therefore corresponds to the quadratic loss of X_i as an estimator of the quantity μ_0. In this way, we again find a result which is analogous to that which we had for the credibility estimator: the weights are proportional to the precisions, which are defined as the inverse of the quadratic loss. From the equation

$$\operatorname{Var}[X_i] = E[\operatorname{Var}[X_i|\,\Theta_i]] + \operatorname{Var}[E[X_i|\,\Theta_i]]$$

it also follows that

$$\begin{aligned}\operatorname{Var}[X_i] &= \tau^2 + \frac{\sigma^2}{w_{i\bullet}} \\ &= \tau^2\left(\frac{w_{i\bullet}\tau^2 + \sigma^2}{w_{i\bullet}\tau^2}\right) \\ &= \tau^2\,\alpha_i^{-1}.\end{aligned} \tag{4.17}$$

From (4.16) and (4.17) it follows that the optimal weights are proportional to the credibility weights α_i, and from (4.14) that

$$\widehat{\mu_0} = \sum\nolimits_i \frac{\alpha_i}{\alpha_\bullet} X_i,$$

where $\alpha_\bullet = \sum_i \alpha_i$.

Lastly, we must show that $\widehat{\mu_0}$ satisfies (4.13), i.e. that

$$\mu_0 - \widehat{\mu_0} \perp X_{kl} - X_{rs} \qquad \text{for all } k, l, r \text{ and } s.$$

Because

$$\mu_0 - \widehat{\mu_0} = \sum\nolimits_i \frac{\alpha_i}{\alpha_\bullet}(\mu_0 - X_i),$$

this is equivalent to

$$\mu_0 - X_i \perp X_{kl} - X_{rs} \qquad \text{for } i, k, l, r, s. \tag{4.18}$$

The right-hand side of (4.18) can be written as

$$(X_{kl} - X_k) + (X_k - X_r) + (X_r - X_{rs}).$$

Because of the independence of $\mathbf{X}_i$ and $\mathbf{X}_k$ for $i \neq k$ we have

$$\mu_0 - X_i \perp X_{kl} - X_k \qquad \text{for } k \neq i.$$

Moreover,

$$\mu_0 - \widehat{\mu_0} \perp (X_k - X_r)$$

because of (4.15). Hence it remains to show that

$$\mu_0 - X_i \perp X_{ij} - X_i.$$

Since X_i is the homogeneous credibility estimator of $\mu(\Theta_i)$ based on $\mathbf{X}_i$ (see (4.5)), we have that

$$\mu(\Theta_i) - X_i \perp X_{ij} - X_i \qquad \text{for } j = 1, 2, \ldots, n.$$

If we write $\mu_0 - X_i = (\mu_0 - \mu(\Theta_i)) + (\mu(\Theta_i) - X_i)$ it remains to show that

$$\mu_0 - \mu(\Theta_i) \perp X_{ij} - X_i \qquad \text{for } j = 1, 2, \ldots, n,$$

which is obvious by conditioning on Θ_i.

Thus we have proved the following result:

Theorem 4.4 (homogeneous credibility estimator). *The homogeneous credibility estimator of $\mu(\Theta_i)$ in the Bühlmann–Straub model (Model Assumptions 4.1) is given by*

$$\widehat{\widehat{\mu(\Theta_i)}}^{\text{hom}} = \alpha_i X_i + (1 - \alpha_i)\,\widehat{\widehat{\mu_0}} = \widehat{\widehat{\mu_0}} + \alpha_i \left(X_i - \widehat{\widehat{\mu_0}}\right), \tag{4.19}$$

where $\widehat{\widehat{\mu_0}} = \sum_{i=1}^{I} \frac{\alpha_i}{\alpha_\bullet} X_i,$

$$\alpha_i = \frac{w_{i\bullet}}{w_{i\bullet} + \frac{\sigma^2}{\tau^2}} \quad \textit{and} \quad \alpha_\bullet = \sum_{i=1}^{I} \alpha_i.$$

Remarks:

- Intuition failed us. As an estimator for μ_0 we should not use the observed average

  $$\overline{X} = \sum\nolimits_{i=1}^{I} \frac{w_{i\bullet}}{w_{\bullet\bullet}} X_i,$$

  but rather the credibility weighted average

  $$\widehat{\widehat{\mu_0}} = \sum\nolimits_{i=1}^{I} \frac{\alpha_i}{\alpha_\bullet} X_i.$$

  If we look at $s_i := w_{i\bullet}/w_{\bullet\bullet}$ as sample weights, then $\alpha_i/\alpha_\bullet$ will be closer to s_i the closer the s_i are to the uniform sample weights. Also, given

$\{s_i : i = 1, 2, \ldots, I\}$, $\alpha_i/\alpha_\bullet \to s_i$ for $w_{\bullet\bullet} \to 0$ and $\alpha_i/\alpha_\bullet \to 1/I$ for $w_{\bullet\bullet} \to \infty$.

It is important to note that it may happen that $\widehat{\widehat{\mu(\Theta_i)}}^{\text{hom}} > X_i$, even though $\overline{X} < X_i$, i.e. the homogeneous credibility estimator is not necessarily between the observed individual and the observed collective mean.

- In contrast to the (inhomogeneous) credibility estimator, we use for the homogeneous estimator the observations from the entire collective (and not only those from the ith contract): they are needed for the estimation of μ_0.
- In practice, the homogeneous estimator is as important (if not more important) than the inhomogeneous estimator, because the estimation of μ_0 is automatically built in to the formula for the homogeneous estimator.
- Notice:
  For the calculation of the homogeneous credibility estimator we proceed as follows (compare with Figure 4.2):
  - first calculate X_i and $\widehat{\mu_0}$ "bottom up" (from the bottom to the top);
  - then calculate $\widehat{\widehat{\mu(\Theta_i)}}^{\text{hom}}$ by inserting the top value $\widehat{\mu_0}$ ("top down").

  This "bottom up" and "to down" procedure will become more apparent in the hierarchical credibility model (see Chapter 6).

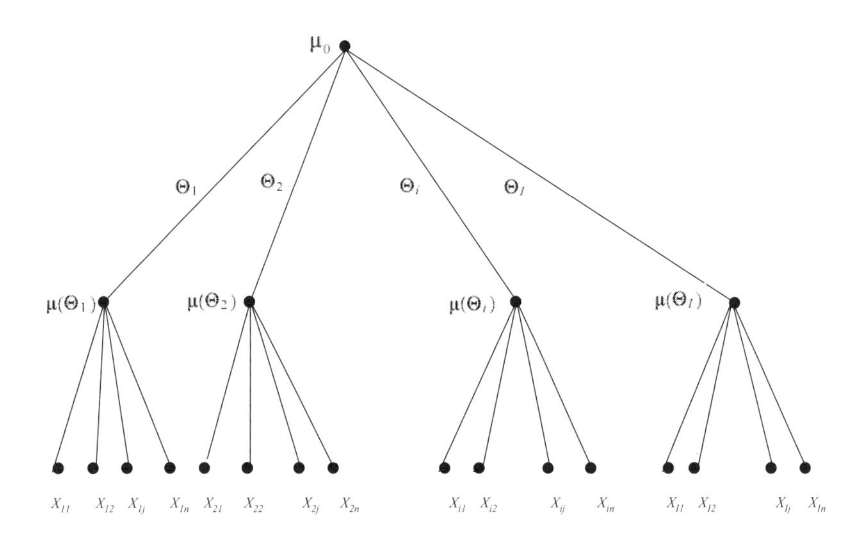

Fig. 4.2. Tree structure of the Bühlmann–Straub model

There is another noteworthy property of the homogeneous credibility estimator, which from the point of view of practitioners is important and which we will now formulate as a theorem.

Theorem 4.5 (balance property). *Given Model Assumptions 4.1 we have for the homogeneous credibility estimator in the Bühlmann–Straub model*

$$\sum_{i,j} w_{ij}\,\widehat{\widehat{\mu(\Theta_i)}}^{\text{hom}} = \sum_{i,j} w_{ij} X_{ij}. \tag{4.20}$$

Remarks:

- In practice, w_{ij} is mostly identical with the underlying tariff volume measure V_{ij}. Then the left-hand side of the above equation corresponds exactly to the (total) credibility premium which would have resulted over the whole portfolio if the credibility premium had been applied over the entire past observation period. On the right-hand side we have the total aggregate claim amount over the past observation period. The equation therefore says that with respect to the past observation period and in total over the whole portfolio, the resulting total credibility premium and the aggregate claim amount are equal. This is true independent of the choice, respectively the estimates, of the structural parameters σ^2 and τ^2.
- Naturally the equality will not be exact for the future period for which we must rate the risks. It does, however, indicate that, under stationary conditions, the premium level over the whole portfolio will be fair even if not all model assumptions are strictly satisfied.
- Under the above conditions, the balance property of the homogeneous credibility estimator may be reinterpreted as a so-called redistribution of the total aggregate claim amount $S_{\bullet\bullet} = \sum_i w_{i\bullet} X_{i\bullet}$. The observed aggregate claim amount $w_{i\bullet} X_i$ for the ith risk is "replaced" by $w_{i\bullet}\,\widehat{\widehat{\mu(\Theta_i)}}^{\text{hom}}$. The resulting total aggregate claim amount over the whole portfolio and the whole observation period remains unchanged by this replacement.

Proof of Theorem 4.5: From $w_{i\bullet}(1-\alpha_i) = \left(\sigma^2/\tau^2\right)\alpha_i$ it follows that

$$\begin{aligned}\sum_i w_{i\bullet}\left(\widehat{\widehat{\mu(\Theta_i)}}^{\text{hom}} - X_i\right) &= \sum_i w_{i\bullet}\left(-(1-\alpha_i)X_i + (1-\alpha_i)\widehat{\widehat{\mu_0}}\right)\\ &= \frac{\sigma^2}{\tau^2}\sum_i \alpha_i(\widehat{\widehat{\mu_0}} - X_i) = 0,\end{aligned}$$

where in the last step we have used the identity $\widehat{\widehat{\mu_0}} = \sum \alpha_i X_i/\alpha_\bullet$. This concludes the proof of Theorem 4.5. □

4.7 Quadratic Loss of the Homogeneous Credibility Estimator

In this section we calculate the quadratic loss of the homogeneous credibility estimator (4.19) in the Bühlmann–Straub model.

Theorem 4.6. *Given Model Assumptions 4.1 we have for the quadratic loss of the homogeneous credibility estimator (4.19):*

$$\mathbf{E}\left[\left(\widehat{\widehat{\mu(\Theta_i)}}^{\mathrm{hom}} - \mu(\Theta_i)\right)^2\right] = \tau^2(1-\alpha_i)\left(1+\frac{1-\alpha_i}{\alpha_\bullet}\right). \tag{4.21}$$

Proof: By iteratively applying the projection operator

$$\widehat{\widehat{\mu(\Theta_i)}}^{\mathrm{hom}} = \operatorname{Pro}\{\operatorname{Pro}(\mu(\Theta_i)|\, L(\mathbf{X},1))|\, L_e(\mathbf{X})\}$$

we get

$$\begin{aligned} & E\left[\left(\widehat{\widehat{\mu(\Theta_i)}}^{\mathrm{hom}} - \mu(\Theta_i)\right)^2\right] \\ &= E\left[\left(\widehat{\widehat{\mu(\Theta_i)}} - \mu(\Theta_i)\right)^2\right] + E\left[\left(\widehat{\widehat{\mu(\Theta_i)}} - \widehat{\widehat{\mu(\Theta_i)}}^{\mathrm{hom}}\right)^2\right]. \end{aligned}$$

From Theorem 4.3 we have that

$$E\left[\left(\widehat{\widehat{\mu(\Theta_i)}} - \mu(\Theta_i)\right)^2\right] = \tau^2(1-\alpha_i).$$

For the second term we have that

$$E\left[\left(\widehat{\widehat{\mu(\Theta_i)}} - \widehat{\widehat{\mu(\Theta_i)}}^{\mathrm{hom}}\right)^2\right] = (1-\alpha_i)^2\, E\left[(\widehat{\mu_0} - \mu_0)^2\right]$$

and

$$\begin{aligned} E\left[(\widehat{\mu_0} - \mu_0)^2\right] &= E\left[\left(\sum_i \frac{\alpha_i}{\alpha_\bullet}(X_i - \mu_0)\right)^2\right] \\ &= \sum_i \left(\frac{\alpha_i}{\alpha_\bullet}\right)^2 \underbrace{E\left[(X_i-\mu_0)^2\right]}_{=\frac{\sigma^2}{w_{i\bullet}}+\tau^2} \\ &= \sum_i \left(\frac{\alpha_i}{\alpha_\bullet}\right)^2 \underbrace{\left(\frac{\sigma^2 + w_{i\bullet}\tau^2}{w_{i\bullet}\tau^2}\right)}_{=\alpha_i^{-1}}\tau^2 = \frac{\tau^2}{\alpha_\bullet}. \end{aligned}$$

Hence we get

$$E\left[\left(\widehat{\widehat{\mu(\Theta_i)}}^{\text{hom}} - \mu(\Theta_i)\right)^2\right] = \tau^2(1-\alpha_i) + (1-\alpha_i)^2\frac{\tau^2}{\alpha_\bullet}$$
$$= \tau^2(1-\alpha_i)\left(1+\frac{1-\alpha_i}{\alpha_\bullet}\right),$$

which concludes the proof of Theorem 4.6. □

4.8 Estimation of the Structural Parameters σ^2 and τ^2

We have seen that the formula for the (inhomogeneous) credibility estimator involves the three structural parameters μ_0, σ^2, τ^2. An estimate for μ_0 is already built in to the formula for the homogeneous credibility estimator, so that only the two structural parameters σ^2 and τ^2 remain to be determined.

In practice, these two parameters are also unknown and must be estimated from the data of the collective. Various articles are to be found in the actuarial literature about the estimation of such parameters (e.g. Dubey and Gisler [DG81], Norberg [Nor82]). However, none of the various considered alternatives has proved to be generally better than that suggested by Bühlmann and Straub in their original work. We will now present that estimator here.

Estimation of σ^2: For the variance within the ith contract we consider

$$S_i := \frac{1}{n-1}\sum_{j=1}^{n} w_{ij}(X_{ij}-X_i)^2.$$

If we write

$$S_i = \frac{1}{n-1}\sum_{j=1}^{n} w_{ij}(X_{ij}-\mu(\Theta_i)+\mu(\Theta_i)-X_i)^2$$
$$= \frac{1}{n-1}\left\{\sum_{j=1}^{n} w_{ij}(X_{ij}-\mu(\Theta_i))^2 - w_{i\bullet}(\mu(\Theta_i)-X_i)^2\right\},$$

we see immediately that

$$E[S_i|\Theta_i] = \sigma^2(\Theta_i),$$

and therefore

$$E[S_i] = E\{E[S_i|\Theta_i]\} = E\left[\sigma^2(\Theta_i)\right] = \sigma^2. \tag{4.22}$$

Each of the random variables S_i is therefore an unbiased estimator for σ^2, and the question is, how should they be weighted? For example, we could simply take the arithmetic mean (each S_i is given the same weight). Or should the larger contracts have a greater weight?

The theoretically optimal weights would be proportional to the precisions $\left(E\left[\left(S_i-\sigma^2\right)^2\right]\right)^{-1}$. This expression, however, depends on moments up to the fourth order which are also unknown and would have to be estimated. The estimation of these would involve new structural parameters, so that we see that this line of thinking is not helpful.

In order to get more insight, we consider the special case, where the X_{ij} are conditionally on Θ_i normally distributed. Then we have

$$\begin{aligned} E\left[\left(S_i-\sigma^2\right)^2\right] &= E\left[\left(S_i-\sigma^2(\Theta_i)+\sigma^2(\Theta_i)-\sigma^2\right)^2\right] \\ &= \frac{2}{n-1}E\left[\sigma^4\left(\Theta_i\right)\right]+E\left[\left(\sigma^2(\Theta_i)-\sigma^2\right)^2\right], \end{aligned} \tag{4.23}$$

which does not depend on i. In this case, it is therefore optimal to weight all S_i's equally and to take their arithmetic mean as estimator of σ^2.

Obviously, most of the insurance data are not normally distributed. On the other hand, other weights which would be better in a multitude of situations are not known. It is therefore reasonable to use as an estimator

$$\widehat{\sigma^2}=\frac{1}{I}\sum_{i=1}^{I}\frac{1}{n-1}\sum_{j=1}^{n}w_{ij}\left(X_{ij}-X_i\right)^2. \tag{4.24}$$

This estimator has the following properties:

- Unbiasedness, i.e. $E\left[\widehat{\sigma}^2\right]=\sigma^2$. This follows directly from (4.22).
- Consistency, i.e. $\widehat{\sigma}^2 \underset{I\to\infty}{\overset{P}{\to}} \sigma^2$. This follows from Chebychev's inequality.

Estimation of τ^2: For didactic reasons, we first consider the case where we have equal aggregate weights for all risks and having done that, we will then consider the more general case.

a) $w_{1\bullet}=w_{2\bullet}=\cdots=w_{I\bullet}$:
Consider

$$T=\frac{1}{I-1}\sum_{i=1}^{I}\left(X_i-\overline{X}\right)^2,$$

$$\text{where } \overline{X}=\frac{1}{I}\sum_{i=1}^{I}X_i.$$

Since

$$E\left[T\right]=\mathrm{Var}\left[X_i\right]=\frac{\sigma^2}{w_{i\bullet}}+\tau^2,$$

we get that

$$\widehat{\widehat{\tau}}^2=\frac{1}{I-1}\sum_{i=1}^{I}\left(X_i-\overline{X}\right)^2-\frac{\widehat{\sigma}^2}{w_{i\bullet}} \tag{4.25}$$

is an unbiased estimator of τ^2.

b) General case:
We define

$$T = \frac{I}{I-1} \sum_{i=1}^{I} \frac{w_{i\bullet}}{w_{\bullet\bullet}} \left(X_i - \overline{X}\right)^2,$$

$$\text{where } \overline{X} = \sum_i \frac{w_{i\bullet}}{w_{\bullet\bullet}}.$$

After some calculation we get the following, unbiased estimator:

$$\widehat{\widehat{\tau}}^2 = c \cdot \left\{ \frac{I}{I-1} \sum_{i=1}^{I} \frac{w_{i\bullet}}{w_{\bullet\bullet}} \left(X_i - \overline{X}\right)^2 - \frac{I\widehat{\sigma^2}}{w_{\bullet\bullet}} \right\}, \tag{4.26}$$

$$\text{where} \quad c = \frac{I-1}{I} \left\{ \sum_{i=1}^{I} \frac{w_{i\bullet}}{w_{\bullet\bullet}} \left(1 - \frac{w_{i\bullet}}{w_{\bullet\bullet}}\right) \right\}^{-1}.$$

Remarks:

–
$$c = \begin{cases} = 1 \text{ if } w_{1\bullet} = w_{2\bullet} = \cdots = w_{I\bullet}, \\ > 1 \text{ otherwise.} \end{cases}$$

– Of course, (4.26) coincides with (4.25) in the case of identical aggregate weights $w_{i\bullet}$.

The estimator $\widehat{\widehat{\tau}}^2$ has the following properties:

- Unbiasedness, i.e. $E\left[\widehat{\widehat{\tau}}^2\right] = \tau^2$. This is easily checked.
- Consistency, i.e. $\widehat{\widehat{\tau}}^2 \underset{I\to\infty}{\overset{P}{\to}} \tau^2$, if no one of the risks is "dominating" (i.e. if $\sum_i \left(\frac{w_{i\bullet}}{w_{\bullet\bullet}}\right)^2 \to 0$ for $I \to \infty$). A proof can be found for instance in [DG81].
- $\widehat{\widehat{\tau}}^2$ can possibly be negative. This means that there is no detectable difference between the risks. This point will become more clear below. Naturally, in this case we put $\widehat{\tau^2} = 0$. Hence, we finally use as estimator

$$\widehat{\tau}^2 = \max\left(\widehat{\widehat{\tau}}^2, 0\right), \tag{4.27}$$

$$\text{where } \widehat{\widehat{\tau}}^2 \text{ is given by (4.26).}$$

Of course, $\widehat{\tau}^2$ is no longer unbiased.

4.9 Empirical Credibility Estimator

The empirical credibility estimator is obtained from the homogeneous credibility formula by replacing the structural parameters σ^2 and τ^2 by their estimators derived in Section 4.8.

$$\widehat{\widehat{\mu(\Theta_i)}}^{emp} = \widehat{\alpha}_i X_i + (1 - \widehat{\alpha}_i) \widehat{\widehat{\mu_0}},$$

$$\text{where} \qquad \widehat{\alpha}_i = \frac{w_{i\bullet}}{w_{i\bullet} + \widehat{\kappa}}, \qquad \widehat{\kappa} = \frac{\widehat{\sigma^2}}{\widehat{\tau^2}}, \qquad \widehat{\widehat{\mu_0}} = \frac{\sum_i \widehat{\alpha}_i X_i}{\sum_i \widehat{\alpha}_i}.$$

Observe that $\widehat{\alpha}_i = 0$, if $\widehat{\tau}^2 = 0$. In this case $\widehat{\widehat{\mu_0}}$ is defined by $\widehat{\widehat{\mu_0}} = \sum_i (w_{i\bullet}/w_{\bullet\bullet}) X_i$.

More insight and a better understanding of the credibility weight $\widehat{\alpha}_i$ can be gained by the following considerations making use of the decomposition of the sum of squares known from classical statistics. We define

$$SS_{tot} = \sum_{i=1}^{I} \sum_{j=1}^{n} w_{ij} \left(X_{ij} - \overline{X}\right)^2 \qquad \text{(sum of squares total)},$$

$$SS_w = \sum_{i=1}^{I} \sum_{j=1}^{n} w_{ij} \left(X_{ij} - X_i\right)^2 \qquad \text{(sum of squares within groups)},$$

$$SS_b = \sum_{i=1}^{I} w_{i\bullet} \left(X_i - \overline{X}\right)^2 \qquad \text{(sum of squares between groups)}.$$

It is well known and easy to show that

$$SS_{tot} = SS_w + SS_b.$$

We now have

$$\begin{aligned} &\text{if } \widehat{\tau}^2 = 0, \text{then } \widehat{\alpha}_i = 0, \\ &\text{if } \widehat{\tau}^2 > 0, \text{ then } \widehat{\alpha}_i = \frac{w_{i\bullet}}{w_{i\bullet} + \frac{\widehat{\sigma}^2}{\widehat{\tau}^2}} = \frac{w_{i\bullet}}{w_{i\bullet} + \widehat{\kappa}}, \end{aligned}$$

where $\widehat{\kappa}$ can be rewritten as follows:

$$\begin{aligned} \widehat{\kappa} &= \frac{\widehat{\sigma}^2}{\widehat{\widehat{\tau}}^2} \\ &= c^{-1} \cdot \frac{w_{\bullet\bullet}}{I} \cdot \left\{ \frac{SS_b}{I-1} \cdot \left(\frac{SS_w}{I \cdot (n-1)} \right)^{-1} - 1 \right\}^{-1} \\ &= c^{-1} \cdot \frac{w_{\bullet\bullet}}{I} \cdot \{F - 1\}^{-1}, \end{aligned} \tag{4.28}$$

where

$$c = \frac{I-1}{I} \left\{ \sum_{i=1}^{I} \frac{w_{i\bullet}}{w_{\bullet\bullet}} \left(1 - \frac{w_{i\bullet}}{w_{\bullet\bullet}} \right) \right\}^{-1},$$

$$F = \left(\frac{SS_b}{I-1} \right) \left(\frac{SS_w}{I\,(n-1)} \right)^{-1}.$$

Remarks:

- Note that F is just the test statistic used in classical statistics for testing whether the means of observations from normally distributed random variables are all the same. Indeed, if the X_{ij} were conditionally, given the Θ_i, normally distributed with the same mean and the same normalized variance $\left(\operatorname{Var}(X_{ij}|\Theta_i, w_{ij}) = \sigma^2/w_{ij}\right)$, then F would have an F-distribution with $I-1$ degrees of freedom in the nominator and $I \cdot (n-1)$ degrees of freedom in the denominator. A negative value of $\widehat{\widehat{\tau}}^2$ is equivalent to $F < 1$, in which case the F-test would never reject the hypothesis of all means being equal. This explains the remark made at the end of Section 4.8, that a negative outcome of $\widehat{\widehat{\tau}}^2$ should be interpreted as there were no detectable differences in the expected value between the risks.
- In general, $\widehat{\widehat{\mu(\Theta_i)}}^{emp.}$ is no longer unbiased. Although the estimators for the structural parameters are "almost" unbiased (i.e. aside from the exception that no negative values for τ^2 are allowed), in general this is not true for the empirical credibility estimator.
- If no contract dominates (i.e. if $\sum_i (w_{i\bullet}/w_{\bullet\bullet})^2 \to 0$ for $I \to \infty$), then we have

$$\widehat{\widehat{\mu(\Theta_i)}}^{emp.} \xrightarrow{P} \widehat{\widehat{\mu(\Theta_i)}} \qquad \text{for } I \to \infty,$$

$$E\left[\widehat{\widehat{\mu(\Theta_i)}}^{emp.}\right] \longrightarrow \mu_0 \qquad \text{for } I \to \infty.$$

- Since Theorem 4.5 holds true independent of the choice of the structural parameters σ^2 and τ^2, it also holds true for the empirical credibility estimator, that is, the premium resulting from the empirical credibility estimator if applied over the past observation period and over the whole portfolio is equal to the observed total claim amount over the same period and over the whole portfolio.
- From this it also follows that

$$E\left[\sum_{i=1}^{I} \frac{w_{i\bullet}}{w_{\bullet\bullet}} \widehat{\widehat{\mu(\Theta_i)}}^{emp.}\right] = \mu_0,$$

i.e. the weighted average of the empirical credibility estimator taken over the whole portfolio is unbiased.

These last two points indicate that the premium resulting from the empirical credibility estimator over the whole portfolio is on the right level.

4.10 Credibility for Claim Frequencies

Claim frequencies play a central role in the calculation of a multivariate tariff as well as for Bonus–Malus systems. A frequently made assumption is that the claim numbers can be modelled as random variables having a conditional Poisson distribution. Thus we have more structure, and this special structure should be taken into account in credibility calculations. This section focuses on this set-up.

Again we consider the situation where we have a portfolio of risks (or risk groups) and associated observations, namely

N_{ij}	number of claims of risk i in year j,
w_{ij}	associated weight, e.g. years at risk,
$F_{ij} = N_{ij}/w_{ij}$	claim frequency.

Analogously to the Bühlmann–Straub model we get the following:

Model Assumptions 4.7 (claim frequency)
The risk i is characterized by its individual risk profile ϑ_i, which is itself the realization of a random variable Θ_i, and we assume that:

Fr1: Given Θ_i the N_{ij} $(j = 1, 2, \dots)$ are independent and Poisson distributed with Poisson parameter $\lambda_{ij}(\Theta_i) = w_{ij}\Theta_i\lambda_0$;
Fr2: The pairs $(\Theta_1, \mathbf{N}_1), (\Theta_2, \mathbf{N}_2), \dots$ are independent, and $\Theta_1, \Theta_2, \dots$ are independent and identically distributed with $E[\Theta_i] = 1$.

Remark:

- The Θ_i reflect the frequency structure between the risks, whereas λ_0 is the overall frequency level, in analogy to the distinction between *tariff structure* and *tariff level*, which is quite common in insurance practice. We will come back to this point below.

Model Assumptions 4.7 imply that

$$\begin{aligned} E[F_{ij} | \Theta_i] &= \Theta_i\lambda_0, \\ \operatorname{Var}[F_{ij} | \Theta_i] &= \frac{\Theta_i\lambda_0}{w_{ij}}. \end{aligned}$$

The conditions of the Bühlmann–Straub model (Model Assumptions 4.1) are therefore satisfied for the claim frequencies with

$$\begin{aligned} \mu(\Theta_i) &= \lambda(\Theta_i) = \Theta_i\lambda_0, \\ \sigma^2(\Theta_i) &= \Theta_i\lambda_0. \end{aligned}$$

The structural parameters are

$$\begin{aligned} \lambda_0 &:= E[\lambda(\Theta_i)], \\ \sigma^2 &:= E\left[\sigma^2(\Theta_i)\right] = \lambda_0, \\ \tau^2 &:= \operatorname{Var}[\lambda(\Theta_i)] = \lambda_0^2 \operatorname{Var}[\Theta_i]. \end{aligned}$$

We therefore have:

Corollary 4.8 (claim frequency). *The credibility estimator and the homogeneous credibility estimator for the (absolute) claim frequency under Model Assumptions 4.7 are given by*

$$\begin{aligned}
\widehat{\widehat{\lambda(\Theta_i)}} &= \alpha_i F_i + (1-\alpha_i)\,\lambda_0 \;=\; \lambda_0 + \alpha_i\,(F_i - \lambda_0)\,, \\
\text{where } F_i &= \sum_i \frac{w_{ij}}{w_{i\bullet}}\,F_{ij}, \\
\alpha_i &= \frac{w_{i\bullet}}{w_{i\bullet}+\kappa}, \\
\kappa &= \frac{\lambda_0}{\tau^2} = \left(\lambda_0 \operatorname{Var}[\Theta_i]\right)^{-1}.
\end{aligned} \tag{4.29}$$

$$\begin{aligned}
\widehat{\widehat{\lambda(\Theta_i)}}^{\text{hom}} &= \alpha_i F_i + (1-\alpha_i)\,\widehat{\widehat{\lambda_0}} \;=\; \widehat{\widehat{\lambda_0}} + \alpha_i\left(F_i - \widehat{\widehat{\lambda_0}}\right), \\
\text{where } \widehat{\widehat{\lambda_0}} &= \sum\nolimits_i \frac{\alpha_i}{\alpha_\bullet}\,F_i.
\end{aligned}$$

Remarks:

- Note that $\sqrt{\operatorname{Var}[\Theta_i]}$ is equal to the coefficient of variation (= standard deviation divided by expected value) of the individual claim frequency $F_i^{\text{ind}} = \lambda(\Theta_i) = \Theta_i \lambda_0$.
- By multiplying the numerator and denominator on the right-hand side of formula (4.29) by λ_0 we see that the credibility weight can also be written as

$$\begin{aligned}
\alpha_i &= \frac{\nu_{i\bullet}}{\nu_{i\bullet}+\widetilde{\kappa}}, \\
\text{where } \nu_{i\bullet} &= \lambda_0 w_{i\bullet} = \text{a priori expected number of claims}, \\
\widetilde{\kappa} &= \frac{1}{\operatorname{Var}[\Theta_i]} = \left[\operatorname{CoVa}\left(F_i^{\text{ind}}\right)\right]^{-2}.
\end{aligned} \tag{4.30}$$

  $\widetilde{\kappa}$ has now a direct interpretation: it is the coefficient of variation of the individual claim frequency $F_i^{\text{ind}} = \lambda(\Theta_i)$ to the power minus two. In many applications in practice, one has some feeling about $\operatorname{CoVa}(F_i^{\text{ind}})$, which is often in the range between 10% and 50% with a resulting $\widetilde{\kappa}$ in the range between 100 and 4.
- If the observed claim numbers in the individual cells are large (several hundred claims), the credibility weights are also very large and lie in the neighbourhood of 1. In such cases one can simply rely on the observed individual claim frequency and there is hardly a need for credibility, since the results of the credibility estimates deviate very little from the observed individual claim frequencies. Example 4.1 on page 103 shows a numerical illustration. However, credibility for claim frequencies makes sense in situations with a small number of claims, for instance if one is interested

in the claim frequency of large claims (large claims are small in number, but are often responsible for half or even more of the claims load). For a numerical illustration, see Example 4.2 on page 104. Credibility for claim frequencies is also used for Bonus–Malus systems where an individual risk in the physical sense is considered or where the risk classes considered are rather small (e.g. a particular fleet of cars).

- From Theorem 4.5 it follows that

$$\sum_{ij} w_{ij} \widehat{\widehat{\lambda(\Theta_i)}}^{\text{hom}} = \sum_{i,j} w_{ij} F_{ij} = N_{\bullet\bullet},$$

respectively

$$\sum_{ij} \frac{w_{ij}}{w_{\bullet\bullet}} \widehat{\widehat{\lambda(\Theta_i)}}^{\text{hom}} = \sum_{ij} \frac{w_{ij}}{w_{\bullet\bullet}} F_{ij} = \overline{F}.$$

Whereas the structural parameter λ_0 represents the overall frequency level and will be crucial for the tariff level, the random variables Θ_i show the structure of the portfolio with regard to claim frequency and are important for the tariff structure. A value of say 1.2 for Θ_i means that the frequency of risk i is 20% higher than the average. Often one is directly interested in this structure and hence in estimating the random variables Θ_i. Of course, because of the linearity property of the credibility estimators (see Exercise 3.1), the credibility estimator of Θ_i is immediately obtained by dividing $\widehat{\widehat{\lambda(\Theta_i)}}$ from Corollary 4.8 by λ_0. Therefore it is natural to consider the "standardized" claim frequencies

$$F_{ij}^{\text{st}} := \frac{F_{ij}}{\lambda_0} = \frac{N_{ij}}{\nu_{ij}},$$

where $\nu_{ij} = w_{ij}\lambda_0 =$ a priori expected claim number.

Considering standardized frequencies has a further big advantage: in Model Assumptions 4.7 it was assumed that the a priori claim frequency is the same for all risks and for all observation years. However, there are situations in practice where this assumption is not meaningful. For instance, if we consider fleets of cars, then the a priori claim frequency will depend on the types of cars in the fleet and possibly on other explanatory variables. Sometimes one also finds trends over the years, such that time homogeneity does not hold true either. But if we consider standardized frequencies, we can easily deal with such situations. Formally, we replace *Fr1* in Model Assumptions 4.7 by the following weaker and more general condition:

- *Fr1'*:Given Θ_i the N_{ij} $(j = 1, 2, \dots)$ are independent and Poisson distributed with Poisson parameter $\lambda_{ij}(\Theta_i) = \Theta_i \nu_{ij}$, where ν_{ij} is the a priori expected claim number of risk i in year j.

Note that the standardized frequencies

$$F_{ij}^{\mathrm{st}} := \frac{N_{ij}}{\nu_{ij}}$$

all have an a priori expected value of one. Note also that we are in the same formal situation as in Model 4.7: we just have to put there $\lambda_0 = 1$ and $w_{ij} = \nu_{ij}$. We therefore have:

Corollary 4.9 (standardized claim frequency). *The credibility estimator for the standardized claim frequency Θ_i under Model Assumptions 4.7 is given by*

$$\widehat{\widehat{\Theta_i}} = 1 + \alpha_i \left(F_i^{\mathrm{st}} - 1 \right),$$

$$\text{where } F_i^{\mathrm{st}} = \sum\nolimits_j \frac{\nu_{ij}}{\nu_{i\bullet}} F_{ij}^{\mathrm{st}},$$

$$\alpha_i = \frac{\nu_{i\bullet}}{\nu_{i\bullet} + \widetilde{\kappa}},$$

$$\widetilde{\kappa} = \left(\mathrm{Var}\left[\Theta_i\right] \right)^{-1}.$$

Remark:

- Under Model Assumptions 4.7 we have

$$\nu_{ij} = w_{ij} \lambda_0,$$

$$\frac{\nu_{ij}}{\nu_{i\bullet}} = \frac{w_{ij}}{w_{i\bullet}},$$

  and $\widetilde{\kappa}$ has a direct interpretation: it is the coefficient of variation of the individual claim frequency F^{ind} to the power -2.

In the homogeneous case, the a priori expected claim numbers ν_{ij} are only known up to a constant factor. For instance, under Model Assumptions 4.7, $\nu_{ij} = w_{ij}\lambda_0$ depends on λ_0, which is unknown and which has to be estimated from the data too. The question arises, how we should then define standardized frequencies?

We suggest standardizing in the homogeneous case in such a way that the resulting numbers $\widetilde{\nu}_{ij}$ satisfy the condition

$$\widetilde{\nu}_{\bullet\bullet} = N_{\bullet\bullet}. \tag{4.31}$$

This means that the unknown multiplicative factor is chosen in such a way that, over the whole portfolio, there is equality between the a priori expected claim number and the observed claim number. Under Model Assumptions 4.7

$$\widetilde{\nu}_{ij} = w_{ij} \overline{F},$$

where $\overline{F}$ is the observed claim frequency over the whole portfolio. Note that in this case

$$F_{ij}^{\mathrm{st}} = \frac{F_{ij}}{\overline{F}},$$

which is very natural to look at.

The rule (4.31) has the advantage that the numbers $\widetilde{\nu}_{ij}$ are known at the very beginning and do not depend on parameters to be estimated. With this rule it also follows that

$$\sum_i \frac{\widetilde{\nu}_{i\bullet}}{\widetilde{\nu}_{\bullet\bullet}} \widehat{\widehat{\Theta}}_i^{\mathrm{hom}} = 1.$$

Corollary 4.8 still holds for the standardized frequencies and corresponding "weights" $\widetilde{\nu}_{ij}$ fulfilling (4.31), and we get:

Corollary 4.10 (hom. estimator of the standardized frequency). *The homogeneous credibility estimator of Θ_i is given by*

$$\begin{aligned}
\widehat{\widehat{\Theta}}_i^{\mathrm{hom}} &= \widehat{\widehat{\vartheta}}_0 + \alpha_i \left(F_i^{\mathrm{st}} - \widehat{\widehat{\vartheta}}_0 \right), \\
\textit{where } \alpha_i &= \frac{\nu_{i\bullet}}{\nu_{i\bullet} + \widetilde{\kappa}}, \\
\widetilde{\kappa} &= \left(\mathrm{Var}\left[\Theta_i\right] \right)^{-1}, \\
\widehat{\widehat{\vartheta}}_0 &= \sum_i \frac{\alpha_i}{\alpha_\bullet} F_i^{\mathrm{st}}.
\end{aligned}$$

Remarks:

- Under Model Assumptions 4.7, we have

$$\widehat{\widehat{\Theta}}_i^{\mathrm{hom}} = \frac{\widehat{\widehat{\lambda(\Theta_i)}}^{\mathrm{hom}}}{\overline{F}},$$

  and hence

$$\widehat{\widehat{\vartheta}}_0 = \frac{\widehat{\widehat{\lambda}}_0}{\overline{F}}.$$

Estimation of structural parameters:
A consequence of the Poisson assumption is that the structural parameters σ^2 and λ_0 are equal. To have this equality also for the estimated structural parameters, we can use the following iterative procedure:

First we define

$$\begin{aligned}
c &:= \frac{I-1}{I} \left\{ \sum_i \frac{w_{i\bullet}}{w_{\bullet\bullet}} \left(1 - \frac{w_{i\bullet}}{w_{\bullet\bullet}} \right) \right\}^{-1}, \\
T &:= \frac{I}{I-1} \cdot \sum_i \frac{w_{i\bullet}}{w_{\bullet\bullet}} \left(F_i - \overline{F} \right)^2, \\
\text{where } \overline{F} &= \sum_i \frac{w_{i\bullet}}{w_{\bullet\bullet}} F_i.
\end{aligned}$$

Then we start with

$$\begin{aligned} {}^{(0)}\widehat{\widehat{\lambda_0}} &= \overline{F}, \\ {}^{(0)}\widehat{\tau}^2 &= c\left\{T - \frac{I^{(0)}\widehat{\lambda}_0}{w_{\bullet\bullet}}\right\}. \end{aligned} \tag{4.32}$$

The iteration from step n to step $n+1$ is

$$\begin{aligned} {}^{(n)}\widehat{\kappa} &= \frac{{}^{(n)}\widehat{\widehat{\lambda_0}}}{{}^{(n)}\widehat{\tau}^2}, \qquad {}^{(n)}\alpha_i = \frac{w_{i\bullet}}{w_{i\bullet} + {}^{(n)}\widehat{\kappa}}, \\ {}^{(n+1)}\widehat{\widehat{\lambda_0}} &= \sum\nolimits_i \frac{{}^{(n)}\alpha_i}{{}^{(n)}\alpha_\bullet} F_i, \\ {}^{(n+1)}\widehat{\tau}^2 &= c \cdot \left\{T - \frac{I^{(n+1)}\widehat{\widehat{\lambda_0}}}{w_{\bullet\bullet}}\right\}. \end{aligned}$$

Remarks:

- The same procedure for estimating the structural parameters also applies for estimating the structural parameters for the "standardized" frequency. Then we have simply to replace the weights w_{ij} by ν_{ij} ($\nu_{ij} = w_{ij}\overline{F}$ under Model Assumptions 4.7), and instead of $\widehat{\widehat{\lambda_0}}$, we estimate $\widehat{\widehat{\vartheta_0}}$.
- Note that it is sufficient to know the multiyear claim frequencies F_i and that we do not need the yearly observations to estimate the structural parameters.

Example 4.1 (frequency of "normal" claims)

The table below shows the number and frequencies of normal claims of a motor liability portfolio for 21 regions.

With the recursive procedure for estimating the structural parameters we obtain after two recursion steps:

$$\begin{aligned} \widehat{\widehat{\lambda_0}} &= 88.3‰, \\ \widehat{\tau}^2 &= 2.390 \cdot 10^{-4}, \end{aligned}$$

with a resulting

$$\widehat{\kappa} = 370 \quad \text{(relating to the number of risks)},$$

and an estimated coefficient of variation

$$\widehat{\mathrm{CoVa}\left(\lambda(\Theta_i)\right)} = \frac{\widehat{\tau}}{\widehat{\widehat{\lambda_0}}} = 17.5\%,$$

or for the standardized frequency Θ_i:

$$\widehat{\widehat{\vartheta_0}} = 0.976,$$
$$\widehat{\widehat{\tau}}^2 = 2.923 \cdot 10^{-2},$$

with a resulting

$$\widehat{\widehat{\kappa}} = 33 \text{ (relating to the a priori expected claim number)},$$

and an estimated coefficient of variation

$$\widehat{\mathrm{CoVa}}\left(\Theta_i\right) = \frac{\widehat{\widehat{\vartheta_0}}}{\widehat{\widehat{\tau}}} = 17.5\%.$$

Of course, $\mathrm{CoVa}\left(\lambda(\Theta_i)\right)$ must be identical to $\mathrm{CoVa}\left(\Theta_i\right)$, which must also hold for the estimated values.

Group i	number of risks	number of claims a priori expected	number of claims observed	Frequency ‰	Frequency standardized	credibility weight %	Cred. Estimator ‰	Cred. Estimator standardized
1	50 061	4 534	3 880	77.5	0.86	99.3	77.6	0.86
2	10 135	918	794	78.3	0.86	96.5	78.7	0.87
3	121 310	10 987	8 941	73.7	0.81	99.7	73.7	0.81
4	35 045	3 174	3 448	98.4	1.09	99.0	98.3	1.09
5	19 720	1 786	1 672	84.8	0.94	98.2	84.9	0.94
6	39 092	3 541	5 186	132.7	1.47	99.1	132.2	1.46
7	4 192	380	314	74.9	0.83	91.9	76.0	0.84
8	19 635	1 778	1 934	98.5	1.09	98.2	98.3	1.09
9	21 618	1 958	2 285	105.7	1.17	98.3	105.4	1.16
10	34 332	3 109	2 689	78.3	0.86	98.9	78.4	0.87
11	11 105	1 006	661	59.5	0.66	96.8	60.5	0.67
12	56 590	5 125	4 878	86.2	0.95	99.4	86.2	0.95
13	13 551	1 227	1 205	88.9	0.98	97.3	88.9	0.98
14	19 139	1 733	1 646	86.0	0.95	98.1	86.0	0.95
15	10 242	928	850	83.0	0.92	96.5	83.2	0.92
16	28 137	2 548	2 229	79.2	0.87	98.7	79.3	0.88
17	33 846	3 065	3 389	100.1	1.11	98.9	100.0	1.10
18	61 573	5 577	5 937	96.4	1.06	99.4	96.4	1.06
19	17 067	1 546	1 530	89.6	0.99	97.9	89.6	0.99
20	8 263	748	671	81.2	0.90	95.7	81.5	0.90
21	148 872	13 483	15 014	100.9	1.11	99.8	100.8	1.11
Total	763 525	69 153	69 153	90.6	1.00		90.6	1.00

Table: *Claim frequency of "normal" claims*

The results show, that the deviations between the credibility estimates and the observed individual frequencies are very small and negligible for practical purposes, such that credibility gives very little additional insight.

Example 4.2 (frequency of large claims)

The table below shows the number of large claims (claims above CHF 50 000) and the corresponding claim frequencies for the same motor liability portfolio as in Example 4.1. The number of large claims is rather small compared to the number of "normal" claims. However, they make more than half of the claims load, and therefore it can be crucial in tariff making, to estimate as precisely as possible the frequency of large claims.

Group i	number of risks	number of xs-claims a priori expected	observed	Frequency %o	standardized	credibility weight %	Cred. Estimator %o	standardized
1	50 061	45	42	0.84	0.93	62.5	0.86	0.95
2	10 136	9	5	0.49	0.55	25.2	0.79	0.88
3	121 310	109	105	0.87	0.96	80.1	0.87	0.97
4	35 045	32	32	0.91	1.01	53.8	0.90	1.00
5	19 720	18	32	1.62	1.80	39.6	1.18	1.31
6	39 092	35	37	0.95	1.05	56.5	0.92	1.02
7	4 192	4	7	1.67	1.85	12.2	0.99	1.10
8	19 635	18	16	0.81	0.90	39.5	0.86	0.96
9	21 618	20	18	0.83	0.92	41.8	0.87	0.96
10	34 332	31	30	0.87	0.97	53.3	0.88	0.98
11	11 105	10	7	0.63	0.70	27.0	0.82	0.91
12	56 590	51	52	0.92	1.02	65.3	0.91	1.01
13	13 551	12	10	0.74	0.82	31.1	0.85	0.94
14	19 139	17	15	0.78	0.87	38.9	0.85	0.94
15	10 242	9	14	1.37	1.51	25.4	1.01	1.12
16	28 137	25	20	0.71	0.79	48.3	0.81	0.89
17	33 846	31	17	0.50	0.56	53.0	0.69	0.76
18	61 573	56	37	0.60	0.67	67.2	0.70	0.77
19	17 067	15	21	1.23	1.36	36.2	1.02	1.13
20	8 263	7	9	1.09	1.21	21.6	0.94	1.04
21	148 872	134	163	1.09	1.21	83.2	1.06	1.18
Total	763 525	689	689	0.90	1.00		0.90	1.00

Table: *Claim frequency of large claims*

With the recursive procedure we obtain (after two recursion steps) the following estimates for the structural parameters:

$$\widehat{\lambda}_0 = 0.90\text{‰}$$
$$\widehat{\tau}^2 = 2.978 \cdot 10^{-8}$$

with a resulting

$$\widehat{\kappa} = 30\,058 \quad \text{(relating to the number of risks),}$$

and an estimated coefficient of variation

$$\widehat{\mathrm{CoVa}\,(\lambda(\Theta_i))} = \frac{\widehat{\tau}}{\widehat{\lambda_0}} = 19.3\%,$$

or for the standardized frequency Θ_i:

$$\widehat{\widehat{\vartheta_0}} = 0.992,$$
$$\widehat{\widehat{\tau}}^2 = 3.657 \cdot 10^{-2},$$

with a resulting

$$\widehat{\widehat{\kappa}} = 27 \quad \text{(relating to the a priori expected claim number),}$$

and an estimated coefficient of variation

$$\widehat{\mathrm{CoVa}\,(\Theta_i)} = \frac{\widehat{\widehat{\vartheta_0}}}{\widehat{\widehat{\tau}}} = 19.3\%.$$

Note that the heterogeneity (in terms of CoVa) is about the same as in Example 4.1. This is reflected in the values of $\widetilde{\kappa}$ (kappa for the standardized frequency), which are very close to each other in Examples 4.1 and 4.2. However, this is not the case for the values of κ (kappa value for the absolute claim frequency). Indeed, the value of κ is more difficult to interpret and is determined by the observed over all claim frequency $\overline{F}$, whereas $\widetilde{\kappa}$ has a very direct meaning, as it denotes the number of expected claims to give a credibility weight of 50%. Note also, that $\widehat{\widehat{\vartheta_0}}$ is close to one in both examples, which is usually the case in practical applications.

From the above table we can see that the deviation between the credibility estimates and the individual observations is far from negligible.

4.11 Credibility for Claim Sizes

Frequently, in the calculation of premiums, the expected values of claim frequencies and claim sizes are determined and modelled separately. Our interest is in the risk premium which is a product of both of these quantities. For this reason we might ask if it really makes sense to consider the claim frequency and the average claim size separately.

For each risk i, we assume that the following amounts over a given observation period (one year or several years) are known:

N_i	claim number,
w_i	associated weight or volume (e.g. number of years at risk),
$F_i = \frac{N_i}{w_i}$	claim frequency,
$Y_i^{(\nu)}, \quad \nu = 1, 2, \ldots, N_i$	claim sizes, claim amounts,
$S_i = \sum_{\nu=1}^{N_i} Y_i^{(\nu)}$	total aggregate claim amount,
$Y_i = \frac{S_i}{N_i}$	average claim size, average claim amount,
$X_i = F_i \cdot Y_i = \frac{S_i}{w_i}$	claims ratio.

Again, we want to calculate the pure risk premium

$$E\left[X_i \middle| \Theta_i\right] = E\left[F_i \cdot Y_i \middle| \Theta_i\right].$$

A "factorization" into frequency and average claim amount is allowable if

$$E\left[X_i \middle| \Theta_i\right] = E\left[F_i \middle| \Theta_i\right] \cdot E\left[Y_i \middle| \Theta_i\right].$$

This is the case if claim number and claim sizes are conditionally independent. Note that only *conditional*, and not unconditional, independence is required. The conditional independence is an assumption which is frequently appropriate in many situations in insurance practice.

We can now estimate $E[F_i|\Theta_i]$ and $E[Y_i|\Theta_i]$ separately and then multiply the two estimators to get an estimate of $E[X_i|\Theta_i]$. $E[F_i|\Theta_i]$ and $E[Y_i|\Theta_i]$ are possibly not independent. If this were the case, this would mean, that we could possibly learn something from the claim frequency with respect to the expected value of the claim size and vice versa. When estimating claim frequency and claim averages separately, we ignore this additional information. In Chapter 7 we will present a technique to show how this additional information could be taken into account in a credibility frame work.

The first factor, the claim frequency, was the subject of the previous section. In this section, we concern ourselves with the second part, the determination of the expected value of the individual claim size.

For claim sizes, we make the following assumptions:

Model Assumptions 4.11 (claim sizes)
Conditional on Θ_i, the claim sizes $Y_i^{(\nu)}$, $\nu = 1, 2, \ldots, N_i$, are independent and identically distributed with the same distribution as the random variable Y and the conditional moments

$$\mu(\Theta_i) = E[Y|\Theta_i],$$
$$\sigma^2(\Theta_i) = \mathrm{Var}(Y|\Theta_i).$$

Note that Model Assumptions 4.11 imply the following conditional moments for the average claim sizes Y_i:

$$E[Y_i|\Theta_i] = \mu(\Theta_i),$$
$$(\mathrm{Var}[Y_i|\Theta_i, N_i = n_i]) = \frac{\sigma^2(\Theta_i)}{n_i}.$$

We need to estimate $\mu(\Theta_i) = E[Y|\Theta_i]$. This situation also fits in very nicely with the Bühlmann–Straub model (Model Assumptions 4.1). We must (mentally) substitute the year index j by the "count index" ν. The weights $w_{i\nu}$ are 1, and instead of the number of observation years n we have the number n_i of observed claims. The point is that both ν and j denote the index numbering the observations of the same risk. This index is often interpreted as time, but of course, as this example shows, other interpretations are possible.

Corollary 4.12. *The credibility estimator of $\mu(\Theta_i)$ is given by*

$$\widehat{\widehat{\mu(\Theta_i)}} = \mu_0 + \alpha_i (Y_i - \mu_0),$$

$$\textit{where } \alpha_i = \frac{n_i}{n_i + \frac{\sigma^2}{\tau^2}}.$$

Remarks:

- We obtain for the credibility coefficient

$$\begin{aligned} \kappa &= \frac{\mu_0^2}{\tau^2} \cdot \frac{\sigma^2}{\mu_0^2}, \\ &= \frac{\mu_0^2}{\tau^2} \cdot \left(\frac{\sigma^2 + \tau^2}{\mu_0^2} \right) - 1, \\ &= \{\mathrm{CoVa}\left(\mu\left(\Theta_i\right)\right)\}^{-2} \cdot \left(\mathrm{CoVa}\left(Y\right)\right)^2 - 1. \end{aligned} \tag{4.33}$$

- The coefficient of variation of the individual claim size $\mu(\Theta_i)$ is often in the range between 10% and 30%, whereas the (unconditional) coefficient of variation of the claim size (CoVa (Y)) depends on the line of business and the claim segments (all claims, normal claims, large claims). These values are often known in insurance practice (see table below).

Line of business	Coefficient of variation (CoVa (Y))		
	all claims	claims	claims
		< CHF 50 000	≧ CHF 50 000
motor liability	9.0	1.7	1.5
motor hull	2.0		
general liability	7.5	2.5	2.0
workmen compensation in case of accident	9.0	2.5	1.2
workmen compensation in case of sickness	2.5		
industry fire	7.0	2.0	2.0
household	2.0		

Table: *Claim sizes: coefficient of variation from a Swiss insurance company*

For instance, in the motor liability line and for CoVa $(\mu(\Theta_i)) = 20\%$, we obtain, using formula (4.33), that $\kappa \cong 2\,000$ for all claims, $\kappa \cong 50$ for normal claims and $\kappa \cong 30$ for large claims. If we consider all claims, we need about 2 000 claims for a credibility weight α_i of 50%. However, if we separate "normal" and "large" claims, we often have a similar situation to the claim frequencies: for the large mass of normal claims, the credibility estimators for the average claim size will often differ little from the observed individual average claim amount.

In contrast to the usual situation in the Bühlmann–Straub model, the modelling of the claim size has to describe a situation where the number n_i can be substantially different from one unit to be rated to another. This requires a modification to the estimation procedure for the structural parameter σ^2.

Again let

$$S_i = \frac{1}{n_i - 1} \sum_{\nu=1}^{n_i} \left(Y_i^{(\nu)} - Y_i \right)^2,$$

$$\text{where} \quad Y_i = \frac{1}{n_i} \sum_{\nu=1}^{n_i} Y_i^{(\nu)}.$$

The random variables S_i $(i = 1, 2, \ldots)$ are all unbiased and independent estimators for σ^2. Again the question arises as to how they should be weighted. The considerations already discussed on page 94 can help us again here. In the case of normally distributed random variables, according to (4.23), we have

$$E\left[\left(S_i - \sigma^2\right)^2\right] = \frac{2}{n_i - 1} E\left[\sigma^4\left(\Theta_i\right)\right] + E\left[\left(\sigma^2(\Theta_i) - \sigma^2\right)^2\right].$$

If we ignore the second summand, we should choose the weights proportional to $n_i - 1$. Alternatives, i.e. weights which would be better in the majority of cases, are not available. On the basis of these considerations, the following estimator is a reasonable choice:

$$\widehat{\sigma^2} = \frac{1}{n_\bullet - I} \sum_{i=1}^{I} \sum_{\nu=1}^{n_i} \left(Y_i^{(\nu)} - Y_i \right)^2. \tag{4.34}$$

Remark:

- It may happen that n_i is equal to one or zero for some risk groups. These risk groups do not contribute to the estimation of the "within" variance component σ^2. To be strict, we should then use I^* = number of risks groups with $n_i \geqslant 2$ instead of I in formula (4.34). However, the difference in the results when using I instead of I^* in such situations will often be so small that it can be ignored.

For estimating τ^2 we can use the usual estimator of Section 4.8, formula (4.27), that is

$$\widehat{\tau}^2 = \max\left(\widehat{\widehat{\tau}}^2, 0\right),$$

$$\text{where } \widehat{\widehat{\tau}}^2 = c \cdot \left\{ \frac{I}{I-1} \sum_{i=1}^{I} \frac{n_i}{n_\bullet} \left(Y_i - \overline{Y}\right)^2 - \frac{I \widehat{\sigma}^2}{n_\bullet} \right\}, \tag{4.35}$$

$$\overline{Y} = \sum_{i=1}^{I} \frac{n_i}{n_\bullet} Y_i,$$

$$c = \frac{I-1}{I} \left\{ \sum_{i=1}^{I} \frac{n_i}{n_\bullet} \left(1 - \frac{n_i}{n_\bullet}\right) \right\}^{-1}.$$

As an alternative, it is sometimes convenient to make use of the decomposition of the sum of squares, which are in this case:

$$SS_{tot} = \sum_{i=1}^{I} \sum_{\nu=1}^{n_i} \left(Y_i^{(\nu)} - \overline{Y}\right)^2 \qquad \text{(sum of squares total)},$$

$$SS_w = \sum_{i=1}^{I} \sum_{\nu=1}^{n_i} \left(Y_i^{(\nu)} - Y_i\right)^2 \qquad \text{(sum of squares within groups)},$$

$$SS_b = \sum_{i=1}^{I} n_i \left(Y_i - \overline{Y}\right)^2 \qquad \text{(sum of squares between groups)}.$$

It holds again that

$$SS_{tot} = SS_w + SS_b.$$

Analogously to (4.28) we can write

$$\begin{aligned} \widehat{\kappa} &= \frac{\widehat{\sigma}^2}{\widehat{\widehat{\tau}}^2} \\ &= c^{-1} \cdot \frac{n_\bullet}{I} \cdot \left\{ \frac{SS_b}{SS_w} \cdot \frac{n_\bullet - I}{I - 1} - 1 \right\}^{-1}. \end{aligned} \tag{4.36}$$

Remarks:

- The coefficient of variation of the claim size is usually estimated by

$$\widehat{\text{CoVa}}(Y) = \frac{\sqrt{\frac{1}{n_\bullet - 1} \sum_{i=1}^{I} \sum_{\nu=1}^{n_i} \left(Y_i^{(\nu)} - \overline{Y}\right)^2}}{\overline{Y}}.$$

  Hence

$$SS_{tot} = (n_\bullet - 1)\, \overline{Y}^2\, \widehat{\text{CoVa}}(Y)^2. \tag{4.37}$$

- Sometimes, there are only available summary statistics with the observed average claim size Y_i, but not the individual claim amounts $Y_i^{(\nu)}$. But on the other hand, one has from other sources an estimation of the coefficient of variation $\widehat{\text{CoVa}}(Y)$. Then one can still use formula (4.36) by setting $SS_w = SS_{tot} - SS_b$ and by estimating the "unknown" SS_{tot} by

$$\widehat{SS_{tot}} = (n_\bullet - 1)\, \overline{Y}^2\, \widehat{\text{CoVa}}(Y)^2.$$

- If one knows the observed average claim amounts Y_{ij} of several years $j = 1, 2, \ldots, n$ as well as the corresponding numbers of claims n_{ij}, then one can of course also estimate the structural parameters by the usual estimators of Section 4.8 and by taking the number of observed claims as weights.

4.12 Credibility for Risk Groups of Known Individual Contracts

Assume that we have a portfolio of risk groups $i = 1, 2, \ldots, I$, each of them consisting of n_i individual contracts with observations $X_i^{(\nu)}$, $\nu = 1, 2, \ldots,$

n_i, and corresponding individual weights $w_{i\nu}$. We further assume that these individual observations fulfil the Bühlmann–Straub model conditions. We then have the following model:

Model Assumptions 4.13

i) Conditional on Θ_i, *the observations* $X_i^{(\nu)}$, $\nu = 1, 2, \ldots, n_i$, *are independent with*

$$E\left[X_i^{(\nu)}\middle|\Theta_i\right] = \mu(\Theta_i),$$

$$\operatorname{Var}\left(X_i^{(\nu)}\middle|\Theta_i\right) = \frac{\sigma^2(\Theta_i)}{w_{i\nu}}.$$

ii) The pairs $(\Theta_1, \mathbf{X}_1)$, $(\Theta_2, \mathbf{X}_2), \ldots$ *are independent, and* Θ_1, $\Theta_2, \ldots$ *are independent and identically distributed.*

Here, the elements of the vectors $\mathbf{X}_i$ are the observations of the individual contracts, that is $\mathbf{X}_i' = \left(X_i^{(1)}, X_i^{(2)}, \ldots, X_i^{(n_i)}\right)$.

Again, this model is in line with the Bühlmann–Straub model, and we just have to substitute the year index j in Model Assumptions 4.1 by the "count index" ν. As for the claim sizes, the only difference to the usual set-up is that the number of contracts (resp. observations) n_i differ between risk groups, which requires a modification of the estimation of σ^2. The estimators of the structural parameters are the same as the ones encountered in Section 4.11. We have simply to replace $Y_i^{(\nu)}, Y_i, \overline{Y}$ in the formulae (4.34) and (4.35) by $X_i^{(\nu)}, X_i$, and $\overline{X}$.

4.13 Modification for the Case of Known a Priori Differences

The model assumption of the Bühlmann–Straub model, that the risks are "a priori equal", is a strong condition that is not fulfilled in many situations in insurance practice. Often, we have available "a priori" information about differences between the expected risk premiums of the individual risks. For example, the current premium could provide such information. The differentiation between risk premiums based on current rates can be used as an a priori differentiation for the new rating of premiums based on newly available statistics and data. Another example is the situation where technical experts make a classification of the risks based on their knowledge and judgement. Examples of such a priori information are the car classification in motor hull insurance or the grouping of businesses into danger classes in collective accident insurance.

In all these cases, the Bühlmann–Straub model is not directly applicable. However, a minor modification of the model assumptions allows us to use this model also in such situations which enormously increases the applicability of the Bühlmann–Straub model in practice.

Model Assumptions 4.14 (a priori differentiation)
The risk i is characterized by its individual risk profile ϑ_i, which is itself the realization of a random variable Θ_i, and:

BS1′:the observations $\{X_{ij} : j = 1, 2, \ldots, n\}$ are, conditional on Θ_i, independent with

$$E\left[X_{ij} \left|\Theta_i\right.\right] = a_i \, \mu(\Theta_i), \tag{4.38}$$

$$\operatorname{Var}\left[X_{ij} \left|\Theta_i\right.\right] = \sigma^2\left(\Theta_i\right) / \left(\frac{w_{ij}}{a_i}\right), \tag{4.39}$$

where the a_i's are known constants.

BS2′:the same condition as in the Bühlmann–Straub model (Model Assumptions 4.1), i.e. the pairs $(\Theta_1, \mathbf{X}_1)$, $(\Theta_2, \mathbf{X}_2), \ldots$ are independent and Θ_1, $\Theta_2, \ldots$ are independent and identically distributed.

For the transformed random variables $Y_{ij} = X_{ij} / a_i$ we then have

$$E\left[Y_{ij} \left|\Theta_i\right.\right] = \mu(\Theta_i), \tag{4.40}$$

$$\operatorname{Var}\left[Y_{ij} \middle| \Theta_i\right] = \sigma^2(\Theta_i) / \left(a_i w_{ij}\right), \tag{4.41}$$

i.e. the transformed variables Y_{ij} together with the transformed weights $w_{ij}^* = a_i w_{ij}$ satisfy the conditions of the Bühlmann-Straub model (Model Assumptions 4.1). Therefore the credibility estimator of $\mu\left(\Theta_i\right)$ based on the Y_{ij}'s is given by

$$\widehat{\widehat{\mu(\Theta_i)}} = \alpha_i Y_i + (1 - \alpha_i)\, \mu_0,$$

$$\text{where } \; Y_i = \sum\nolimits_j \frac{w_{ij}}{w_{i\bullet}} Y_{ij} \quad \text{and} \quad \alpha_i = \frac{a_i w_{i\bullet}}{a_i w_{i\bullet} + \frac{\sigma^2}{\tau^2}}.$$

Applying the linearity property of projections we immediately obtain the credibility estimator for $\mu_X\left(\Theta_i\right) := a_i \mu\left(\Theta_i\right)$, namely:

Corollary 4.15 *Given Model Assumptions 4.14, the (inhomogeneous) credibility estimator is given by*

$$\widehat{\widehat{\mu_X(\Theta_i)}} = a_i \, \widehat{\widehat{\mu(\Theta_i)}} = \alpha_i X_i + (1 - \alpha_i)\, \mu_i,$$

$$\textit{where} \quad \mu_i = a_i \mu_0,$$

$$\alpha_i = \frac{a_i w_{i\bullet}}{a_i w_{i\bullet} + \frac{\sigma^2}{\tau^2}}.$$

Remarks:

- Note that the a_i must be known only up to a constant factor; only "relative" differences between the risks play a role.
- The whole argument also works when, instead of, or in addition to the a priori differentiation between the risks, we have an a priori differentiation which is year dependent (instead of the constants a_i, we have constants a_j or a_{ij}).
- The balance property stated in Theorem 4.5 (balance between aggregated premiums and claim amounts) continues to be valid, because

$$\sum_{i,j} w_{ij} \widehat{\widehat{\mu_X(\Theta_j)}}^{\text{hom}} = \sum_{i,j} a_j w_{ij} \widehat{\widehat{\mu(\Theta_j)}}^{\text{hom}} = \sum_{i,j} a_j w_{ij} Y_{ij} = \sum_{i,j} w_{ij} X_{ij}. \tag{4.42}$$

  Hence, if the weights w_{ij} are the volume measure of the underlying tariff, which is the standard case, then the resulting credibility premium aggregated over the whole portfolio and applied to the past observation period would exactly match the observed total aggregate claim amount.
- The variance condition (4.41) is motivated by the following considerations:
  - If $X_{ij} = S_{ij}/w_{ij}$ and the S_{ij} are conditionally Poisson distributed with Poisson parameter $\lambda_{ij}(\Theta_i) = a_i w_{ij} \lambda(\Theta_i)$, then we have directly (4.39).
  - The form of the conditional variance as in (4.41) is necessary to retain the above-stated convenient practical property (4.42).

4.14 Example of Another Kind of Application

Credibility and the Bühlmann–Straub model can also be used in situations other than the direct calculation of the pure risk premium. In this section we show an interesting application of the Bühlmann–Straub model to estimate the parameter of the Pareto distribution (see [Ryt90]).

In excess of loss reinsurance, the primary insurer pays claims up to an agreed threshold (called retention) $c > 0$, and the excess part (often cut off at an upper limit) is paid by the reinsurer. As usual we assume that the claim amounts are i.i.d.. In addition we assume that the claims exceeding the retention limit c have a Pareto distribution, i.e.

$$P(Y > y) = \left(\frac{y}{c}\right)^{-\vartheta}.$$

Observe that under the Pareto law, the distribution of claim amounts exceeding any higher limit c' is also Pareto with parameters ϑ and c'. Hence ϑ is independent of the retention limit.

The problem is to estimate the parameter ϑ.

We consider the reinsurance contract i, and let $Y_i^{(\nu)}$ $(\nu = 1, 2, \ldots, n_i)$ be the claim amounts (reindexed) of all claims which have exceeded the threshold c_i. It is easy to see that

$$\ln\left(\frac{Y_i^{(\nu)}}{c_i}\right) \text{ is exponentially distributed with parameter } \vartheta_i. \tag{4.43}$$

From this we get the maximum-likelihood estimator

$$\widehat{\vartheta}_i^{MLE} = \left(\frac{1}{n_i}\sum_{\nu=1}^{n_i} \ln\left(\frac{Y_i^{(\nu)}}{c_i}\right)\right)^{-1}.$$

From (4.43) and the (conditional) independence of the individual claim amounts it follows that

$$\sum_{\nu=1}^{n_i} \ln\left(\frac{Y_i^{(\nu)}}{c_i}\right) \text{ is Gamma distributed with density } \frac{\vartheta_i^{n_i}}{(n_i-1)!}\, y^{n_i-1}\, e^{-\vartheta_i y}.$$

After some calculation we get

$$E\left[\widehat{\vartheta}_i^{MLE}\right] = \frac{n_i}{n_i-1}\vartheta_i,$$

$$\operatorname{Var}\left[\widehat{\vartheta}_i^{MLE}\right] = \left(\frac{n_i}{n_i-1}\right)^2 \frac{\vartheta_i^2}{n_i-2}.$$

With appropriate normalization we derive from the maximum-likelihood estimator the unbiased estimator

$$\widehat{\vartheta}_i^{MLE^*} = \frac{n_i-1}{n_i}\widehat{\vartheta}_i^{MLE},$$

with

$$E\left[\widehat{\vartheta}_i^{MLE^*}\right] = \vartheta_i, \tag{4.44}$$

$$\operatorname{Var}\left[\widehat{\vartheta}_i^{MLE^*}\right] = \frac{\vartheta_i^2}{n_i-2}, \tag{4.45}$$

$$\operatorname{CoVa}\left(\widehat{\vartheta}_i^{MLE^*}\right) = \frac{1}{\sqrt{n_i-2}}. \tag{4.46}$$

The number of observed "excess claims" n_i is often very small. This means that the estimator $\widehat{\vartheta}_i^{MLE^*}$ has a large coefficient of variation and is therefore rather unreliable. We can get more reliable information when studying a portfolio of similar excess of loss contracts. This calls for credibility methods and brings us to the following:

Model Assumptions 4.16 (credibility model for Pareto)
The risk i is characterized by its individual risk profile ϑ_i, which is itself the realization of a random variable Θ_i, and it holds that:

- $Y_i^{(\nu)}$ $(\nu = 1, 2, \ldots, n_i)$ *are conditionally (given $\Theta_i = \vartheta_i$) independent and Pareto(c_i, ϑ_i) distributed;*
- $\Theta_1, \Theta_2, \ldots, \Theta_I$ *are independent and identically distributed with $E[\Theta_i] = \vartheta_0$ and $\mathrm{Var}[\Theta_i] = \tau^2$.*

According to (4.44) and (4.45) it holds that

$$E\left[\widehat{\vartheta}_i^{MLE^*} \middle| \Theta_i\right] = \Theta_i,$$
$$\mathrm{Var}\left[\widehat{\vartheta}_i^{MLE^*} \middle| \Theta_i\right] = \frac{\Theta_i^2}{(n_i - 2)}.$$

From this it is obvious that the modified (unbiased) ML-estimators $\left\{\widehat{\vartheta}_i^{MLE^*} : i = 1, 2, \ldots, I\right\}$ satisfy Model Assumptions 4.1 of the Bühlmann–Straub model with

$$\mu(\Theta_i) = \Theta_i,$$
$$\sigma^2(\Theta_i) = \Theta_i^2,$$
$$\text{weight } w_i = n_i - 2.$$

The structural parameters in this Pareto credibility model are

$$\vartheta_0 = E[\Theta_i],$$
$$\sigma^2 = E\left[\Theta_i^2\right] = \tau^2 + \vartheta_0^2,$$
$$\tau^2 = \mathrm{Var}[\Theta_i].$$

With this we have derived the following result:

Corollary 4.17. *Under Model Assumptions 4.16, the credibility estimator (based on the statistics of the modified ML-estimators $\widehat{\vartheta}_i^{MLE^*} : i = 1, \ldots, I$) is given by*

$$\widehat{\Theta}_i = \alpha_i \widehat{\vartheta}_i^{MLE^*} + (1 - \alpha_i)\,\vartheta_0,$$

$$\textit{where} \quad \alpha_i = \frac{n_i - 2}{n_i - 1 + \frac{\vartheta_0^2}{\tau^2}}.$$

Remarks:

- Note that $\vartheta_0^2/\tau^2 = \mathrm{CoVa}^{-2}(\Theta_i)$.

- From the experience with many excess-of-loss portfolios, an experienced underwriter often has an a priori idea about the values of ϑ_0 and $\mathrm{CoVa}^{-2}(\Theta_i)$, which he can use in the application of the credibility formula (pure Bayesian procedure). We can, however, also estimate these structural parameters from the data (empirical credibility).
- Compare the result of Corollary 4.17 with the Bayes estimator (2.40).

In the following we show how we can use the data from the collective to estimate the structural parameters ϑ_0 and τ. To simplify the notation we write

$$X_i := \widehat{\vartheta}_i^{MLE^*}.$$

We also assume that $n_i > 2$ for $i = 1, 2, \ldots, I$.

First, we consider the random variable

$$T = \frac{1}{I-1} \sum_i \alpha_i \left(X_i - \widehat{\vartheta}_0\right)^2, \tag{4.47}$$

$$\text{where} \quad \widehat{\vartheta}_0 = \sum_i \frac{\alpha_i}{\alpha_\bullet} X_i. \tag{4.48}$$

Observe that, contrary to the usual volume weights, the credibility weights are taken in (4.47) and that these credibility weights depend on the structural parameters we want to estimate. Hence T is a pseudo estimator.

Lemma 4.18. *If we knew the correct credibility weights α_i, then*

$$E[T] = \tau^2.$$

Proof: We rewrite T and get

$$\begin{aligned} E[T] &= E\left[\frac{1}{I-1}\left\{\sum_i \alpha_i (X_i - \vartheta_0)^2 - \alpha_\bullet \left(\widehat{\vartheta}_0 - \vartheta_0\right)^2\right\}\right] \\ &= \frac{1}{I-1}\left\{\sum_i \alpha_i (\tau^2 + \frac{\sigma^2}{w_{i\bullet}}) - \alpha_\bullet \mathrm{Var}\left[\widehat{\vartheta}_0\right]\right\} \\ &= \frac{1}{I-1}\{I\tau^2 - \tau^2\} \\ &= \tau^2. \end{aligned}$$

The lemma is thus proved. □

We estimate now the structural parameters ϑ_0 and τ^2 so that

$$T\left(\widehat{\vartheta}_0, \widehat{\tau}^2\right) = \widehat{\tau}^2. \tag{4.49}$$

These estimators, as is the case for many maximum-likelihood estimators, are defined by an implicit equation. This can be most easily solved by a recursive procedure. Let $\vartheta_0^{(0)}$ and $\left(\tau^2\right)^{(0)}$ be sensibly chosen starting values and let $\alpha_i^{(0)}$ be the credibility weights based on these values. (4.47) and (4.48) then give new estimators $\vartheta_0^{(1)}$and $\left(\tau^2\right)^{(1)}$. This is repeated until (4.49) is satisfied with the desired accuracy. As a simple starting value we could use, e.g.

$$\vartheta_0^{(0)} = \frac{1}{I}\sum_i X_i$$

$$\left(\tau^2\right)^{(0)} = \frac{1}{I-1}\sum_i \left(X_i - \vartheta_0^{(0)}\right)^2 .$$

4.15 Exercises

Exercise 4.1

The following table shows the observations of a fire portfolio consisting of five risk groups, where

w = weight = sum insured in millions,
S = total claim amount in thousands,
$X = S/w$ = claims ratio in ‰
(= total claim amount per unit sum insured).

risk groups / years		1	2	3	4	5	Total
group 1	w	729	786	872	951	1 019	4 357
	S	583	1 100	262	837	1 630	4 412
	X	0.80	1.40	0.30	0.88	1.60	1.01
group 2	w	1 631	1 802	2 090	2 300	2 368	10 191
	S	99	1 298	326	463	895	3 081
	X	0.06	0.72	0.16	0.20	0.38	0.30
group 3	w	796	827	874	917	944	4 358
	S	1 433	496	699	1 742	1 038	5 408
	X	1.80	0.60	0.80	1.90	1.10	1.24
group 4	w	3 152	3 454	3 715	3 859	4 198	18 378
	S	1 765	4 145	3 121	4 129	3 358	16 518
	X	0.56	1.20	0.84	1.07	0.80	0.90
group 5	w	400	420	422	424	440	2 106
	S	40		169	1 018	44	1 271
	X	0.10	0.00	0.40	2.40	0.10	0.60
Total	w	6 708	7 289	7 973	8 451	8 969	39 390
	S	3 920	7 039	4 577	8 189	6 965	30 690
	X	0.58	0.97	0.57	0.97	0.78	0.78

We want to estimate for each risk group $i = 1, 2, \ldots, 5$ the corresponding individual claims ratio $\mu(\Theta_i)$.

a) Calculate the homogeneous credibility estimates based on the Bühlmann–Straub model (Model Assumptions 4.1) with $\kappa = 3\,000$ mio.
b) As a) but with $\kappa = 6\,000$ mio.
c) As a) but estimate κ (resp. σ^2 and τ^2) from the data (empirical credibility estimator).
d) Check that (4.20) of Theorem 4.5 is fulfilled, i.e. show, that for all homogeneous credibility estimators obtained in a)–c) it holds that the credibility premium obtained by applying the credibility rates over the past observation period coincides in total over the whole portfolio with the observed claim amount.
e) The old tariff (based on earlier observation periods and a priori knowledge) had the following relative structure:

risk group	1	2	3	4	5
a priori premium factor	0.70	0.80	1.50	1.00	0.50

Calculate the pure risk premiums based on the (empirical) homogeneous credibility estimators by using the above factors as a priori differences between risks (Model Assumptions 4.14).

Compare the relative structure of the old and the new tariff obtained with e) and c) (normalizing factors such that group 4 has a factor equal to one).

Exercise 4.2

The following table shows the risk exposure, the total claim amount and the average claim amount per risk of a motor portfolio, both for normal and large claims (below/above a certain threshold).

region	# of years at risk (yr)	claim amount: all claims total in 1 000	all claims average per yr	normal claims total in 1 000	normal claims average per yr	large claims total in 1 000	large claims average per yr
1	41 706	17 603	422	7 498	180	10 105	242
2	9 301	5 822	626	1 900	204	3 922	422
3	33 678	15 033	446	7 507	223	7 526	223
4	3 082	758	246	560	182	198	64
5	27 040	11 158	413	4 142	153	7 016	259
6	44 188	25 029	566	7 928	179	17 101	387
7	10 049	5 720	569	1 559	155	4 161	414
8	30 591	14 459	473	6 568	215	7 891	258
9	2 275	329	145	329	145	0	0
10	109 831	54 276	494	21 859	199	32 417	295
Total	311 741	150 187	482	59 850	192	90 337	290

a) Calculate the pure risk premium, i.e. the values of the (empirical) homogeneous credibility estimator based on the figures in column 4 (all claims, average claims cost per year at risk).
b) Calculate the pure risk premium for the claims cost of normal claims, i.e. the values of the (empirical) homogeneous credibility estimator based on the figures in column 6 (normal claims, average claims cost per year at risk).
c) As b) for the large claims.
d) Compare the pure risk premium obtained by adding the results of b) and c) with those obtained in a).

Exercise 4.3

The following table shows the number and the claim frequencies of normal and large claims of the same motor portfolio as in Exercise 4.2. Calculate the values of the credibility estimator for the normal and the large claim frequency (absolute and relative values) under Model Assumptions 4.7.

region	# of years at risk	normal claims		large claims	
		number	frequency ‰	number	frequency ‰
1	41 706	2 595	62.2	21	0.50
2	9 301	763	82.0	8	0.86
3	33 678	3 302	98.0	26	0.77
4	3 082	233	75.6	1	0.32
5	27 040	1 670	61.8	15	0.55
6	44 188	3 099	70.1	33	0.75
7	10 049	631	62.8	12	1.19
8	30 591	2 428	79.4	25	0.82
9	2 275	116	51.0	0	0.00
10	109 831	8 246	75.1	84	0.76
Total	311 741	23 083	74.0	225	0.72

Exercise 4.4

The following table shows for each region i the observed average claim amount X_i, the observed empirical standard deviation σ_i of the claim amounts and the observed empirical coefficient of variation $\mathrm{CoVa}_i = \sigma_i / X_i$, both for normal and large claims of the same motor portfolio as in Exercises 4.2 and 4.3.

a) Calculate the values of the (empirical) homogeneous credibility estimator of the severity of claims under Model Assumptions 4.11.
b) Dito for the large claims.

region	normal claims				large claims			
	claim number	claim amounts in CHF			claim number	claim amounts in CHF 1 000		
		$X_i^{(n)}$	$\sigma_i^{(n)}$	$CoVa_i^{(n)}$		$X_i^{(l)}$	$\sigma_i^{(l)}$	$CoVa_i^{(l)}$
1	2 595	2 889	4 867	1.7	21	481	523	1.1
2	763	2 490	4 704	1.9	8	490	522	1.1
3	3 302	2 273	3 681	1.6	26	289	457	1.6
4	233	2 403	4 143	1.7	1	198	-	-
5	1 670	2 480	3 885	1.6	15	468	478	1.0
6	3 099	2 558	4 232	1.7	33	518	743	1.4
7	631	2 471	4 493	1.8	12	347	309	0.9
8	2 428	2 705	4 671	1.7	25	316	363	1.1
9	116	2 836	4 724	1.7	0	-	-	-
10	8 246	2 651	4 152	1.6	84	386	505	1.3
Total	23 083	2 593	4 255	1.6	225	401	522	1.3

Exercise 4.5

Assume that the number of claims and the claim sizes are conditionally independent given Θ_i. Estimate the individual pure risk premium $\mu(\Theta_i)$ by multiplying the estimates of the claim frequency (Exercise 4.3) and the estimates of the severity (Exercise 4.4). Compare the results with those obtained in Exercise 4.2.

Exercise 4.6

In motor liability, the risks in a specific subportfolio are cross-classified according to two tariff criteria: use of car (A and B) and type of car (five different car types numbered 1,..., 5).

The following data and summary statistics, which will also be used in later exercises in connection with other credibility models, are available:

risk class	A1	A2	A3	A4	A5	B1	B2	B3	B4	B5	Total
year					**years at risk**						
1	3 767	9 113	683	1 288	4 923	57	1 234	84	352	3 939	25 440
2	3 877	9 305	751	1 158	4 995	61	1 133	81	317	4 121	25 799
3	3 492	8 322	619	920	3 874	50	909	63	229	3 108	21 586
4	3 243	7 743	514	742	3 170	51	733	52	164	2 445	18 857
5	2 930	6 963	389	583	2 374	41	663	46	154	1 793	15 936
Total	17 309	41 446	2 956	4 691	19 336	260	4 672	326	1 216	15 406	107 618

risk class	A1	A2	A3	A4	A5	B1	B2	B3	B4	B5	Total
year					**average claims cost per year at risk**						
1	507	599	207	419	1 171	538	617	206	3 499	1 206	810
2	465	631	172	1 401	1 384	293	1 448	332	859	1 307	918
3	173	520	201	456	953	898	1 441	215	436	1 719	740
4	627	512	475	2 339	1 104	1 216	1 825	590	3 209	2 035	976
5	1 041	752	166	304	1 004	306	1 444	278	1 041	1 898	970
Total	543	600	238	958	1 151	646	1 286	310	1 884	1 549	875

claim number and claim amounts, years 1-5
N_i = number of claims, Y_i = observed average claim amount,
σ_i = empirical standard deviation of the claim amounts, $CoVa_i = \sigma_i / Y_i$=coefficient of variation

risk class	A1	A2	A3	A4	A5	B1	B2	B3	B4	B5	Total
all claims											
N_i	1 058	3 146	238	434	2 481	35	825	32	180	2 577	11 006
Y_i	8 885	7 902	2 959	10 355	8 971	4 801	7 281	3 163	12 725	9 259	8 554
σ_i	65 747	59 267	2 959	72 482	52 931	6 241	47 326	3 163	75 077	54 631	59 022
$CoVa_i$	7.4	7.5	1.0	7.0	5.9	1.3	6.5	1.0	5.9	5.9	6.9
normal claims											
N_i	1 043	3 106	238	429	2 446	35	815	32	174	2 547	10 865
Y_i (in CHF)	3 179	3 481	2 959	4 143	4 804	4 801	3 518	3 163	3 786	5 579	4 267
σ_i (in CHF)	3 814	4 177	2 959	4 972	5 765	6 241	4 222	3 163	3 786	6 695	5 121
$CoVa_i$	1.2	1.2	1.0	1.2	1.2	1.3	1.2	1.0	1.0	1.2	1.2
large claims											
N_i	15	40	-	5	35	-	10	-	6	30	141
Y_i (in 1000 CHF)	406	351	-	543	300	-	314	-	272	322	339
σ_i (in 1000 CHF)	365	386	-	435	360	-	314	-	218	386	373
$CoVa_i$	0.9	1.1	-	0.8	1.2	-	1.0	-	0.8	1.2	1.1

Assume that each of the risk classes $i = \text{A1}, \text{A2}, \ldots, \text{B1}, \ldots, \text{B5}$ is characterized by its own risk profile Θ_i and denote by $\mu(\Theta_i)$ the individual premium given Θ_i, by $\mu_F(\Theta_i)$ the claim frequency and by $\mu_Y(\Theta_i)$ the average claim amount.

Assume that the conditions of the Bühlmann–Straub model are fulfilled for the observed average claims cost per year at risk as well as for the claim frequencies and the average claim amounts.

a) Estimate $\mu(\Theta_i)$ (in absolute and relative terms) based on the statistics of the observed claims cost per year at risk.
To get an idea about the heterogeneity of the portfolio, give also an estimate of the heterogeneity parameter $\text{CoVa}(\mu(\Theta_i))$.
b) Make the usual assumption that the number of claims and the claim sizes are conditionally independent. Then

$$\mu(\Theta_i) = \mu_F(\Theta_i) \cdot \mu_Y(\Theta_i). \tag{4.50}$$

For estimating $\mu(\Theta_i)$, use (4.50) and determine the credibility estimator of the claims frequency $\mu_F(\Theta_i)$ based on the observed claim numbers (all claims) under Model Assumptions 4.7 and the credibility estimator of $\mu_D(\Theta_i)$ based on the observed average claim amounts (all claims) under Model Assumptions 4.11.
To get an idea about the heterogeneity of the portfolio with respect to the claim frequency as well as the average claim size, give also an estimate of the heterogeneity parameters $\text{CoVa}(\mu_F(\Theta_i))$ and $\text{CoVa}(\mu_Y(\Theta_i))$.

c) As b) but now separately for normal and large claims. Sum up the results for normal and large claims to determine the total pure risk premiums.
d) Compare the results and comment.

Exercise 4.7

region	# of claims	total claim amount	average claim amount
1	5'424	24'914'365	4'593
2	160	723'410	4'514
3	885	3'729'472	4'216
4	10'049	43'242'683	4'303
5	2'469	10'739'070	4'350
6	1'175	5'402'705	4'598
7	2'496	12'221'317	4'897
8	7'514	34'956'886	4'652
9	482	2'424'411	5'030
10	2'892	13'570'966	4'693
11	839	4'276'629	5'097
12	3'736	15'787'934	4'226
13	2'832	13'820'842	4'879
14	548	2'521'255	4'603
15	406	1'847'533	4'555
16	6'691	31'168'608	4'658
17	1'187	5'616'656	4'734
18	2'105	9'396'649	4'463
19	1'533	7'193'857	4'693
20	2'769	13'594'889	4'909
21	5'602	28'446'646	5'078
22	283	1'256'937	4'443
23	9'737	49'778'011	5'112
24	2'839	14'722'732	5'187
25	1'409	6'357'640	4'513
25	20'267	93'827'437	4'630
26	6'977	32'464'942	4'653
Total	103'303	484'004'484	4'685

The table gives summary statistics for the average claim amount in motor hull insurance. We also know that for this line of business the coefficient of

variation of the claim amounts is about 2.0. Estimate the expected value of the claims' severity for each region based on this information.
Hint: consider the decomposition of the sum of squares as described at the end of Section 4.11 and use formula (4.37) for estimating SS_{tot}.

Exercise 4.8

For car fleets, a company uses experience rating based on the observed number of claims. The starting point is a basic premium calculated on the basis of estimated (a priori) claim frequencies and average claim amounts for the different types of vehicles. The a priori values are taken from suitable claims statistics (containing the whole motor portfolio, not just fleets). Depending on the composition of the fleet, one can then calculate for each fleet i the a priori expected claim number ν_i. It is then assumed that the individual expected claim number of the fleet i is $\nu_i \cdot \Theta_i$ with $E[\Theta_i] = 1$. Estimate for each fleet $i = 1, \ldots, 10$ the experience factor Θ_i based on the statistics given below.

fleet	# of years at risks	# of claims observed	# of claims a priori expected
1	103	18	23
2	257	27	16
3	25	2	4
4	84	20	17
5	523	109	120
6	2 147	518	543
7	72	11	10
8	93	16	13
9	154	116	129
10	391	156	118
Total	3 849	993	993

Exercise 4.9

Show that formula (4.26),

$$\widehat{\widehat{\tau}}^2 = c \cdot \left\{ \frac{I}{I-1} \sum_{i=1}^{I} \frac{w_{i\bullet}}{w_{\bullet\bullet}} \left(X_i - \overline{X}\right)^2 - \frac{I\widehat{\sigma^2}}{w_{\bullet\bullet}} \right\},$$

$$\text{where} \quad c = \frac{I-1}{I} \left\{ \sum_{i=1}^{I} \frac{w_{i\bullet}}{w_{\bullet\bullet}} \left(1 - \frac{w_{i\bullet}}{w_{\bullet\bullet}}\right) \right\}^{-1},$$

is an unbiased estimator for τ^2.

Exercise 4.10

In the *general liability line* we assume that the claims exceeding 200 000 CHF are Pareto distributed. In order to calculate the premiums for different upper limits of the insurance cover, we want to estimate the Pareto parameter for each of the six risk groups based on the data given below and assuming that Model Assumptions 4.16 are fulfilled.

group	1	2	3	4	5	6
	214 266	201 015	208 415	201 232	201 991	210 000
	215 877	215 670	208 791	260 000	290 190	210 150
	220 377	221 316	210 232	272 856	339 656	225 041
	230 501	241 750	213 618	395 807	349 910	226 312
	241 872	249 730	216 678	736 065	385 545	253 700
	245 992	258 311	227 952	776 355	443 095	280 000
	260 000	260 000	237 824	890 728	480 712	315 055
	260 000	265 696	240 500	1 005 186	714 988	335 000
	263 214	268 880	247 000		800 000	340 000
	263 497	273 000	255 879		1 280 928	354 929
	278 934	296 245	279 284		1 361 490	382 500
	290 000	299 000	279 376		2 112 000	450 000
	290 452	363 744	300 398			500 000
	292 500	382 581	319 141			500 000
	301 636	400 000	321 628			500 000
	312 920	400 000	400 465			500 000
	319 757	477 581	401 461			700 000
	340 733	640 131	404 027			729 000
	384 489	656 086	405 576			810 000
	392 437	800 000	472 138			900 000
	397 649	1014 262	632 161			900 000
	400 000	1046 388	671 758			1 500 000
	400 000		1 231 272			2 000 000
	400 000		1 303 129			3 000 000
	400 000		1 500 929			
	400 000		2 400 644			
	415 178		2 485 646			
	433 144					
	451 256					
	464 000					
	478 146					
	640 000					
	720 000					
	880 000					
# of claims	34	22	27	8	12	24

Table: *observed claim sizes above 200 000 CHF of a liability portfolio*

5

Treatment of Large Claims in Credibility

5.1 Motivation

In applying the Bühlmann–Straub model to calculate risk premiums in collective health insurance, practitioners noticed the following points:

- Because of the possible occurrence of large claims, the credibility weights are rather small for small and medium-sized insurance contracts.
- Despite the small credibility weights, a single large claim during the observation period can cause a considerable jump in the credibility premium. This is perceived by the insured as unfair, and is indeed questionable, since occasionally large claims are usually not very informative about the true risk profile.
- Due to the small credibility weights, credibility premiums vary little from one insurance contract to another if there are no large claims during the observation period.

The reason for these difficulties results from a high within risk variance component σ^2. Indeed, the potential occurrence of big claims contributes heavily to σ^2 due to the chosen quadratic loss function.

In order to avoid the difficulties mentioned, we look for transformations of the data. The credibility estimator is then applied to the transformed data. One approach is to truncate either the aggregate or the individual claims. This is the topic of the next two sections. In a further section, we will indicate other ways of dealing with large claims, but without going into details.

5.2 Semi-Linear Credibility with Truncation in the Simple Bühlmann Model

We again consider the simple Bühlmann model without volume measures (see Model Assumptions 3.6 or 4.11 for claim sizes). This could for instance mean that we consider aggregate claims in situations where the risks considered have all the same volume, or that we want to estimate the expected value of the claim size and that we consider the individual claim sizes. In the latter case, the number of observations n_i varies between risks.

The basic idea then is not to consider the original data X_{ij}, but rather first to transform this data (e.g. by truncation). De Vylder [DV76b] was the first to have this idea, and he named this credibility estimation technique based on transformed data *semi-linear credibility.*

We consider then the transformed random variables $Y_{ij} = f(X_{ij})$, where f is some arbitrary transformation. However, as before, our goal is to estimate the individual premium

$$P^{\text{ind}} = \mu_X(\Theta_i) = E[X_{ij}|\Theta_i]$$

for the random variables X_{ij}.

Definition 5.1. *The semi-linear credibility estimator of* $\mu_X(\Theta_i)$ *based on the transformed data* $Y_{ij} = f(X_{ij})$ *is the best estimator of the form* $\widehat{\mu_X(\Theta_i)} = a_0 + \sum_{ij} a_{ij} Y_{ij}$. *In order to indicate the dependence of the estimator on the transformation* f, *we use the notation* $\widehat{\widehat{\mu}}_X^{(f)}(\Theta_i)$.

Remark:

- The random variables Y_{ij} also satisfy the conditions of the simple Bühlmann model. The formula for the credibility estimator for $\mu_Y(\Theta_i) := E[Y_{ij}|\Theta_i]$ is therefore given by Theorem 3.7. However, we want to estimate $\mu_X(\Theta_i)$ (and not $\mu_Y(\Theta_i)$) from the observations Y_{ij}.

Theorem 5.2. *The semi-linear credibility estimator of* $\mu_X(\Theta_i)$ *in the simple Bühlmann model, based on the transformed data* $Y_{ij} = f(X_{ij})$, *is given by*

$$\begin{aligned}
\widehat{\widehat{\mu}}_X^{(f)}(\Theta_i) &= \mu_X + \frac{\tau_{XY}}{\tau_Y^2}\alpha_Y^{(i)}(Y_i - \mu_Y) \\
&= \mu_X + \frac{\tau_X}{\tau_Y}\rho_{XY}\,\alpha_Y^{(i)}(Y_i - \mu_Y),
\end{aligned} \tag{5.1}$$

where

$$\begin{aligned}
\mu_X &= E[\mu_X(\Theta_i)], \quad \mu_Y = E[\mu_Y(\Theta_i)], \\
\tau_Y^2 &= \text{Var}[\mu_Y(\Theta_i)], \quad \tau_X^2 = \text{Var}[\mu_X(\Theta_i)], \\
\tau_{XY} &= \text{Cov}(\mu_X(\Theta_i),\, \mu_Y(\Theta_i)), \\
\rho_{XY} &= \text{Corr}(\mu_X(\Theta_i),\, \mu_Y(\Theta_i)) = \frac{\tau_{XY}}{\tau_X\tau_Y}, \\
\sigma_Y^2 &= E[\text{Var}[Y_{ij}|\Theta_i]], \\
\alpha_Y^{(i)} &= \frac{n_i}{n_i + \kappa_Y}, \text{ where } \kappa_Y = \frac{\sigma_Y^2}{\tau_Y^2}, \\
Y_i &= \frac{1}{n_i}\sum_{j=1}^{n_i} Y_{ij}.
\end{aligned}$$

Proof: As in the Bühlmann–Straub model, the credibility estimator

$$\widehat{\widehat{\mu}}_X^{(f)}(\Theta_i) = \mathrm{Pro}\left(\mu_X(\Theta_i) \middle| L(1, \mathbf{Y}_1, \ldots, \mathbf{Y}_I)\right)$$

does not depend on the data from other risks, i.e.

$$\widehat{\widehat{\mu}}_X^{(f)}(\Theta_i) = \mathrm{Pro}\left(\mu_X(\Theta_i) \middle| L(1, \mathbf{Y}_i)\right). \tag{5.2}$$

We write (5.2) in a slightly more complicated way:

$$\widehat{\widehat{\mu}}_X^{(f)}(\Theta_i) = \mathrm{Pro}(\mathrm{Pro}(\mu_X(\Theta_i)| L(1, \mathbf{Y}_i, \mu_Y(\Theta_i))) \,|\, L(1, \mathbf{Y}_i)). \tag{5.3}$$

The inner projection $\mu_X'(\Theta_i) := \mathrm{Pro}\left(\mu_X(\Theta_i)\middle| L(1, \mathbf{Y}_i, \mu_Y(\Theta_i))\right)$ must satisfy the normal equations (Corollary 3.17), i.e.

$$E\left[\mu_X'(\Theta_i)\right] = E\left[\mu_X(\Theta_i)\right] = \mu_X, \tag{5.4}$$

$$\mathrm{Cov}\left(\mu_X(\Theta_i), Y_{ij}\right) = \mathrm{Cov}\left(\mu_X'(\Theta_i), Y_{ij}\right), \tag{5.5}$$

$$\mathrm{Cov}\left(\mu_X(\Theta_i), \mu_Y(\Theta_i)\right) = \mathrm{Cov}\left(\mu_X'(\Theta_i), \mu_Y(\Theta_i)\right). \tag{5.6}$$

It is intuitively clear, that the inner projection $\mu_X'(\Theta_i)$ does not depend on the variables Y_{ij}, since the Y_{ij} merely contain information on $\mu_Y(\Theta_i)$ and since $\mu_Y(\Theta_i)$ is already included as an explanatory variable. Indeed, if

$$\mu_X'(\Theta_i) = a_0' + a_1' \, \mu_Y(\Theta_i)$$

satisfies(5.6), then (5.5) is automatically satisfied, since $E\left[\mu_X(\Theta_i) \cdot Y_{ij}\right] = E\left[E\left\{\mu_X(\Theta_i) \cdot Y_{ij}\middle|\Theta_i\right\}\right] = E\left[\mu_X(\Theta_i) \cdot \mu_Y(\Theta_i)\right]$. This implies that $\mu_X'(\Theta_i)$ does not depend on the observations Y_{ij}, but only on $\mu_Y(\Theta_i)$. This is easy to understand, since the Y_{ij} merely contain information on the $\mu_Y(\Theta_i)$, and $\mu_Y(\Theta_i)$ is already included as an explanatory variable in $\mu_X'(\Theta_i)$. Thus, we have

$$\mu_X'(\Theta_i) = a_0' + a_1' \, \mu_Y(\Theta_i).$$

From (5.6) and (5.4) it follows that

$$\mathrm{Cov}\left(\mu_X(\Theta_i), \mu_Y(\Theta_i)\right) = a_1' \, \mathrm{Var}\left[\mu_Y(\Theta_i)\right],$$

$$\mu_X'(\Theta_i) = \mu_X + \frac{\mathrm{Cov}\left(\mu_X(\Theta_i), \mu_Y(\Theta_i)\right)}{\mathrm{Var}\left[\mu_Y(\Theta_i)\right]} \left(\mu_Y(\Theta_i) - \mu_Y\right).$$

Using the iterative procedure (5.3) and denoting by $\widehat{\widehat{\mu_Y(\Theta_i)}}$ the credibility estimator for $\mu_Y(\Theta_i)$ based on the simple Bühlmann model (Model Assumptions 3.6), we obtain

$$\widehat{\widehat{\mu}}_X^{(f)}(\Theta_i) = \mu_X + \frac{\mathrm{Cov}\left(\mu_X(\Theta_i), \mu_Y(\Theta_i)\right)}{\mathrm{Var}\left[\mu_Y(\Theta_i)\right]}\left(\widehat{\widehat{\mu_Y(\Theta_i)}} - \mu_Y\right)$$
$$= \mu_X + \frac{\tau_{XY}}{\tau_Y^2}\alpha_Y^{(i)}\left(Y_i - \mu_Y\right),$$

which ends the proof of Theorem 5.2. □

The quadratic loss is also easy to calculate and has a similar form to that of formula (4.11).

Theorem 5.3. *The quadratic loss of the semi-linear credibility estimator is given by*

$$E\left[\left(\widehat{\widehat{\mu}}_X^{(f)}(\Theta_i) - \mu_X(\Theta_i)\right)^2\right] = \tau_X^2\left(1 - \rho_{XY}^2\,\alpha_Y^{(i)}\right), \tag{5.7}$$

where $\rho_{XY} = \mathrm{Corr}\left(\mu_X(\Theta_i), \mu_Y(\Theta_i)\right).$

Proof:

$$E\left[\left(\widehat{\widehat{\mu}}_X^{(f)}(\Theta_i) - \mu_X(\Theta_i)\right)^2\right]$$
$$= E\left[\left(\mu_X(\Theta_i) - \mu_X'(\Theta_i)\right)^2\right] + E\left[\left(\mu_X'(\Theta_i) - \widehat{\widehat{\mu}}_X^{(f)}(\Theta_i)\right)^2\right]$$
$$= \left(\tau_X^2 - 2\frac{\tau_{XY}}{\tau_Y^2}\tau_{XY} + \frac{\tau_{XY}^2}{\left(\tau_Y^2\right)^2}\tau_Y^2\right) + \frac{\tau_{XY}^2}{\left(\tau_Y^2\right)^2}\tau_Y^2\left(1 - \alpha_Y^{(i)}\right)$$
$$= \tau_X^2 - \frac{\tau_{XY}^2}{\tau_Y^2}\alpha_Y^{(i)}.$$

This proves Theorem 5.3. □

An obvious idea when dealing with the problem of large claims is to truncate the claims at some suitable chosen truncation point m (see [Gis80]), i.e. to choose the function

$$f(X) = \min(X, m)$$

as the transformation. The semi-linear credibility estimator with truncation depends on the selected point of truncation and is given by

$$\widehat{\widehat{\mu}}_X^{(m)}(\Theta_i) = \mu_X + \frac{\tau_X}{\tau_Y}\rho_{XY}\,\alpha_Y^{(i)}\left(Y_i - \mu_Y\right).$$

In this formula, the observed average Y_i of the truncated claims appears. Notice that the excess part of the claim amount above the truncated point is taken into account by the semilinear credibility premium, since the starting point is given by μ_X, the *mean over the whole collective* of the expected claim amount *before truncation.* Only the "experience correction" $(\tau_X/\tau_Y)\,\rho_{XY}\,\alpha_Y^{(i)}\,(Y_i-\mu_Y)$ is based, not on the original data, but rather on the truncated data. Notice also that the weight $(\tau_X/\tau_Y)\,\rho_{XY}\,\alpha_Y^{(i)}$ associated with the observed deviation of the observed truncated claim amount from the expected truncated claim amount (the term $(Y_i-\mu_Y)$) can in principle be greater than 1, because of the term $(\tau_X/\tau_Y)\,\rho_{XY}$.

When using credibility with truncation, we must decide how to choose the truncation point. The *optimal truncation point m* is that which minimizes the quadratic loss, respectively maximizes $R(m)=\rho_{XY}^2\,\alpha_Y^{(i)}$. This optimal truncation point may be at infinity, which means that no truncation is needed. How to calculate the theoretically optimal truncation point can be found in Gisler [Gis80]. However, in practice the choice of the truncation point is often based on some priori knowledge of the claim distribution and is not calculated theoretically. This is justifiable because the quadratic loss as a function of the truncation point is relatively flat over a large range, and therefore one doesn't lose much if the truncation point is not fully optimal (as an example see Exercise 5.1).

As in the Bühlmann–Straub model, the semi-linear credibility estimator also depends on structural parameters, in this case on μ_X, μ_Y, τ_{XY}, τ_Y^2, σ_Y^2. These can again be estimated on the basis of data from the collective. Below we list the estimators analogous to the Bühlmann–Straub model.

We introduce the following quantities:

$$S_X=\frac{1}{n_\bullet-I}\sum_i\sum_j(X_{ij}-X_i)^2,\qquad \text{where } n_\bullet=\sum_i n_i,$$

$$S_{XY}=\frac{1}{n_\bullet-I}\sum_i\sum_j(X_{ij}-X_i)(Y_{ij}-Y_i),$$

$$S_Y=\frac{1}{n_\bullet-I}\sum_i\sum_j(Y_{ij}-Y_i)^2,$$

$$T_X=\frac{I}{I-1}\sum_i\frac{n_i}{n_\bullet}(X_i-\overline{X})^2,$$

$$T_{XY}=\frac{I}{I-1}\sum_i\frac{n_i}{n_\bullet}(X_i-\overline{X})(Y_i-\overline{Y}),$$

$$T_Y=\frac{I}{I-1}\sum_i\frac{n_i}{n_\bullet}(Y_i-\overline{Y})^2,$$

$$c=\frac{I-1}{I}\left\{\sum_i\frac{n_i}{n_\bullet}\left(1-\frac{n_i}{n_\bullet}\right)\right\}^{-1}.$$

The following parameter estimators are analogous to the parameter estimators in the Bühlmann–Straub model:

$$
\begin{aligned}
\widehat{\sigma_Y^2} &= S_Y, \quad \widehat{\sigma_X^2} = S_X, \\
\widehat{\tau_Y^2} &= \max\left(c \cdot \left(T_Y - \frac{I \cdot \widehat{\sigma_Y^2}}{n_\bullet}\right), 0\right), \\
\widehat{\tau_X^2} &= \max\left(c \cdot \left(T_X - \frac{I \cdot \widehat{\sigma_X^2}}{n_\bullet}\right), 0\right), \\
\widehat{\tau_{XY}} &= \operatorname{signum}\left(T_{XY} - \frac{I \cdot S_{XY}}{n_\bullet}\right) \\
&\quad \cdot \sqrt{\min\left\{c^2 \cdot (T_{XY} - \frac{I \cdot S_{XY}}{n_\bullet})^2, \left(\widehat{\tau_Y^2} \cdot \widehat{\tau_X^2}\right)\right\}}, \\
\widehat{\mu_X} &= \left(\sum\nolimits_i \alpha_X^{(i)} X_i\right) \left(\sum\nolimits_i \alpha_X^{(i)}\right)^{-1}, \\
\widehat{\mu_Y} &= \left(\sum\nolimits_i \alpha_Y^{(i)} Y_i\right) \left(\sum\nolimits_i \alpha_Y^{(i)}\right)^{-1}.
\end{aligned}
$$

5.3 Semi-Linear Credibility with Truncation in a Model with Weights

We turn next to the question of whether the techniques which were discussed in Section 5.2 can be carried over to the Bühlmann–Straub model with arbitrary volume measures. The answer is no, and the reason for this is that the transformed random variables $Z_{ij} = \min(X_{ij}, m)$, where X_{ij} is the ratio of the *aggregate claim amount* and the corresponding volume, do not satisfy the assumptions of the Bühlmann–Straub model. For example, the quantity $E[Z_{ij}|\Theta_i]$ depends on the weight w_{ij}. In most cases, the bigger w_{ij}, the bigger is $E[Z_{ij}|\Theta_i]$ (since the expected value of the part which has been cut off decreases with increasing volume).

It would, however, be useful in many cases in insurance practice, if the idea of the *truncation of single claims* could also be carried over to models with volume measures. Group accident insurance, or group health insurance are two such cases. In both cases, the policy holder is a firm, and all employees of that firm are covered by the insurance contract. The insurance provides benefits if any one of these employees has an accident or becomes sick. The volume or weight w_{ij} is then typically the size of the payroll of the ith firm in year j. Without truncation, we can consider the conditions for the Bühlmann–Straub model as being satisfied. In order to be able to apply credibility with truncation, however, we need an extended model with somewhat more structure.

Let

$S_{ij} = \sum_{\upsilon=1}^{N_{ij}} Y_{ij}^{(\nu)}$ where S_{ij} = aggregate claim amount, N_{ij} = number of claims, $Y_{ij}^{(\nu)}$ = individual claim sizes,

w_{ij} weights (e.g. amount of payroll),

$X_{ij} = S_{ij}/w_{ij}$ claims ratios.

Model Assumptions 5.4 (volume-dependent model with truncation)
The ith risk is characterized by its individual risk profile ϑ_i, which is itself the realization of a random variable Θ_i, and it holds that:

Gi1: Conditionally, given Θ_i, the random variables

$$N_{ij}, Y_{ij}^{(\nu)} \quad (j = 1, 2, \ldots, n;\ \nu = 1, 2, \ldots) \text{ are independent,} \tag{5.8}$$

$$N_{ij} \sim \text{Poisson } (w_{ij}\, \eta(\Theta_i)), \tag{5.9}$$

$$Y_{ij}^{(\nu)} \quad (j = 1, 2, \ldots, n;\ \nu = 1, 2, \ldots) \text{ are independent and identically distributed with distribution } F_{\Theta_i}(y). \tag{5.10}$$

Gi2: The pairs $(\Theta_1, \mathbf{X}_1)$, $(\Theta_2, \mathbf{X}_2)$, ... are independent, and Θ_1, Θ_2, ... are independent and identically distributed.

Under the above model assumptions we have:

$$E[X_{ij}|\,\Theta_i] = \eta(\Theta_i)\, E\left[Y_{ij}^{(\nu)}\middle|\,\Theta_i\right] =: \mu_X(\Theta_i), \tag{5.11}$$

$$\text{Var}[X_{ij}|\,\Theta_i] = \frac{1}{w_{ij}}\eta(\Theta_i)\, E\left[Y_{ij}^{(\nu)2}\middle|\,\Theta_i\right] =: \frac{1}{w_{ij}}\sigma_X^2(\Theta_i), \tag{5.12}$$

i.e. the claims ratios X_{ij} satisfy the model assumptions of the Bühlmann–Straub model (Model Assumptions 4.1).

Remarks:

- It is only assumed that the number of claims and the claim sizes are *conditionally* independent given Θ_i. However, $\eta(\Theta_i)$ and $\mu_X(\Theta_i)$ and/or $\mu_Y(\Theta_i)$ may be correlated.
- All the results of this section also hold, when the Poisson assumption for the number of claims is replaced by the following weaker assumption:

$$E[N_{ij}|\,\Theta_i] = w_{ij}\,\eta(\Theta_i), \qquad \text{Var}[N_{ij}|\,\Theta_i] \propto w_{ij}\,\eta(\Theta_i).$$

  This weaker assumption is satisfied, in particular, if the claim numbers are Negative Binomial distributed, under the additional assumption that the quotient of the expected value and the variance does not depend on Θ_i.

We use the notation

$G_{ij}^{(\nu)} = \min\left(Y_{ij}^{(\nu)}, m\right)$ the claim sizes truncated at m,

$T_{ij} = \sum_{\nu=1}^{N_{ij}} G_{ij}^{(\nu)}$ aggregate claim amounts after truncation of the individual claims,

$Z_{ij} = T_{ij}/w_{ij}$ claims ratios after truncation of the individual claim sizes.

Clearly $G_{ij}^{(\nu)}$, T_{ij} and Z_{ij} depend on the selected truncation point m, and are therefore also functions of this truncation point. However, in order to keep the notation simple, we do not explicitly indicate this dependence.

The first two conditional moments of the claims ratios Z_{ij} after truncation are given by

$$E\left[Z_{ij}|\Theta_i\right] =: \mu_Z(\Theta_i) = \eta(\Theta_i) \cdot E\left[\left.G_{ij}^{(\nu)}\right|\Theta_i\right], \tag{5.13}$$

$$\operatorname{Var}\left[Z_{ij}|\Theta_i\right] =: \frac{\sigma_Z^2(\Theta_i)}{w_{ij}}, \tag{5.14}$$

$$\text{where } \sigma_Z^2(\Theta_i) := \eta(\Theta_i)\, E\left[\left.G_{ij}^{(\nu)2}\right|\Theta_i\right]. \tag{5.15}$$

The claims ratios Z_{ij} also satisfy (just as the original claims ratios X_{ij}) the model assumptions of the Bühlmann–Straub model. In the following only the structure given by (5.13)–(5.15) is relevant.

The *structural parameters* in this model are:

original data	transformed data
$\mu_X = E\left[\mu_X(\Theta_i)\right]$,	$\mu_Z = E\left[\mu_Z(\Theta_i)\right]$,
$\sigma_X^2 = E\left[\sigma_X^2(\Theta_i)\right]$,	$\sigma_Z^2 = E\left[\sigma_Z^2(\Theta_i)\right]$,
$\tau_X^2 = \operatorname{Var}\left[\mu_X(\Theta_i)\right]$,	$\tau_Z^2 = \operatorname{Var}\left[\mu_Z(\Theta_i)\right]$,
	$\tau_{XZ} = \operatorname{Cov}\left(\mu_X(\Theta_i), \mu_Z(\Theta_i)\right)$.

By the semi-linear credibility estimator of $\mu_X(\Theta_i)$ with truncation we mean the credibility estimator of $\mu_X(\Theta_i)$ based on the statistics $\{\mathbf{Z}_1, \mathbf{Z}_2, \ldots, \mathbf{Z}_I\}$, which corresponds exactly to the content of the following definition.

Definition 5.5. *The semi-linear credibility estimator of $\mu_X(\Theta_i)$ with truncation of the individual claims at the point m is the best estimator of the form $\widehat{\mu_X(\Theta_i)} = a_0 + \sum_{ij} a_{ij} Z_{ij}$. In order to indicate its dependence on the truncation point m, we use the notation $\widehat{\widehat{\mu}}_X^{(m)}(\Theta_i)$.*

With exactly the same proof as in Theorem 5.2 we get the following result.

Theorem 5.6. *Under Model Assumptions 5.4, the semi-linear credibility estimator with truncation of the individual claims at the point m is given by:*

$$\widehat{\widehat{\mu}}_X^{(m)}(\Theta_i) = \mu_X + \frac{\tau_X}{\tau_Z}\rho_{XZ}\alpha_Z^{(i)}\left(Z_i - \mu_Z\right), \tag{5.16}$$

$$\text{where } \alpha_Z^{(i)} = \frac{w_{i\bullet}}{w_{i\bullet} + \frac{\sigma_Z^2}{\tau_Z^2}},$$
$$\rho_{XZ} = \text{Corr}\left(\mu_X(\Theta_i), \mu_Z(\Theta_i)\right),$$
$$Z_i = \sum_j \frac{w_{ij}}{w_{i\bullet}} Z_{ij}.$$

Remarks:

- To apply formula (5.16) one does not need to know all individual claim sizes. It is sufficient to know, for each contract i, the total aggregate claim amount plus the individual claim sizes of the few claims exceeding the truncation point.
- The above result also holds when, instead of truncation, any other transformation of the individual claim amounts is used. The random variables Z_{ij} are then simply otherwise defined.
- Notice the different kind of truncation used in each of the models:
  - truncation of the original X-variable in the model without volumes,
  - truncation of the individual claim amounts and construction of a claims ratio from the truncated data in the model with volumes.

With the same proof as in Theorem 5.3 we get the following result for the quadratic loss.

Theorem 5.7. *The quadratic loss is given by*

$$E\left[\left(\widehat{\widehat{\mu}}_X^{(m)}(\Theta_i) - \mu_X(\Theta_i)\right)^2\right] = \tau_X^2\left(1 - \rho_{XZ}^2\alpha_Z^{(i)}\right). \tag{5.17}$$

The truncation point m is a "parameter", the value of which we can freely choose. The *optimal truncation point* is again that which minimizes the quadratic loss, respectively maximizes the function $R(m) = (\tau_X/\tau_Z)\,\rho_{XZ}^2\alpha_Z^{(i)}$.

Remarks:

- One can show that in the special case where only the number of claims but not the claim sizes $Y_{ij}^{(v)}$ depend on the risk profile Θ_i, the optimal truncation point m^* is such that all truncated claim sizes become equal. The resulting semi-linear credibility formula is then given by

$$\widehat{\widehat{\mu}}_X^{(m^*)}(\Theta_i) = \mu_Y \cdot \widehat{\widehat{\mu_F(\Theta_i)}},$$

  where $\widehat{\widehat{\mu_F(\Theta_i)}}$ is the credibility estimator for the claim frequency based on the statistic of the number of claims.

- In general, the optimal truncation point depends on the volumes $w_{i\bullet}$. As a rule, the bigger these volumes are, the greater is the optimal truncation point. In insurance practice it is usually enough to form a few volume categories and to choose the truncation points for these categories sensibly, e.g. based on the estimated function $R(m)$.

The following estimators for the structural parameters are quite analogous to the ones given on page 130, but now with volume measures and with the same number n of observation years for all risks. We introduce the following quantities:

$$
\begin{aligned}
S_X &= \frac{1}{I}\sum_i \frac{1}{n-1}\sum_j w_{ij}\,(X_{ij}-X_i)^2, \quad \text{where } X_i=\sum_j \frac{w_{ij}}{w_{i\bullet}}X_{ij},\\
S_{XZ} &= \frac{1}{I}\sum_i \frac{1}{n-1}\sum_j w_{ij}\,(X_{ij}-X_i)(Z_{ij}-Z_i),\\
S_Z &= \frac{1}{I}\sum_i \frac{1}{n-1}\sum_j w_{ij}\,(Z_{ij}-Z_i)^2,\\
T_X &= \frac{I}{I-1}\sum_i \frac{w_{i\bullet}}{w_{\bullet\bullet}}(X_i-\overline{X})^2, \qquad \text{where } \overline{X}=\sum_i \frac{w_{i\bullet}}{w_{\bullet\bullet}}X_i,\\
T_{XZ} &= \frac{I}{I-1}\sum_i \frac{w_{i\bullet}}{w_{\bullet\bullet}}(X_i-\overline{X})(Z_i-\overline{Z}),\\
T_Z &= \frac{I}{I-1}\sum_i \frac{w_{i\bullet}}{w_{\bullet\bullet}}(Z_i-\overline{Z})^2,\\
c &= \frac{I-1}{I}\left\{\sum_{i=1}^{I}\frac{w_{i\bullet}}{w_{\bullet\bullet}}\left(1-\frac{w_{i\bullet}}{w_{\bullet\bullet}}\right)\right\}^{-1}.
\end{aligned}
$$

As in Section 5.2 we can define the following estimators for the structural parameters.

$$
\begin{aligned}
\widehat{\sigma_Z^2} &= S_Z, \qquad \widehat{\sigma_X^2} = S_X,\\
\widehat{\tau_Z^2} &= \max\left(c\cdot\left(T_Z-\frac{I\cdot\widehat{\sigma_Z^2}}{w_{\bullet\bullet}}\right),0\right),\\
\widehat{\tau_X^2} &= \max\left(c\cdot\left(T_X-\frac{I\cdot\widehat{\sigma_X^2}}{w_{\bullet\bullet}}\right),0\right),\\
\widehat{\tau_{XZ}} &= \operatorname{signum}\left(T_{XZ}-\frac{I\cdot S_{XZ}}{w_{\bullet\bullet}}\right)\\
&\quad\cdot\sqrt{\min\left\{c^2\cdot\left(T_{XZ}-\frac{I\cdot S_{XZ}}{w_{\bullet\bullet}}\right)^2,\left(\widehat{\tau_Z^2}\cdot\widehat{\tau_X^2}\right)^{1/2}\right\}},
\end{aligned}
$$

$$\widehat{\mu_X} = \left(\sum_i \alpha_X^{(i)} X_i\right) / \left(\sum_i \alpha_X^{(i)}\right),$$
$$\widehat{\mu_Z} = \left(\sum_i \alpha_Z^{(i)} Z_i\right) / \left(\sum_i \alpha_Z^{(i)}\right).$$

5.4 Further Methods for Treating Large Claims

In this section we want to mention other methods for the treatment of large claims within the credibility framework. It is beyond the scope of this book to treat these approaches in detail and so we will merely sketch the basic ideas here and point to the corresponding literature (or later chapters in this book).

As discussed in the previous sections, in semi-linear credibility, the original data is first transformed, and because we were treating the problem of large claims, we considered truncation transformations. The truncated part of the claim is implicitly included by the semi-linear credibility formula in the term μ_X. But the "experience correction" at the level of the individual contract, respectively the dependence of the credibility premium on the individual claim experience, is only calculated from the claims experience *after* truncation, hence the part of the claims experience which has been cut off, the observed "excess claim amount", has no influence on the credibility premium. Another possibility that suggests itself, is not to completely discard the observed "excess" or "large claim amounts", and to take account of both the large claims as well as the normal claims, in parallel. This is a special application of multi-dimensional credibility, and will later be discussed in greater detail in Chapter 7.

By definition, credibility theory is limited to linear estimators. This has the big advantage that the credibility premiums have a simple form and depend only on the first two moments of the data, which can be easily estimated. However, this linearity can also have disadvantages, as we already saw in our treatment of large claims. Since credibility theory leaves open which random variables are considered (with respect to which the estimator must be linear), we can use this freedom to our advantage. In the previous sections the approach consisted of truncating the original data. In Section 4.14 the random variables under consideration were maximum-likelihood estimators.

It was in considering the treatment of large claims that Künsch [Kün92] first had the idea of applying credibility to robust estimators. The basic idea is to replace the observed individual mean by a robust estimator (robust in the sense of classical robust statistics) and then to use credibility estimators based on the statistics of the robust means. This idea was further developed by Gisler and Reinhard [GR93]. An advantage of this approach is that it can be applied directly to claims ratios. For example, this might be a useful approach for fire insurance.

A further approach to the treatment of large claims is given by "robust Bayesian statistics". This has been propagated in the actuarial literature by

Schnieper [Sch04] in connection with large claims. Although the name is very similar to "robust credibility", the two concepts have little in common. Robust Bayesian statistics has nothing to do with classical robust statistics. It belongs to the framework of Bayesian statistics. Robust indicates merely that the conditional distributions and their associated a priori distributions are chosen in such a way that the resulting Bayes estimator is automatically robust. As a consequence the treatment of outliers is driven by the observations themselves. The disadvantage is that the distributions and the estimators (in contrast to the credibility approach) need to be determined numerically and do not have a simple form.

5.5 Exercises

Exercise 5.1

trunc.point	estimates of						
m	σ_Z^2	τ_Z^2	$\kappa_Z=\sigma_Z^2/\tau_Z^2$	τ_X/τ_Z	τ_{XZ}	ρ_{XZ}	μ_Z
∞	15 531	0.0895	173 531	1.0	0.0895	100.0%	1.000
100 000	14 465	0.0891	162 346	1.0	0.0888	99.4%	0.994
95 000	14 339	0.0894	160 391	1.0	0.0889	99.4%	0.992
90 000	14 171	0.0896	158 158	1.0	0.0889	99.3%	0.991
85 000	13 995	0.0900	155 500	1.0	0.0890	99.2%	0.989
80 000	13 828	0.0904	152 965	1.0	0.0891	99.1%	0.988
75 000	13 672	0.0908	150 573	1.0	0.0892	98.9%	0.987
70 000	13 489	0.0912	147 906	1.0	0.0893	98.8%	0.985
65 000	13 222	0.0914	144 661	1.0	0.0893	98.7%	0.983
60 000	12 834	0.0908	141 344	1.0	0.0889	98.6%	0.977
55 000	12 370	0.0904	136 836	1.0	0.0886	98.5%	0.972
50 000	11 758	0.0893	131 669	1.0	0.0880	98.4%	0.964
45 000	10 917	0.0879	124 198	1.0	0.0872	98.3%	0.954
40 000	9 944	0.0854	116 440	1.0	0.0858	98.1%	0.939
35 000	8 970	0.0823	108 991	1.0	0.0838	97.6%	0.921
30 000	7 993	0.0761	105 033	1.1	0.0808	97.9%	0.900
25 000	7 047	0.0708	99 534	1.1	0.0778	97.7%	0.876
20 000	6 089	0.0645	94 403	1.2	0.0734	96.6%	0.848
15 000	5 002	0.0589	84 924	1.2	0.0690	95.0%	0.810
10 000	3 874	0.0523	74 073	1.3	0.0625	91.4%	0.755
5 000	2 570	0.0429	59 907	1.4	0.0525	84.7%	0.657

Consider the group health insurance for daily allowances. In this line of business, the policy holders are firms which buy insurance for workmen compensation in case of sickness. We consider the portfolio of policies with the following coverage: if an employee can't go to work because of sickness, the insurance company pays, after a waiting period of three days, 80% of the salary for the duration of the sickness up to a maximum of two years. The basic tariff premium is calculated by a premium rate times the total salary of all employees in the company.

The claims experience varies substantially between firms, which is not surprising, since individual hidden factors like working climate and control of absences have a considerable influence on the claims load. For that reason an insurance company introduces an experience rating scheme based on the model introduced in Section 5.3, Model Assumptions 5.4. The pure risk premiums of the basic tariff denoted by P_i are taken as weights. At every renewal of a contract, the observed claims experience over the last three years is taken into account by a correction factor applied to the basic premium rate. We assume that the basic premium rate is at the right level, i.e. that $E[X_i] = \mu_X = 1$. The goal is to estimate the correction factor $\mu_X(\Theta_i)$ by applying the semilinear credibility formula with truncation of the individual claims.

The structural parameters were estimated from the observations of the whole portfolio. The table below shows the estimated values $\widehat{\mu}_Z$, $\widehat{\sigma}_Z^2$, $\widehat{\tau}_Z^2$, $\widehat{\kappa}_Z = \widehat{\sigma}_Z^2/\widehat{\tau}_Z^2$, $\widehat{\rho}_{XZ} = \widehat{\tau}_{XZ}/(\widehat{\tau}_X \cdot \widehat{\tau}_Z)$ in dependence of the truncation point m.

a) To get an idea where to choose the truncation point depending on the size of the contract, display the function $\widehat{R}(m) = \widehat{\rho}_{XZ}^2 \cdot \widehat{\alpha}_Z^{(i)}$ (see formula (5.17) on page 133) for contracts with $P = 60\,000,\ 300\,000,\ 500\,000$ and where P denotes the pure risk premium obtained with the basic tariff over the observation period of three years.
b) The company chooses the following truncation points m:

$$
\begin{aligned}
m &= 15\,000 \text{ if } 20\,000 \leq P < 100\,000, \\
m &= 30\,000 \text{ if } 100\,000 \leq P < 1\,000\,000, \\
m &= \infty \quad\ \ \text{ if } P \geq 1\,000\,000.
\end{aligned}
$$

policy 1: $P = 20\,000$,
$S = 0$

policy 2: $P = 20\,000$,
$S = 36\,000$,
whereof no claim $> 15\,000$.

policy 3: $P = 20\,000$,
$S = 36\,000$,
whereof one claim $> 15\,000$: $32\,000$.

policy 4: $P = 300\,000$,
$S = 300\,000$,
whereof no claim $> 30\,000$.

policy 5: $P = 300\,000$,
$S = 360\,000$,
whereof two claims $> 30\,000$: $38\,000$, $162\,000$.

Calculate the resulting experience correction factors for the above contracts, where P is defined as above and where S denotes the total aggregate claim amount over the last three years:

Exercise 5.2

region	# of claims	original claims data X_i	$\sigma_X^{(i)}$	$CoVa_X^{(i)}$	trunc. claims data Y_i	$\sigma_Y^{(i)}$	$CoVa_Y^{(i)}$	Correlation $Corr(X,Y)^{(i)}$
1	1 785	11 224	83 141	7.4	4 760	7 760	1.6	54.7%
2	37	6 348	16 893	2.7	4 887	8 420	1.7	97.5%
3	280	21 273	175 132	8.2	4 932	8 351	1.7	55.3%
4	3 880	11 754	91 977	7.8	4 631	8 008	1.7	52.6%
5	846	6 559	30 280	4.6	4 275	7 285	1.7	71.5%
6	464	10 045	71 253	7.1	5 071	8 521	1.7	48.8%
7	136	5 968	24 890	4.2	4 174	6 945	1.7	75.8%
8	853	5 407	24 070	4.5	3 974	6 155	1.5	70.2%
9	2 151	5 483	21 465	3.9	4 135	6 470	1.6	74.7%
10	155	5 133	16 294	3.2	4 139	6 746	1.6	83.1%
11	653	6 815	30 508	4.5	4 504	7 693	1.7	70.1%
12	252	13 161	74 342	5.6	4 877	8 494	1.7	70.9%
13	1 228	8 670	47 897	5.5	4 849	7 827	1.6	62.4%
14	865	6 780	32 290	4.8	4 482	7 326	1.6	67.0%
15	145	6 273	18 310	2.9	4 949	7 958	1.6	84.7%
16	127	6 694	33 779	5.0	4 070	7 144	1.8	71.5%
17	2 151	8 757	46 631	5.3	4 624	7 573	1.6	69.4%
18	449	6 535	27 296	4.2	4 514	7 013	1.6	73.8%
19	784	7 068	51 113	7.2	4 191	6 954	1.7	50.7%
20	438	15 453	126 781	8.2	5 243	8 447	1.6	49.4%
21	1 011	9 010	41 405	4.6	5 224	8 722	1.7	68.0%
22	1 692	7 666	48 000	6.3	4 671	7 443	1.6	53.5%
23	82	8 856	51 853	5.9	3 710	5 900	1.6	90.2%
24	2 813	5 967	62 570	10.5	4 094	6 001	1.5	32.5%
25	729	9 213	77 005	8.4	4 523	7 520	1.7	46.7%
26	329	6 727	20 080	3.0	5 157	8 444	1.6	83.7%
27	5 743	8 279	49 575	6.0	4 686	7 231	1.5	60.0%
28	3 002	8 149	57 186	7.0	4 491	7 370	1.6	52.4%
Total	33 080	8 486	61 189	7.2	4 555	7 389	1.6	

Table: *average claim amounts of a motor liability portfolio*

The table above shows, for a motor liability portfolio and for different regions, the number n_i of claims, the average claim amount X_i, the empirical standard deviations $\sigma_X^{(i)}$ and the empirical coefficients of variation $\mathrm{CoVa}_X^{(i)} = \sigma_X^{(i)}/X_i$ and analogously Y_i, $\sigma_Y^{(i)}$, $\mathrm{CoVa}_Y^{(i)}$ for the truncated claim amounts (truncation at 50 000). In addition the empirical correlations $\mathrm{Corr}(X,Y)^{(i)}$ between the original claim amounts $X_{i\nu}$ and the truncated claim amounts $Y_{i\nu}$ ($\nu = 1, 2,..., n_i$) are given. The aim is to estimate for each region its individual expected value of the claim size $\mu(\Theta_i) = E[X_i | \Theta_i]$.

a) Estimate $\mu(\Theta_i)$ by applying the Bühlmann–Straub model based on the original data X_i (see Section 4.11).
b) Estimate $\mu(\Theta_i)$ using the credibility model with truncation at 50 000 (see Section 5.2) based on the Y_i.

Exercise 5.3

The table below shows for 10 risk groups the number of claims n_i, the average claim amounts X_i and the coefficients of variation CoVa_i of the claim amounts.

The actuary knows from other sources that a lognormal distribution fits fairly well as a claim size distribution in this line of business. Therefore, he also considers besides the original claim sizes the logarithms $Y_{i\nu} = \ln(X_{i\nu})$. In the table below, the same summary statistics are given for the original claim sizes and for the logarithms $Y_{i\nu}$.

risk group	# of claims	claim amounts			
		original data		logarithms	
i	n_i	X_i	$\text{CoVa}_X^{(i)}$	Y_i	$\text{CoVa}_Y^{(i)}$
1	92	3 478	7 079	6.47	0.31
2	542	5 204	17 179	6.78	0.28
3	57	2 092	3 340	6.45	0.29
4	317	4 437	13 419	6.73	0.28
5	902	4 676	20 391	6.59	0.29
6	137	6 469	19 147	6.60	0.35
7	48	12 285	50 875	7.30	0.27
8	528	3 717	12 349	6.40	0.31
9	183	14 062	87 300	7.25	0.28
10	365	9 799	136 098	6.33	0.31
Total	3 171	5 825	53 294	6.62	0.30

a) Estimate $\mu(\Theta_i)$ by applying the Bühlmann–Straub model based on the original data X_i.
b) Make use of the a priori knowledge that a lognormal distribution fits well as a claim size distribution in this line of business.
Denote by $\mu_Y(\Theta_i)$ and by $\sigma_Y^2(\Theta_i)$ the parameters of the lognormal distribution given Θ_i. We further assume that $\sigma_Y^2(\Theta_i) = \sigma_Y^2$ for all i, which is equivalent to the assumption that the coefficient of variation of the original claim sizes $X_{i\nu}$, $\nu = 1, 2, \ldots$ is the same for all risk groups i. Estimate

$\mu_Y(\Theta_i)$ and σ_Y^2 with the technique described in Section 4.11. Then estimate the expected value of the original claim size $\mu(\Theta_i)$ by taking the first moment of the Lognormal distribution, i.e. use the estimator

$$\widehat{\mu(\Theta_i)} = \exp\left(\widehat{\mu_Z(\Theta_i)} + \frac{\widehat{\sigma}_Z^2}{2}\right).$$

c) Make a table showing the observed average claim sizes and the estimates of the expected values obtained with the two methods. Make also a table with the relative values (ratio of $\widehat{\mu(\Theta_i)}$ divided by the resulting average over the whole portfolio).

Exercise 5.4

The table below shows, for each risk group i, the number of claims n_i, the average claim size X_i and the coefficient of variation CoVa_i of the individual claim sizes. These figures are given for the category of all claims as well as for normal claims (pure physical damage) and large claims (bodily injury claims). The data are the same as those considered in Exercise 4.4. The aim is again to estimate for each risk group i the expected value $\mu(\Theta_i) = E[X_i|\Theta_i]$ of the claim size.

risk group	all claims n_i	X_i	CoVa_i	normal claims $n_i^{(n)}$	$X_i^{(n)}$	$\mathrm{CoVa}_i^{(n)}$	large claims $n_i^{(l)}$	$X_i^{(l)}$ in 1'000	$\mathrm{CoVa}_i^{(l)}$
1	2 616	6 727	9.5	2 595	2 889	1.7	21	481	1.09
2	771	7 548	9.5	763	2 490	1.9	8	490	1.06
3	3 328	4 513	10.5	3 302	2 273	1.6	26	289	1.58
4	234	3 239	4.2	233	2 403	1.7	1	198	0.00
5	1 685	6 624	9.5	1 670	2 480	1.6	15	468	1.02
6	3 132	7 989	11.6	3 099	2 558	1.7	33	518	1.43
7	643	8 901	7.6	631	2 471	1.8	12	347	0.89
8	2 453	5 898	8.3	2 428	2 705	1.7	25	316	1.15
9	116	2 836	1.7	116	2 836	1.7	0	0	0.00
10	8 330	6 517	9.9	8 246	2 651	1.6	84	386	1.31
Total	23 308	6 443	10.1	23 083	2 593	1.6	225	401	1.30

a) Estimate $\mu(\Theta_i)$ by applying the Bühlmann–Straub model to the observed average claim amounts of the category of all claims.
b) Another idea for estimating $\mu(\Theta_i)$ is to estimate the expected value of the claim severities of the normal and large claims $\left(\mu^{(n)}(\Theta_i) \text{ and } \mu^{(l)}(\Theta_i)\right)$ and the percentage of the large claims.

For estimating $\mu^{(n)}(\Theta_i)$ and $\mu^{(l)}(\Theta_i)$, proceed as in a).

For estimating the percentage of large claims for each risk group i, denote by Π_i the probability that a claim is a large claim (bodily injury). Then the number of large claims $N_i^{(l)}$ is conditionally, given Π_i and n_i, binomial distributed with parameter $n_i \Pi_i$. Approximate the Binomial distribution with the Poisson distribution with parameter $\Lambda_i = n_i \Pi_i$ and estimate Π_i with the technique and the model of Section 4.10.
Finally estimate $\mu(\Theta_i)$ by

$$\widehat{\mu(\Theta_i)} = \left(1 - \widehat{\Pi}_i\right) \widehat{\mu^{(n)}(\Theta_i)} + \widehat{\Pi}_i \widehat{\mu^{(l)}(\Theta_i)}.$$

6

Hierarchical Credibility

6.1 Motivation

In practice, we often encounter hierarchical structures, both in statistical data analysis and in the calculation of premiums.

Insurance data frequently have a hierarchical structure. The individual risks are classified according to their "tariff positions", tariff positions are grouped together into "subgroups", subgroups into "groups", groups into "main groups", which together make the total of a line of business. In the statistical interpretation, this structure will be constructed in reverse order. One is first interested in the development of a line of business as a whole, then in the development of the main groups, and so on.

Not infrequently, a hierarchical procedure is also used in the calculation of premiums, whereby the reasoning follows a hierarchical tree. By this is meant a "top down" procedure, in that first the expected aggregate claim amount for the whole line of business is ascertained and then this amount is successively "distributed" over the lower levels. Consider the following two examples:

- Example 1: Group accident insurance in Switzerland, premium for accidents at work (Figure 6.1).
  In Switzerland, each firm has to buy a workers' compensation insurance. The premium calculation scheme used by the private insurers is basically as follows: the individual firms are grouped together according to their type of business (e.g. banks, hospitals), and these types of business are grouped into danger classes (groups of types of business). The premium calculation is carried out by first fixing the "tariff level" for the various danger classes, then within the danger classes for the types of business.

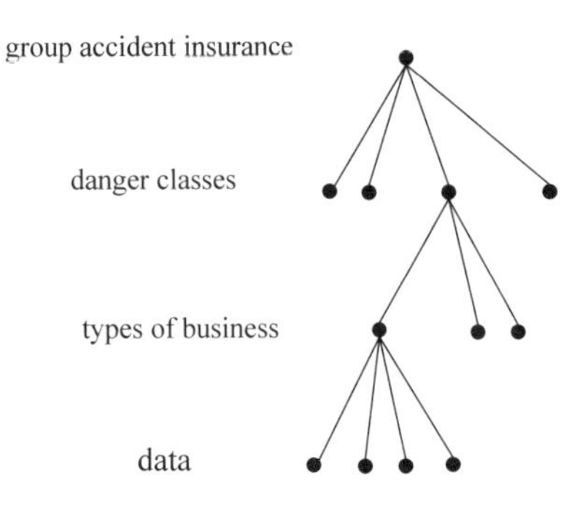

Fig. 6.1.

- Example 2: Industrial fire insurance in Germany (Figure 6.2).
  For a long time, the premium calculation of industrial fire insurance in Germany followed a hierarchical procedure: first the expected aggregate claim amount for the whole line of business is calculated and this is then successively broken down into "books", within the "books" into "groups of statistical accounts", and then within the groups into the individual accounts.

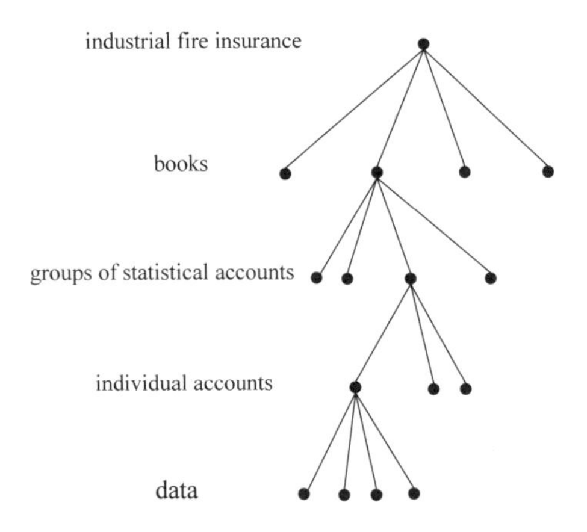

Fig. 6.2.

Such a hierarchical system has the advantage of leading to a well-founded, properly balanced distribution of the burden of claims, in particular of large claims, within the collective. In the following we will incorporate the idea of a hierarchical structure into the credibility framework along the lines presented in the paper by Bühlmann and Jewell [BJ87].

6.2 The Hierarchical Credibility Model

For didactic reasons, we will consider in the following a model with five levels. However, the generalization of this model to an arbitrary order is obvious and straightforward.

The structure of the model can be visualized most easily using Figure 6.3.

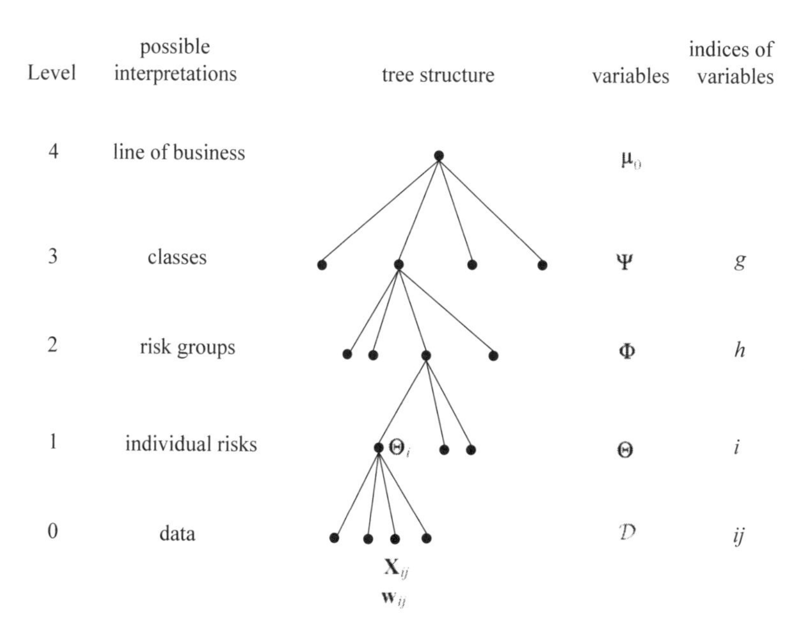

Fig. 6.3. Tree structure of the hierarchical credibility model

In order to keep the number of indices small, we will use the following notation:

$\Phi(\Psi_g)$ = set of Φ's, that stem from Ψ_g,
$\mathcal{D}(\Phi_h)$ = set of observations X_{ij}, that stem from Φ_h,
etc.

Model Assumptions 6.1 (hierarchical credibility)
The probability structure in the hierarchical model is obtained by "drawing" the individual variables top down:

- ***Level 3***
  The random variables Ψ_g $(g = 1, 2, \ldots, |G|)$ *are independent and identically distributed (i.i.d.) with density* $r_3(\psi)$.

- ***Level 2***
  Given Ψ_g the random variables $\Phi_h \in \Phi(\Psi_g)$ are i.i.d. with the ***conditional*** *density $r_2(\varphi|\Psi_g)$.*
- ***Level 1***
  Given Φ_h the random variables $\Theta_i \in \Theta(\Phi_h)$ are i.i.d. with the ***conditional*** *density $r_1(\vartheta|\Phi_h)$.*
- ***Level 0***
  Given Θ_i the observations $X_{ij} \in \mathcal{D}(\Theta_i)$ are ***conditionally independent*** *with densities $r_0(x|\Theta_i, w_{ij})$, for which*

$$\begin{aligned} E[X_{ij}|\Theta_i] &= \mu(\Theta_i), \\ \mathrm{Var}[X_{ij}|\Theta_i] &= \sigma^2(\Theta_i)/w_{ij}, \\ \textit{where } w_{ij} &= \textit{known weights}. \end{aligned}$$

The similarity to the structure of the Bühlmann–Straub model is obvious. The Bühlmann–Straub model also has a tree structure (see Figure 4.2 on page 90), though only with levels zero, one and two. Hierarchical models of higher orders are therefore nothing more than generalizations of the Bühlmann–Straub model to an increased number of levels. In this connection, it is useful to review the calculation of the homogeneous credibility estimator in the Bühlmann–Straub model. As demonstrated in Theorem 4.4, this is done by a bottom-up, followed by a top-down procedure: First the quantities X_i (data compression, linear sufficient statistic) and $\widehat{\widehat{\mu_0}}$ are calculated bottom up, and then the homogeneous credibility estimator $\widehat{\widehat{\mu(\Theta_i)}}^{\text{hom}}$ is calculated top down. As we will see, this method is "inherited" by hierarchical models of higher order.

Notice also that the conditional densities depend only on the variables in the next higher level and not on those from levels higher than that. Mathematically we express this by saying that the random variables Ψ_g, Φ_h, Θ_i, X_{ij} (listed in decreasing order of the tree) possess the Markov property.

6.3 Relevant Quantities and Notation

As before, the goal is to estimate the correct individual premium $\mu(\Theta_i)$ for each of the contracts, or more specifically to find the credibility estimator for $\mu(\Theta_i)$.

It is, however, useful to define the *analogous quantities for the higher levels of the hierarchical tree,* where we have on top, at the level of the whole collective, the collective premium. We use the following notation:

$$\begin{aligned} \mu_0 &:= E[X_{ij}], \quad \text{(collective premium)} \\ \mu(\Psi_g) &:= E[X_{ij}|\Psi_g], \qquad \text{where } X_{ij} \in \mathcal{D}(\Psi_g), \\ \mu(\Phi_h) &:= E[X_{ij}|\Phi_h], \qquad \text{where } X_{ij} \in \mathcal{D}(\Phi_h), \\ \mu(\Theta_i) &:= E[X_{ij}|\Theta_i], \qquad \text{where } X_{ij} \in \mathcal{D}(\Theta_i). \end{aligned} \tag{6.1}$$

It holds that

$$\begin{aligned}
\mu_0 &= E\left[\mu\left(\Psi_g\right)\right], \\
\mu\left(\Psi_g\right) &= E\left[\mu\left(\Phi_h\right)\middle|\,\Psi_g\right], \qquad \text{where } \Phi_h \in \Phi\left(\Psi_g\right), \\
\mu\left(\Phi_h\right) &= E\left[\mu\left(\Theta_i\right)\middle|\,\Phi_h\right], \qquad \text{where } \Theta_i \in \Theta\left(\Phi_h\right).
\end{aligned}$$

This follows easily from the model assumptions and properties of the conditional expectation, as is illustrated in the following for $\mu\left(\Phi_h\right)$:

$$\begin{aligned}
\mu\left(\Phi_h\right) = E\left[X_{ij}\middle|\,\Phi_h\right] &= E\left[E\left[X_{ij}\middle|\,\Theta_i, \Phi_h\right]\middle|\,\Phi_h\right] &\quad (6.2) \\
&= E\left[E\left[X_{ij}\middle|\,\Theta_i\right]\middle|\,\Phi_h\right] &\quad (6.3) \\
&= E\left[\mu\left(\Theta_i\right)\middle|\,\Phi_h\right],
\end{aligned}$$

where (6.2) follows from properties of the conditional expectation and equation (6.3) follows from the model assumptions (Markov property).

Next we want to introduce the *structural parameters of the hierarchical credibility model.* These are the a priori expected value $\mu_0 = E\left[X_{ij}\right]$ (cf. (6.1)) and the variance components

$$\begin{aligned}
&\text{at level 0} \qquad \sigma^2 := E\left[\sigma^2\left(\Theta_i\right)\right], \\
&\text{at level 1} \qquad \tau_1^2 := E\left[\operatorname{Var}\left[\mu\left(\Theta_i\right)\middle|\,\Phi_h\right]\right] = E\left[\tau_1^2\left(\Phi_h\right)\right], \\
&\text{at level 2} \qquad \tau_2^2 := E\left[\operatorname{Var}\left[\mu\left(\Phi_h\right)\middle|\,\Psi_g\right]\right] = E\left[\tau_2^2\left(\Psi_g\right)\right], \\
&\text{at level 3} \qquad \tau_3^2 := \operatorname{Var}\left[\mu\left(\Psi_g\right)\right].
\end{aligned}$$

These quantities can also be visualized by a tree (see Figure 6.4).

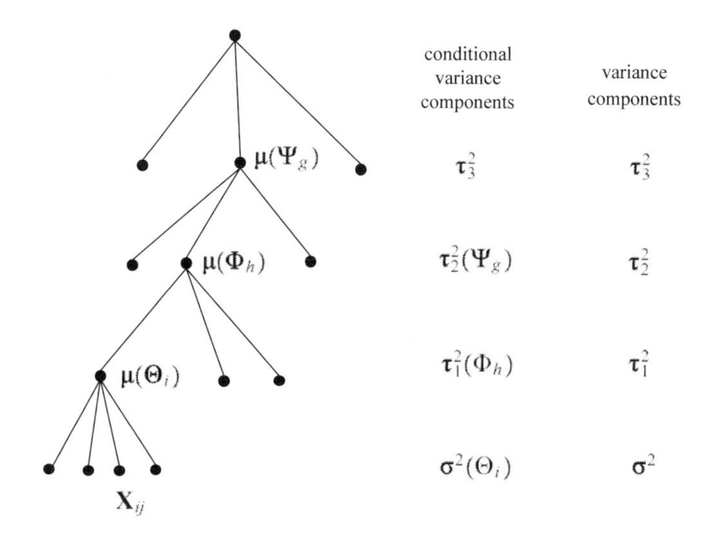

Fig. 6.4. Tree structure with structural parameters

It follows directly from the properties of the conditional expectation that

$$\begin{aligned}\sigma^2 &= E\left[w_{ij}(X_{ij} - \mu(\Theta_i))^2\right],\\ \tau_1^2 &= E\left[(\mu(\Theta_i) - \mu(\Phi_h))^2\right],\\ \tau_2^2 &= E\left[(\mu(\Phi_h) - \mu(\Psi_g))^2\right],\\ \tau_3^2 &= E\left[(\mu(\Psi_g) - \mu_0)^2\right].\end{aligned}$$

Notice further that

$$\mu_0 = E[\mu(\Psi_g)] = E[\mu(\Phi_h)] = E[\mu(\Theta_i)].$$

It is also easy to show that the following equations hold for the unconditional variance

$$\begin{aligned}\text{Var}[\mu(\Theta_i)] &= \tau_1^2 + \tau_2^2 + \tau_3^2,\\ \text{Var}[\mu(\Phi_h)] &= \tau_2^2 + \tau_3^2.\end{aligned}$$

6.4 Credibility Estimator in the Hierarchical Model

Our goal is to find the credibility estimators $\widehat{\widehat{\mu(\Theta_i)}}$ for the individual premium $\mu(\Theta_i)$ for $i = 1, 2, \ldots I$. We shall see that this will also necessitate that we find the credibility estimators $\widehat{\widehat{\mu(\Phi_h)}}$, $h = 1, 2, \ldots, H$ and $\widehat{\widehat{\mu(\Psi_g)}}$, $g = 1, 2, \ldots, G$ and (in the homogeneous case) also $\widehat{\widehat{\mu_0}}$.

Let us start by representing both the quantities to be estimated and the data by a tree (see Figure 6.5).

In Figure 6.5 we have located all quantities in such a way that all possible conditional expected values of them can be read as ancestors. All the credibility estimators can then be understood as linear combinations of the "tree father" μ_0 and all "descendent data" X_{ij}. In the hierarchical set-up, the Hilbert space technique (i.e. our understanding of the estimators as projections) is particularly useful. Especially, we shall use the iterativity property of the projection operator.

By definition

$$\widehat{\widehat{\mu(\Theta_i)}} = \text{Pro}(\mu(\Theta_i)|\, L(\mathcal{D}, 1)).$$

We now use the trick of first projecting onto a bigger space, i.e. we write

$$\begin{aligned}\widehat{\widehat{\mu(\Theta_i)}} &= \text{Pro}(\text{Pro}(\mu(\Theta_i)|\, L(\mathcal{D}, \mu(\Phi_h), 1))|\, L(\mathcal{D}, 1)),\\ &\quad \text{where } \Theta_i \in \Theta(\Phi_h).\end{aligned} \tag{6.4}$$

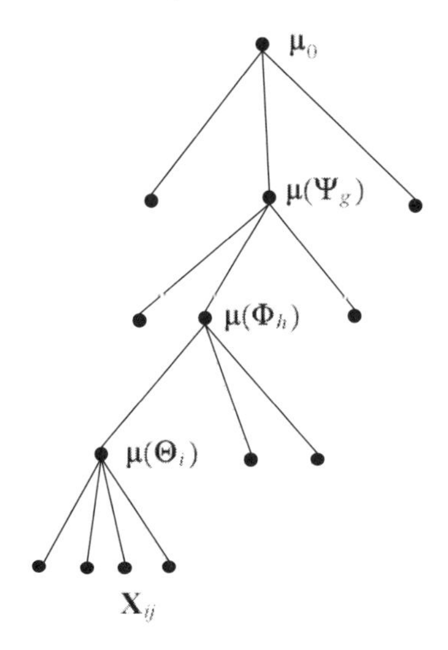

Fig. 6.5. Tree structure with conditional expected values

Note that, for the inner projection, $\mu(\Phi_h)$ is known. Intuitively, it suggests that

$$\widehat{\mu(\Theta_i)}' := \operatorname{Pro}(\mu(\Theta_i)|\, L(\mathcal{D}, \mu(\Phi_h), 1))$$

must be of the form

$$\widehat{\mu(\Theta_i)}' = \alpha_i^{(1)} B_i^{(1)} + (1 - \alpha_i^{(1)})\mu(\Phi_h), \tag{6.5}$$

where $B_i^{(1)}$ are the compressed data from $\mathcal{D}(\Theta_i)$ and $\alpha_i^{(1)}$ are suitable credibility weights. Below we will prove that our intuition is correct and that (6.5) holds true. From (6.4) and (6.5) we then get that

$$\widehat{\widehat{\mu(\Theta_i)}} = \alpha_i^{(1)} B_i^{(1)} + (1 - \alpha_i^{(1)})\widehat{\widehat{\mu(\Phi_h)}}.$$

In order to determine $\widehat{\widehat{\mu(\Phi_h)}}$ we can repeat the same calculation at the next higher level (iteration of the idea above):

$$\begin{aligned}\widehat{\widehat{\mu(\Phi_h)}} &= \operatorname{Pro}(\operatorname{Pro}(\mu(\Phi_h)|\, L(\mathcal{D}, \mu(\Psi_g), 1))|\, L(\mathcal{D}, 1)),\\ &\quad \text{where } \Phi_h \in \Phi(\Psi_g),\end{aligned}$$

$$\begin{aligned}\widehat{\mu(\Phi_h)}' :=& \quad \operatorname{Pro}(\mu(\Phi_h)|\, L(\mathcal{D}, \mu(\Psi_g), 1))\\ &= \alpha_h^{(2)} B_h^{(2)} + (1 - \alpha_h^{(2)})\mu(\Psi_g),\end{aligned} \tag{6.6}$$

where $B_h^{(2)}$ are the compressed data from $\mathcal{D}(\Phi_h)$, and therefore

$$\widehat{\widehat{\mu(\Phi_h)}} = \alpha_h^{(2)} B_h^{(2)} + (1 - \alpha_h^{(2)})\, \widehat{\widehat{\mu(\Psi_g)}}.$$

Finally, for the credibility estimator of $\mu(\Psi_g)$ it must be true that

$$\widehat{\widehat{\mu(\Psi_g)}} = \alpha_g^{(3)} B_g^{(3)} + (1 - \alpha_g^{(3)})\mu_0.$$

Hence, in order to calculate the credibility premium for $\mu(\Theta_i)$ (at the risk level) we must also find the credibility premium for the higher levels. In order to do this we need to find formulae for the compressed data $B_i^{(1)}$, $B_h^{(2)}$, $B_g^{(3)}$ and the corresponding credibility weights $\alpha_i^{(1)}$, $\alpha_h^{(2)}$ and $\alpha_g^{(3)}$.

As to the question of the data compression, we have to ask at each level and at each node, what is the best we can do based only on the data stemming from that node. The obvious candidates are the best linear, individually unbiased estimators, which are again orthogonal projections on affine subspaces. These estimators, which we choose as data compressions, are given by

$$\begin{aligned} B_i^{(1)} &:= \operatorname{Pro}(\mu(\Theta_i)|\, L_e^{ind}(\mathcal{D}(\Theta_i))), \\ B_h^{(2)} &:= \operatorname{Pro}(\mu(\Phi_h)|\, L_e^{ind}(\mathcal{D}(\Phi_h))), \\ B_g^{(3)} &:= \operatorname{Pro}(\mu(\Psi_g))|\, L_e^{ind}(\mathcal{D}(\Psi_g))), \end{aligned} \tag{6.7}$$

where for instance $L_e^{ind}(\mathcal{D}(\Phi_h))$ is defined by

$$L_e^{ind}(\mathcal{D}(\Phi_h)) := \Big\{ \widehat{\mu(\Phi_h)} : \widehat{\mu(\Phi_h)} = \sum_{\{i,j:\ X_{ij} \in \mathcal{D}(\Phi_h)\}} a_{ij} X_{ij}, \\ a_{ij} \in \mathbb{R},\ E\Big[\widehat{\mu(\Phi_h)}\Big|\, \Phi_h\Big] = \mu(\Phi_h) \Big\}.$$

Remarks:

- Note the difference between $L_e^{ind}(.)$ (e.g. $L_e^{ind}(\mathcal{D}(\Phi_h))$) and $L_e(.)$ (e.g. $L_e(\mathbf{X})$ in (3.18) on page 69). $L_e^{ind}(.)$ is an affine subspace of *individually unbiased* estimators, whereas $L_e(.)$ is an affine subspace of *collectively unbiased* estimators (see also Definition 3.9 on page 65).
- $L_e^{ind}(.)$ does not explicitly reflect what quantity we want to estimate and hence what the unbiasedness condition exactly means. However, whenever $L_e^{ind}(.)$ appears, this becomes clear from the context. For instance, it is clear, how $L_e^{ind}(\mathcal{D}(\Phi_h))$ in $\operatorname{Pro}(\mu(\Phi_h)|\, L_e^{ind}(\mathcal{D}(\Phi_h)))$ has to be interpreted: it is the affine subspace of the estimators linear in the data $\mathcal{D}(\Phi_h)$ and whose conditional expected value given Φ_h is equal to $\mu(\Phi_h)$.

We now prove – as an example – formula (6.6). The analogous results hold for all $\widehat{\mu(.)}'$ on the various hierarchical levels.

Lemma 6.2. *It holds that*

$$\widehat{\widehat{\mu(\Phi_h)}}' = \alpha_h^{(2)} B_h^{(2)} + (1 - \alpha_h^{(2)})\mu(\Psi_g), \tag{6.8}$$

$$\textit{where } \alpha_h^{(2)} = \frac{\tau_2^2}{\tau_2^2 + E\left[\left(\mu(\Phi_h) - B_h^{(2)}\right)^2\right]}. \tag{6.9}$$

Remarks:

- An easy calculation shows that the right-hand side of (6.9) can also be written as

$$\alpha_h^{(2)} = \frac{\left\{E\left[\left(B_h^{(2)} - \mu(\Phi_h)\right)^2\right]\right\}^{-1}}{\left\{E\left[\left(B_h^{(2)} - \mu(\Phi_h)\right)^2\right]\right\}^{-1} + \left\{E\left[\left(\mu(\Phi_h) - \mu(\Psi_g)\right)^2\right]\right\}^{-1}}.$$

  Hence $\widehat{\widehat{\mu(\Phi_h)}}'$ is again a weighted mean with the precisions as weights (cf. General Intuitive Principle 3.3).

Proof of Lemma 6.2: We show that:

i) $\widehat{\widehat{\mu(\Phi_h)}}'$ is the credibility estimator based on the "compressed observation value" $B_h^{(2)}$ and $\mu(\Psi_g)$,

ii) $\widehat{\widehat{\mu(\Phi_h)}}'$ is the credibility estimator based on the set $\mathcal{D}$ of all observations and $\mu(\Psi_g)$.

Proof of i): We must show that the normal equations are satisfied, i.e. that

$$E\left[\widehat{\widehat{\mu(\Phi_h)}}'\right] = E\left[\mu(\Phi_h)\right], \tag{6.10}$$

$$\operatorname{Cov}\left(\widehat{\widehat{\mu(\Phi_h)}}', \mu(\Psi_g)\right) = \operatorname{Cov}\left(\mu(\Phi_h), \mu(\Psi_g)\right), \tag{6.11}$$

$$\operatorname{Cov}\left(\widehat{\widehat{\mu(\Phi_h)}}', B_h^{(2)}\right) = \operatorname{Cov}\left(\mu(\Phi_h), B_h^{(2)}\right). \tag{6.12}$$

Obviously, equation (6.10) is satisfied since

$$E\left[B_h^{(2)}\right] = E\left[\mu(\Phi_h)\right] = E\left[\mu(\Psi_g)\right] = \mu_0.$$

If we use the identity $\operatorname{Cov}(X,Y) = E\left[\operatorname{Cov}(X,Y|\,Z)\right] + \operatorname{Cov}(E\left[X|\,Z\right], E\left[Y|\,Z\right])$, and if we condition on the random variable Ψ_g in (6.11), then we get, both on the left-hand side as well as on the right-hand side, the value $\operatorname{Var}\left[\mu(\Psi_g)\right]$. Analogously, we find that equation (6.12) is equivalent to

$$\alpha_h^{(2)} E\left[\operatorname{Var}\left[B_h^{(2)}\middle|\Psi_g\right]\right] = E\left[\operatorname{Var}\left[\mu(\Phi_h)|\Psi_g\right]\right],$$

and hence

$$\begin{aligned}\alpha_h^{(2)} &= \frac{E\left[\operatorname{Var}\left[\mu(\Phi_h)|\Psi_g\right]\right]}{E\left[\operatorname{Var}\left[B_h^{(2)}\middle|\Psi_g\right]\right]} = \frac{E\left[\left(\mu\left(\Phi_h\right)-\mu\left(\Psi_g\right)\right)^2\right]}{E\left[\left(B_h^{(2)}-\mu\left(\Psi_g\right)\right)^2\right]} \\ &= \frac{E\left[\left(\mu\left(\Phi_h\right)-\mu\left(\Psi_g\right)\right)^2\right]}{E\left[\left(B_h^{(2)}-\mu\left(\Phi_h\right)\right)^2\right]+E\left[\left(\mu\left(\Phi_h\right)-\mu\left(\Psi_g\right)\right)^2\right]} \\ &= \frac{\tau_2^2}{\tau_2^2+E\left[\left(\mu(\Phi_h)-B_h^{(2)}\right)^2\right]}.\end{aligned}$$

Thus we have shown i).

Proof of ii): We must show that $\mu(\Phi_h)-\widehat{\mu(\Phi_h)}' \perp X_{ij}$ for all observations in the lowest level.

- $X_{ij} \notin \mathcal{D}(\Phi_h)$:

$$E\left[\left\{\mu(\Phi_h)-\widehat{\mu(\Phi_h)}'\right\}X_{ij}\right] = \alpha_h^{(2)} E\left[\left(\mu(\Phi_h)-B_h^{(2)}\right)X_{ij}\right] + (1-\alpha_h^{(2)})E\left[\left(\mu(\Phi_h)-\mu\left(\Psi_g\right)\right)\cdot X_{ij}\right].$$

  In the first term on the right-hand side of the equation, we condition on the set of all Φ_h and in the second term we condition on the set of all Φ_h and on the set of all Ψ_g. Using the conditional independence assumed in Model Assumptions 6.1 we see immediately that both terms are equal to zero.
- $X_{ij} \in \mathcal{D}(\Phi_h)$:
  We must show that $\mu(\Phi_h)-\widehat{\mu(\Phi_h)}' \perp X_{ij} = \left(X_{ij}-B_h^{(2)}\right)+B_h^{(2)}$.
  Since $\widehat{\mu(\Phi_h)}'$ is the credibility estimator based on $B_h^{(2)}$ and $\mu\left(\Psi_g\right)$, it follows that $\mu(\Phi_h)-\widehat{\mu(\Phi_h)}' \perp B_h^{(2)}$.
  To show orthogonality with respect to $X_{ij}-B_h^{(2)}$ we write

$$\mu(\Phi_h)-\widehat{\mu(\Phi_h)}' = \alpha_h^{(2)}\cdot(\mu(\Phi_h)-B_h^{(2)}) + (1-\alpha_h^{(2)})\cdot(\mu(\Phi_h)-\mu\left(\Psi_g\right)).$$

  The definition of $B_h^{(2)}$ (see (6.7)) as optimal compressed data implies that $\mu(\Phi_h)-B_h^{(2)} \perp X_{ij}-B_h^{(2)}$. Conditioning on Φ_h and Ψ_g, we see immediately that $E\left[\left(\mu(\Phi_h)-\mu\left(\Psi_g\right)\right)\cdot\left(X_{ij}-B_h^{(2)}\right)\right]=0$. Thus we have $\mu(\Phi_h)-\widehat{\mu(\Phi_h)}' \perp X_{ij}-B_h^{(2)}$.

This concludes the proof of Lemma 6.2. □

Lemma 6.2 describes the relevant elements for evaluating a hierarchical credibility model. Still there are two questions to be answered before we can numerically proceed in this evaluation:

a) Formula (6.9) contains the quadratic error $E\left[\left(\mu(\Phi_h) - B_h^{(2)}\right)^2\right]$. How do we find the value for this error term?
b) We understand $B_h^{(2)}$ as the projection on an appropriate affine subspace. How do we numerically find the value of this estimator?

Both these questions are answered by the following lemma, which is formulated and proved for the data compression on level 3, but which holds analogously for the other levels.

Lemma 6.3. *It holds that*

$$i) \quad B_g^{(3)} = \sum_{h \in H_g} \frac{\alpha_h^{(2)}}{w_g^{(3)}} B_h^{(2)}, \tag{6.13}$$

$$\text{where } H_g = \{h : \Phi_h \in \Phi(\Psi_g)\},$$

$$w_g^{(3)} = \sum_{h \in H_g} \alpha_h^{(2)},$$

$$ii) \quad E\left[\left(B_g^{(3)} - \mu(\Psi_g)\right)^2\right] = \frac{\tau_2^2}{w_g^{(3)}}. \tag{6.14}$$

Proof:
First we note that

$$B_g^{(3)} = \operatorname{Pro}\left[\operatorname{Pro}\left[\mu(\Psi_g) \middle| L_e^{ind}\left(\mathcal{D}(\Psi_g), \{\mu(\Phi_h) : h \in H_g\}\right)\right] \middle| L_e^{ind}\left(\mathcal{D}(\Psi_g)\right)\right]. \tag{6.15}$$

It is intuitively clear that the inner projection does not depend on $L_e(\mathcal{D}(\Psi_g))$. Indeed, given $\{\mu(\Phi_h) : h \in H_g\}$, the data $\mathcal{D}(\Psi_g)$ cannot provide more information with regard to $\mu(\Psi_g)$ than that already contained in $\{\mu(\Phi_h) : h \in H_g\}$. Formally this can be checked using the orthogonality condition of projection. Conditionally, given Ψ_g, $\{\mu(\Phi_h) : h \in H_g\}$ are independent with the same conditional variance. Hence we have for the inner projection

$$\operatorname{Pro}\left[\mu(\Psi_g) \middle| L_e^{ind}\left(\mathcal{D}(\Psi_g)\right), \{\mu(\Phi_h) : h \in H_g\}\right] = \frac{1}{|H_g|} \sum\nolimits_{h \in H_g} \mu(\Phi_h), \tag{6.16}$$

where $|H_g|$ denotes the number of nodes stemming from Ψ_g.

From (6.15) and (6.16) we get

$$B_g^{(3)} = \text{Pro}\left[\frac{1}{|H_g|} \sum\nolimits_{h \in H_g} \mu(\Phi_h) \middle| L_e^{ind}(\mathcal{D}(\Psi_g)) \right]. \tag{6.17}$$

We also have that

$$\begin{aligned} \text{Pro}\left[\mu(\Phi_h) | L_e^{ind}(\mathcal{D}(\Psi_g)) \right] &= \text{Pro}\left[\mu(\Phi_h)' | L_e^{ind}(\mathcal{D}(\Psi_g)) \right] \\ &= \alpha_h^{(2)} B_h^{(2)} + \left(1 - \alpha_h^{(2)}\right) B_g^{(3)}. \end{aligned} \tag{6.18}$$

Combining (6.17) and (6.18) we get

$$B_g^{(3)} = \frac{1}{|H_g|} \sum_{h \in H_g} \left\{ \alpha_h^{(2)} B_h^{(2)} + \left(1 - \alpha_h^{(2)}\right) B_g^{(3)} \right\},$$

and

$$0 = \frac{1}{|H_g|} \sum_{h \in H_g} \left\{ \alpha_h^{(2)} B_h^{(2)} - w_g^{(3)} B_g^{(3)} \right\},$$

respectively

$$B_g^{(3)} = \sum_{h \in H_g} \frac{\alpha_h^{(2)}}{w_g^{(3)}} B_h^{(2)},$$

which proves (6.13).

For (6.14) we calculate

$$E\left[\left(B_g^{(3)} - \mu(\Psi_g) \right)^2 \right] = \sum_{h \in H_g} \left(\frac{\left(\alpha_h^{(2)}\right)^2}{\left(w_g^{(3)}\right)^2} E\left[\left(B_h^{(2)} - \mu(\Psi_g) \right)^2 \right] \right), \tag{6.19}$$

$$\begin{aligned} E\left[\left(B_h^{(2)} - \mu(\Psi_g) \right)^2 \right] &= E\left[\left(B_h^{(2)} - \mu(\Phi_h) + \mu(\Phi_h) - \mu(\Psi_g) \right)^2 \right], \\ &= E\left[\left(B_h^{(2)} - \mu(\Phi_h) \right)^2 \right] + E\left[\left(\mu(\Phi_h) - \mu(\Psi_g) \right)^2 \right], \\ &= E\left[\left(B_h^{(2)} - \mu(\Phi_h) \right)^2 \right] + \tau_2^2. \end{aligned} \tag{6.20}$$

From Lemma 6.2, respectively formula (6.9), it holds that

$$\alpha_h^{(2)} = \frac{\tau_2^2}{\tau_2^2 + E\left[\left(\mu(\Phi_h) - B_h^{(2)} \right)^2 \right]}. \tag{6.21}$$

(6.21) and (6.20) plugged into (6.19) yields

$$E\left[\left(B_g^{(3)} - \mu(\Psi_g) \right)^2 \right] = \frac{\tau_2^2}{w_g^{(3)}}. \tag{6.22}$$

This concludes the proof of Lemma 6.3. □

Analogously we have

$$E\left[\left(B_h^{(2)} - \mu\left(\Phi_h\right)\right)^2\right] = \frac{\tau_1^2}{w_h^{(2)}},$$

$$E\left[\left(B_i^{(1)} - \mu\left(\Theta_i\right)\right)^2\right] = \frac{\sigma^2}{w_{i\bullet}}.$$

Note that the last equation is the same as in the Bühlmann–Straub model.

On the basis of Lemmas 6.2 and 6.3 we can now proceed to the numerical evaluation of the data compression and of the credibility weights at the various levels. This is summarized in the following theorem.

Theorem 6.4 (data compression and credibility weights). *The best, individually unbiased estimator (and the credibility weights for the various levels) can be calculated from bottom to top as follows:*

$$B_i^{(1)} = \sum\nolimits_j \frac{w_{ij}}{w_{i\bullet}} X_{ij},$$

$$\text{where } w_{i\bullet} = \sum\nolimits_j w_{ij},$$

$$\alpha_i^{(1)} = \frac{w_{i\bullet}}{w_{i\bullet} + \frac{\sigma^2}{\tau_1^2}},$$

$$B_h^{(2)} = \sum_{i \in I_h} \frac{\alpha_i^{(1)}}{w_h^{(2)}} B_i^{(1)},$$

$$\text{where } I_h = \left\{i : \Theta_i \in \Theta\left(\Phi_h\right)\right\}, \qquad w_h^{(2)} = \sum\nolimits_{i \in I_h} \alpha_i^{(1)},$$

$$\alpha_h^{(2)} = \frac{w_h^{(2)}}{w_h^{(2)} + \frac{\tau_1^2}{\tau_2^2}},$$

$$B_g^{(3)} = \sum_{h \in H_g} \frac{\alpha_h^{(2)}}{w_g^{(3)}} B_h^{(2)},$$

$$\text{where } H_g = \left\{h : \Phi_h \in \Phi\left(\Psi_g\right)\right\}, \qquad w_g^{(3)} = \sum\nolimits_{h \in H_g} \alpha_h^{(2)},$$

$$\alpha_g^{(3)} = \frac{w_g^{(3)}}{w_g^{(3)} + \frac{(\tau_2)^2}{(\tau_3)^2}},$$

$$\widehat{\widehat{\mu_0}} = \sum_g \frac{\alpha_g^{(3)}}{w^{(4)}} B_g^{(3)},$$

$$\text{where } w^{(4)} = \sum\nolimits_g \alpha_g^{(3)}.$$

Remarks:

i) Since the credibility estimator depends only on these "optimally compressed" data $\{B_i^{(1)}, B_h^{(2)}, B_g^{(3)}\}$, we also call them a *linear sufficient statistics*.

ii) Note: the estimator of the next higher level is always the weighted average (weighted with the credibility weights) of the current level, e.g. $B_h^{(2)}$ is the weighted average of the $B_i^{(1)}$'s, where the weights are the $\alpha_i^{(1)}$'s.

iii) The following formulae for the error terms have already been proved:

$$E\left[\left(B_i^{(1)} - \mu(\Theta_i)\right)^2\right] = \frac{\sigma^2}{w_{i\bullet}},$$

$$E\left[\left(B_h^{(2)} - \mu(\Phi_h)\right)^2\right] = \frac{\tau_1^2}{w_h^{(2)}},$$

$$E\left[\left(B_g^{(3)} - \mu(\Psi_g)\right)^2\right] = \frac{\tau_2^2}{w_g^{(3)}}.$$

iv) We also have that

$$E\left[\left(B_i^{(1)} - \mu(\Phi_h)\right)^2\right] = \frac{\tau_1^2}{\alpha_i^{(1)}}, \tag{6.23}$$

$$E\left[\left(B_h^{(2)} - \mu(\Psi_g)\right)^2\right] = \frac{\tau_2^2}{\alpha_h^{(2)}}, \tag{6.24}$$

$$E\left[\left(B_g^{(3)} - \mu_0\right)^2\right] = \frac{\tau_3^2}{\alpha_g^{(3)}}, \tag{6.25}$$

i.e. the credibility weights are proportional to the "precisions" with respect to the quantities to be estimated on the next higher level. From this it becomes intuitively clear that, e.g. $B_h^{(2)}$ is the credibility weighted mean of $\{B_i^{(1)} : i \in I_h\}$.

The validity of (6.23)–(6.25) is easy to check. For instance

$$\begin{aligned} E[(B_h^{(2)} - \mu(\Psi_g))^2] &= E[(B_h^{(2)} - \mu(\Phi_h) + \mu(\Phi_h) - \mu(\Psi_g))^2], \\ &= \frac{\tau_1^2}{w_h^{(2)}} + \tau_2^2, \\ &= \frac{\tau_2^2}{\alpha_h^{(2)}}. \end{aligned}$$

The discussion at the beginning of this section together with Theorem 6.4 provide the basis for the calculation of the credibility estimator at the various hierarchical levels. The result is summarized in the following theorem.

Theorem 6.5 (credibility estimator). *The (inhomogeneous) credibility estimators* $\widehat{\widehat{\mu(\cdot)}}$ *can be determined top down as follows:*

$$\widehat{\widehat{\mu(\Psi_g)}} = \alpha_g^{(3)} B_g^{(3)} + \left(1 - \alpha_g^{(3)}\right) \mu_0,$$

$$\widehat{\widehat{\mu(\Phi_h)}} = \alpha_h^{(2)} B_h^{(2)} + \left(1 - \alpha_h^{(2)}\right) \widehat{\widehat{\mu(\Psi_g)}}, \qquad \Phi_h \in \Phi(\Psi_g),$$

$$\widehat{\widehat{\mu(\Theta_i)}} = \alpha_i^{(1)} B_i^{(1)} + \left(1 - \alpha_i^{(1)}\right) \widehat{\widehat{\mu(\Phi_h)}}, \qquad \Theta_i \in \Theta(\Phi_h).$$

Remarks:

- The credibility weights (the α's) as well as the best, individually unbiased estimators (the B's), were defined and calculated in Theorem 6.4.
- The credibility estimators depend on the data only through the B-values. These are *linear sufficient statistics.* The method for compressing the data is the same as that which we met earlier, namely the best, individually unbiased estimator based on the underlying data.
- Notice also that the procedure for calculating the credibility estimator is exactly analogous to that for the homogeneous credibility estimator in the Bühlmann–Straub model. This is a two-step procedure, where first the linear sufficient statistic (data compression) is calculated from bottom to top and then the credibility estimator is determined top down.

Proof of Theorem 6.5:
The theorem follows directly from the discussion at the beginning of the section as well as from Theorem 6.4. □

With exactly the same arguments we also get ("for free", so to speak) the homogeneous credibility estimator in the hierarchical model. The results in the following theorem are obvious.

Theorem 6.6 (homogeneous credibility estimator). *The homogeneous credibility estimator* $\widehat{\widehat{\mu(\cdot)}}^{\text{hom}}$ *can be calculated top down as follows:*

$$\widehat{\widehat{\mu(\Psi_g)}}^{\text{hom}} = \alpha_g^{(3)} B_g^{(3)} + \left(1 - \alpha_g^{(3)}\right) \widehat{\widehat{\mu_0}},$$

$$\widehat{\widehat{\mu(\Phi_h)}}^{\text{hom}} = \alpha_h^{(2)} B_h^{(2)} + \left(1 - \alpha_h^{(2)}\right) \widehat{\widehat{\mu(\Psi_g)}}^{\text{hom}}, \qquad \Phi_h \in \Phi(\Psi_g),$$

$$\widehat{\widehat{\mu(\Theta_i)}}^{\text{hom}} = \alpha_i^{(1)} B_i^{(1)} + \left(1 - \alpha_i^{(1)}\right) \widehat{\widehat{\mu(\Phi_h)}}^{\text{hom}}, \qquad \Theta_i \in \Theta(\Phi_h).$$

Remarks:

- As in the Bühlmann–Straub model, the homogeneous estimator is found by replacing μ_0 with the best linear estimator $\widehat{\widehat{\mu_0}}$.
- The calculation and definition of $\widehat{\widehat{\mu_0}}$ is given in Theorem 6.4.

The remarkable property of the homogeneous credibility estimator, namely to define a tariff which can be seen as a redistribution of total observed claims, which was already found for the Bühlmann–Straub model (cf. Theorem 4.5) also holds in the hierarchical model and is stated in the following theorem.

Theorem 6.7 (balance property). *For the homogeneous credibility estimator in the hierarchical model it holds that*

$$\sum_{i,j} w_{ij} \widehat{\widehat{\mu(\Theta_i)}}^{\text{hom}} = \sum_{i,j} w_{ij} X_{ij}.$$

Remark:

- This property does *not*, however, hold for the credibility estimator for higher levels. For example, in general we have

$$\sum_h w_{h\bullet} \widehat{\widehat{\mu(\Phi_h)}}^{\text{hom}} \neq \sum_{i,j} w_{ij} X_{ij}, \qquad \text{where } w_{h\bullet} = \sum_{\{i,j:\ X_{ij} \in \mathcal{D}(\Phi_h)\}} w_{ij}.$$

  We must therefore be careful in interpreting $\widehat{\widehat{\mu(\cdot)}}$ for higher levels: It may not be interpreted as the "average premium amount" resulting for the higher levels.

Proof of Theorem 6.7: With Theorem 6.4 we find that

$$\begin{aligned}
\sum_{i,j} w_{ij} \left(\widehat{\widehat{\mu(\Theta_i)}}^{\text{hom}} - X_{ij} \right) &= \sum_i w_{i\bullet} \left(\widehat{\widehat{\mu(\Theta_i)}}^{\text{hom}} - B_i^{(1)} \right) \\
&= \sum_i w_{i\bullet} \left(1 - \alpha_i^{(1)}\right) \left(\widehat{\widehat{\mu(\Phi_h)}}^{\text{hom}} - B_i^{(1)} \right) \\
&= \frac{\sigma^2}{\tau_1^2} \sum_h w_h^{(2)} \left(\widehat{\widehat{\mu(\Phi_h)}}^{\text{hom}} - B_h^{(2)} \right) \\
&= \frac{\sigma^2}{\tau_1^2} \sum_h w_h^{(2)} \left(1 - \alpha_h^{(2)}\right) \left(\widehat{\widehat{\mu(\Psi_g)}}^{\text{hom}} - B_h^{(2)} \right)
\end{aligned}$$

$$
\begin{aligned}
&= \frac{\sigma^2}{\tau_2^2} \sum_g w_g^{(3)} \left(\widehat{\widehat{\mu(\Psi_g)}}^{\text{hom}} - B_g^{(3)} \right) \\
&= \frac{\sigma^2}{\tau_2^2} \sum_g w_g^{(3)} \left(1 - \alpha_g^{(3)}\right) \left(\widehat{\widehat{\mu_0}} - B_g^{(3)} \right) \\
&= \frac{\sigma^2}{\tau_3^2} \sum_g \alpha_g^{(3)} \left(\widehat{\widehat{\mu_0}} - B_g^{(3)} \right) \\
&= 0.
\end{aligned}
$$

In the third, fifth and seventh equations we use the fact that

$$
\begin{aligned}
w_{i\bullet}(1 - \alpha_i^{(1)}) &= \frac{\sigma^2}{\tau_1^2} \alpha_i^{(1)}, \\
w_h^{(2)}(1 - \alpha_h^{(2)}) &= \frac{\tau_1^2}{\tau_2^2} \alpha_h^{(2)}, \\
w_g^{(3)}(1 - \alpha_g^{(3)}) &= \frac{\tau_2^2}{\tau_3^2} \alpha_g^{(3)}.
\end{aligned}
$$

This completes the proof of Theorem 6.7. □

6.5 Quadratic Loss in the Hierarchical Model

The quadratic loss in the hierarchical credibility model can also be best calculated using a recursive procedure.

Theorem 6.8 (quadratic loss of the credibility estimator). *The quadratic loss of the credibility estimator can be determined top down as follows:*

$$
q_g^{(3)} := E\left[\left(\widehat{\widehat{\mu(\Psi_g)}} - \mu(\Psi_g) \right)^2 \right] = \tau_3^2 \left(1 - \alpha_g^{(3)}\right), \tag{6.26}
$$

$$
\begin{aligned}
q_h^{(2)} :=& E\left[\left(\widehat{\widehat{\mu(\Phi_h)}} - \mu(\Phi_h) \right)^2 \right] \\
&= \tau_2^2 \left(1 - \alpha_h^{(2)}\right) + q_g^{(3)} \left(1 - \alpha_h^{(2)}\right)^2,
\end{aligned} \tag{6.27}
$$

$$
\begin{aligned}
q_i^{(1)} :=& E\left[\left(\widehat{\widehat{\mu(\Theta_i)}} - \mu(\Theta_i) \right)^2 \right] \\
&= \tau_1^2 \left(1 - \alpha_i^{(1)}\right) + q_h^{(2)} \left(1 - \alpha_i^{(1)}\right)^2.
\end{aligned} \tag{6.28}
$$

Proof: At the top level we find

$$
\begin{aligned}
q_g^{(3)} &= E\left[\left\{\alpha_g^{(3)}(B_g^{(3)} - \mu(\Psi_g)) + (1-\alpha_g^{(3)})(\mu_0 - \mu(\Psi_g))\right\}^2\right] \\
&= \left(\frac{w_g^{(3)}}{w_g^{(3)} + \frac{\tau_2^2}{\tau_3^2}}\right)^2 \frac{\tau_2^2}{w_g^{(3)}} + \left(\frac{\frac{\tau_2^2}{\tau_3^2}}{w_g^{(3)} + \frac{\tau_2^2}{\tau_3^2}}\right)^2 \tau_3^2 \qquad (6.29) \\
&= \tau_3^2 \frac{\frac{\tau_2^2}{\tau_3^2}}{w_g^{(3)} + \frac{\tau_2^2}{\tau_3^2}} = \tau_3^2(1-\alpha_g^{(3)}).
\end{aligned}
$$

Because of the iterative property of projections we have that

$$
\begin{aligned}
q_h^{(2)} &= E\left[\left(\widehat{\widehat{\mu(\Phi_h)}} - \mu(\Phi_h)\right)^2\right] \\
&= E\left[\left(\widehat{\mu(\Phi_h)}' - \mu(\Phi_h)\right)^2\right] + E\left[\left(\widehat{\mu(\Phi_h)}' - \widehat{\widehat{\mu(\Phi_h)}}\right)^2\right]. \qquad (6.30)
\end{aligned}
$$

A calculation analogous to (6.29) shows that

$$
\begin{aligned}
& E\left[\left(\widehat{\mu(\Phi_h)}' - \mu(\Phi_h)\right)^2\right] \\
&= E\left[\left\{\alpha_h^{(2)}\left(B_h^{(2)} - \mu(\Phi_h)\right) + (1-\alpha_h^{(2)})(\mu(\Psi_g) - \mu(\Phi_h))\right\}^2\right] \\
&= \tau_2^2\,(1-\alpha_h^{(2)}).
\end{aligned}
$$

For the last term in (6.30) we get

$$
\begin{aligned}
E\left[\left(\widehat{\mu(\Phi_h)}' - \widehat{\widehat{\mu(\Phi_h)}}\right)^2\right] &= \left(1-\alpha_h^{(2)}\right)^2 E\left[\left(\widehat{\widehat{\mu(\Psi_g)}} - \mu(\Psi_g)\right)^2\right], \\
&= \left(1-\alpha_h^{(2)}\right)^2 q_g^{(3)}.
\end{aligned}
$$

Thus (6.27) is proved. The proof of (6.28) is analogous. This ends the proof of Theorem 6.8. □

Theorem 6.9 (quadratic loss of the hom. credibility estimator). *The quadratic loss of the homogeneous credibility estimator can be calculated from top down as follows:*

$$
\begin{aligned}
q_{\text{hom}}^{(4)} &:= E\left[\left(\widehat{\widehat{\mu_0}} - \mu_0\right)^2\right] = \frac{\tau_3^2}{\alpha_\bullet^{(3)}}, \\
q_{g,\text{hom}}^{(3)} &:= E\left[\left(\widehat{\widehat{\mu(\Psi_g)}}^{\text{hom}} - \mu(\Psi_g)\right)^2\right] \\
&= \tau_3^2\left(1-\alpha_g^{(3)}\right) + q_{\text{hom}}^{(4)}\left(1-\alpha_g^{(3)}\right)^2, \qquad (6.31)
\end{aligned}
$$

$$q_{h,\text{hom}}^{(2)} := E\left[\left(\widehat{\widehat{\mu(\Phi_h)}}^{\text{hom}} - \mu(\Phi_h)\right)^2\right]$$
$$= \tau_2^2\left(1 - \alpha_h^{(2)}\right) + q_{g,\text{hom}}^{(3)}\left(1 - \alpha_h^{(2)}\right)^2, \tag{6.32}$$
$$q_{i,\text{hom}}^{(1)} := E\left[\left(\widehat{\widehat{\mu(\Theta_i)}}^{\text{hom}} - \mu(\Theta_i)\right)^2\right]$$
$$= \tau_1^2\left(1 - \alpha_i^{(1)}\right) + q_{h,\text{hom}}^{(2)}\left(1 - \alpha_i^{(1)}\right)^2. \tag{6.33}$$

Proof:

$$E\left[\left(\widehat{\widehat{\mu_0}} - \mu_0\right)^2\right] = \sum_g \left(\frac{\alpha_g^{(3)}}{\alpha_\bullet^{(3)}}\right)^2 E\left[\left(B_g^{(3)} - \mu_0\right)^2\right]$$
$$= \sum_g \left(\frac{\alpha_g^{(3)}}{\alpha_\bullet^{(3)}}\right)^2 \frac{\tau_3^2}{\alpha_g^{(3)}}$$
$$= \frac{\tau_3^2}{\alpha_\bullet^{(3)}}. \tag{6.34}$$

Remark:

- Note that (6.34) is of the same structure as formulae (6.23)–(6.25).

It follows from Theorem 3.14 that

$$E\left[\left(\widehat{\widehat{\mu(\Psi_g)}}^{\text{hom}} - \mu(\Psi_g)\right)^2\right] = E\left[\left(\widehat{\widehat{\mu(\Psi_g)}}^{\text{hom}} - \widehat{\widehat{\mu(\Psi_g)}}\right)^2\right] + E\left[\left(\widehat{\widehat{\mu(\Psi_g)}} - \mu(\Psi_g)\right)^2\right].$$

(6.31) then follows from

$$E\left[\left(\widehat{\widehat{\mu(\Psi_g)}}^{\text{hom}} - \widehat{\widehat{\mu(\Psi_g)}}\right)^2\right] = \left(1 - \alpha_g^{(3)}\right)^2 E\left[\left(\widehat{\widehat{\mu_0}} - \mu_0\right)^2\right]$$
$$= \left(1 - \alpha_g^{(3)}\right)^2 \frac{\tau_3^2}{\alpha_\bullet^{(3)}}$$

and (6.26).

(6.32) and (6.33) can be proved in exactly the same way as (6.27) and (6.28) by just replacing everywhere $\widehat{\widehat{\mu(.)}}$ by $\widehat{\widehat{\mu(.)}}^{\text{hom}}$. This completes the proof of Theorem 6.9. □

6.6 Estimation of the Structural Parameters in the Hierarchical Model

The estimator for μ_0 has been given by the homogeneous credibility estimator. In the following, we give estimators for the components of variance.

For the two lowest levels, i.e. for the estimation of σ^2 and τ_1^2, we could use estimators analogous to those found for the Bühlmann–Straub model (see Section 4.8). For the higher levels of the hierarchy, i.e. for the estimation of τ_2^2 and τ_3^2 in the model of order 3, another approach will be recommended.

We have already introduced the sets of indices I_h and H_g (Theorem 6.4). For the following it is useful to introduce the sets of nodes in the hierarchical tree:

$$\begin{aligned} G &:= \{g : \Psi_g \in \Psi(\mu_0)\}, \\ H &:= \bigcup_{g \in G} H_g, \\ I &:= \bigcup_{h \in H} I_h. \end{aligned}$$

Further, we denote the number of elements in a set with $| \cdot |$.

Examples:

- $|I_h|$ = Number of nodes at the Θ-level, which stem from Θ_h.
- $|I|$ = Total number of nodes at the Θ-level.
- $|G|$ = Total number of nodes at level 3.

We denote the number of observations for the ith risk by n_i.

i) **Estimation of σ^2:**
As in the Bühlmann–Straub model, we consider

$$S_i = \frac{1}{n_i - 1} \sum_j w_{ij} \left(X_{ij} - B_i^{(1)} \right)^2 .$$

$$S = \sum_{i \in I} \frac{n_i - 1}{n_\bullet - |I|} S_i = \frac{1}{n_\bullet - |I|} \sum_{i,j} w_{ij} \left(X_{ij} - B_i^{(1)} \right)^2 .$$

Note that $E[S_i | \Theta_i] = \sigma^2(\Theta_i)$ and hence $E[S] = \sigma^2$.
We choose

$$\widehat{\sigma^2} = S$$

as an estimator for σ^2.

ii) **Estimation of τ_1^2:**
We also use here estimators which are analogous to those used in the Bühlmann–Straub model. Consider the following random variables:

$$\widehat{T_h^{(1)}} = c_h \cdot \left\{ \frac{|I_h|}{|I_h|-1} \sum_{i \in I_h} \frac{w_{i\bullet}}{z_h^{(1)}} \left(B_i^{(1)} - \overline{B}_h^{(1)} \right)^2 - \frac{|I_h| \cdot \widehat{\sigma^2}}{z_h^{(1)}} \right\},$$

where

$$z_h^{(1)} = \sum_{i \in I_h} w_{i\bullet},$$

$$\overline{B}_h^{(1)} = \sum_{i \in I_h} \frac{w_{i\bullet}}{z_h^{(1)}} B_i^{(1)},$$

$$c_h = \frac{|I_h|-1}{|I_h|} \left\{ \sum_{i \in I_h} \frac{w_{i\bullet}}{z_h^{(1)}} \left(1 - \frac{w_{i\bullet}}{z_h^{(1)}} \right) \right\}^{-1}.$$

It holds again that

$$E\left[\widehat{T_h^{(1)}} \middle| \Phi_h \right] = \tau_1^2 (\Phi_h),$$

$$E\left[T_h^{(1)} \right] = \tau_1^2.$$

We choose

$$\widehat{\tau_1}^2 = \frac{1}{|H|} \sum_{h \in H} \max\{\widehat{T_h^{(1)}}, 0\}$$

as estimator for τ_1^2.

iii) **Estimation of τ_2^2:**
We define the following random variables:

$$T_g^{(2)} = c_g \cdot \left\{ \frac{|H_g|}{|H_g|-1} \sum_{h \in H_g} \frac{w_h^{(2)}}{z_g^{(2)}} \left(B_h^{(2)} - \overline{B}_g^{(2)} \right)^2 - \frac{|H_g| \, \tau_1^2}{z_g^{(2)}} \right\},$$

where

$$z_g^{(2)} = \sum_{h \in H_g} w_h^{(2)}, \qquad w_h^{(2)} \text{ was defined in Theorem 6.4,}$$

$$\overline{B}_g^{(2)} = \sum_{h \in H_g} \frac{w_h^{(2)}}{z_g^{(2)}} \overline{B}_h^{(2)},$$

$$c_g = \frac{|H_g|-1}{|H_g|} \left\{ \sum_{h \in H_g} \frac{w_h^{(2)}}{z_g^{(2)}} \left(1 - \frac{w_h^{(2)}}{z_g^{(2)}} \right) \right\}^{-1}.$$

It is easy to show that the random variables $T_g^{(2)}$, $g = 1, 2, \ldots, |G|$, are again unbiased estimators for τ_2^2. However, $T_g^{(2)}$ depends on the structural parameters σ^2 and τ_1^2 through $w_h^{(2)}$ and for τ_1^2 also directly. But these were already estimated in steps i) and ii) and so we substitute the structural parameters in the formula for $T_g^{(2)}$ by their estimates. Of course the resulting random variables $\widehat{T_g^{(2)}}$ are no longer unbiased. This procedure leads then to the following estimator:

$$\widehat{\tau_2}^2 = \frac{1}{|G|} \sum_{g \in H_g} \max\{\widehat{T_g^{(2)}}, 0\}.$$

iv) **Estimation of τ_3^2:**
This follows the same scheme as the estimation of τ_2^2.

$$T^{(3)} = c \cdot \left\{ \frac{|G|}{|G|-1} \sum_{g \in G} \frac{w_g^{(3)}}{z^{(3)}} \left(B_g^{(3)} - \overline{B}^{(3)} \right)^2 - \frac{|G| \cdot \tau_2^2}{z^{(3)}} \right\},$$

where

$$z^{(3)} = \sum_{g \in G} w_g^{(3)}, \quad w_g^{(3)} \text{ was defined in Theorem 6.4,}$$

$$\overline{B}^{(3)} = \sum_{g \in G} \frac{w_g^{(3)}}{z^{(3)}} B_g^{(3)},$$

$$c = \frac{|G|-1}{|G|} \left\{ \sum_{g \in G} \frac{w_g^{(3)}}{z^{(3)}} \left(1 - \frac{w_g^{(3)}}{z^{(3)}} \right) \right\}^{-1}.$$

Again, $T^{(3)}$ is an unbiased estimator and depends on the structural parameters σ^2, τ_1^2 and τ_2^2. These have been estimated in steps i) to iii) and so we replace these values by their estimators in the formula for $T^{(3)}$. We denote the resulting random variable by $\widehat{T^{(3)}}$. This leads to the following estimator:

$$\widehat{\tau_3}^2 = \max\{\widehat{T^{(3)}}, 0\}.$$

General remarks:

- The estimation of the structural parameters follows a bottom-up procedure in parallel with the estimation of the sufficient statistics (see Theorem 6.4).
- The extension of this procedure to models of higher order is obvious.
- Observe that the number of parameters to be estimated increases with the number of levels of the hierarchical model. This means that in practice one should be parsimonious when choosing the number of levels in a hierarchical model.
- If we modified Model Assumptions 6.1, modelling all variances as constants, i.e. $\operatorname{Var}[X_{ij} | w_{ij}] = \sigma_i^2 / w_{ij}$, $\operatorname{Var}[\mu(\Theta_i)] = \tau_{1h}^2$ for $\Theta_i \in \Theta(\Phi_h)$,

$\mathrm{Var}\left[\mu\left(\Phi_h\right)\right]=\tau_{2g}^2$ for $\Phi_h \in \Psi_g$, these constants would appear in the credibility formulae instead of the "averaged" variance components $\sigma^2, \tau_1^2, \tau_2^2$. If for instance a specific risk group showed more stable results than another, this specific risk group would be given a bigger credibility weight. The situation is the same as already discussed in the Bühlmann–Straub model in the last point of the remarks on pages 85–86. The point is that these constants would have to be estimated from the data and the imposed estimation error would typically offset the "theoretical" improvement of the credibility estimator if the constants were known. This effect of estimation errors is inherent in each level of the hierarchical model.

6.7 Exercises

Exercise 6.1

An insurance company wants to calculate the expected value of the claims ratio (claim amount divided by a suitable weight) for different risk categories in a certain line of business. The company's own data are available as well as "industry-wide" data (i.e. pooled data delivered to a statistics bureau by several companies) for an observation period of n years. The table below gives the following data:

- the observed average of the claims ratio X_i,
- the empirical within-risk variance σ_i^2 of the yearly claims ratios X_{ij} given by

$$\sigma_i^2=\frac{1}{n-1}\sum_{j=1}^{n} w_{ij}\left(X_{ij}-X_i\right)^2,$$

- the weight $w_{i\bullet}$ over the whole observation period.

risk	own company			other companies		
group	X_i	σ_i^2	$w_{i.}$	X_i	σ_i^2	$w_{i.}$
1	0.81	42.30	3 847	0.75	43.60	14 608
2	0.94	31.60	1 150	0.86	27.10	3 081
3	0.93	27.20	2 843	1.06	49.30	4 644
4	1.14	49.25	1 123	0.89	17.30	3 487
5	0.73	16.40	532	1.01	23.90	9 004
6	1.04	37.65	1 309	0.86	27.45	4 467
7	1.48	39.60	1 332	1.19	46.60	4 536
8	1.10	21.60	923	0.79	23.35	2 718
total	0.98		13 059	0.90		46 545

a) Calculate the values of the (empirical) homogeneous credibility estimator using the hierarchical model with the different risk groups on level 2 and the distinction between the data of the own and the data of the other

companies on level 1.
Beside the point estimates, give also an indication of the precisions of these estimates. As a measure of the precision, take the square root of the quadratic loss obtained by inserting the estimated structural parameters into the formula of the quadratic loss.
Remark: Be aware that the estimated quadratic loss obtained by this procedure does not contain the estimation error of the structural parameters.

b) Do the same but with the distinction between own/other companies on level 2 and the different risk groups on level 1.
c) Discuss shortly the difference between the two models a) and b).

Remark: We will use the same data set in Exercise 7.1 in connection with multidimensional credibility.

Exercise 6.2

Consider the same data set as in Exercise 4.6 and treat them within a hierarchical framework of order two, where the five different types of car are on level one and the use of car (A or B) are on level two.

a) Calculate the values of the (empirical) hierarchical credibility estimator of the claim frequencies, both for normal and large claims. Compare the results with the results obtained in Exercise 4.6 and comment briefly on the results.
b) The same for the claim severities.
c) The same for the average claims cost per risk directly. Discuss the differences in the pure risk premiums obtained by combining a) and b) on the one hand and the results obtained by c) on the other.

7

Multidimensional Credibility

7.1 Motivation

The following three examples are a selection of typical situations which pricing actuaries working in practice are confronted with. The list could easily be extended.

- "Normal" claims load, "large" claims load.
  In many lines of business (e.g. motor liability, general liability, industry fire), one or two percent of the largest claims are responsible for more than half of the total claims load. This means that more than half of the pure risk premium stems from a rather small number of claims, or in other words, there is relatively little statistical information concerning half of the pure risk premium. The question arises, whether and how much we can learn from the bulk of normal claims, where we have a huge number of observations and a lot of statistical information, with respect to the large claims load. The distinction between the two categories of "normal" and "large" claims can be either the size of the claims or any other criterion, for instance "pure physical damage" and "bodily injury" claims in motor liability.
- Claim frequencies, claim sizes.
  It is standard in the pricing of insurance products to analyse claim frequencies and claim amounts separately and then to derive from these two quantities the expected average claims load per risk. However, there might be dependencies between frequencies and claim sizes. For instance the motor accidents in towns are often greater in number but smaller in the average claim size compared to rural areas.
- Own data, pool data.
  In many countries there are summary statistics from a data pool (industry-wide or several companies) available. A company actuary, faced with the problem of calculating a tariff for the own company, has two sources of statistical information: the company's own data and the pool data. He

is then confronted with the question of how much he should rely on the company's own statistics and how much he should take into account the information from the pool statistics.

- Group accident insurance (workers' compensation) in Switzerland.
  In Switzerland, employers are obliged to insure all their employees against accidents both at work and "not at work" by means of group accident insurance policies. In the resulting statistics we can differentiate between accidents occurring at work and those occurring "not at work" and, within each of these classifications, between long-term benefits (for disability or dependents) and short-term benefits (medical costs, wage payment interruptions). Again the question arises, whether and how much we can learn from the observations of the one category of claims with respect to the other categories.

As an example let us consider the problem of "pure physical damage (normal)" and "bodily injury (large) claims" in motor liability rate-making. Assume that, as a first step, we are interested in the corresponding claim frequencies. For each risk i and each year we have a two-dimensional vector of observations, namely the frequency of "normal" claims and the frequency of "large" claims. The expected value of the observation vector, given the risk characteristic Θ, is the vector

$$\boldsymbol{\mu}'(\Theta) = (\mu_1(\Theta), \mu_2(\Theta)).$$

Primarily we are interested in the pure risk premium. We make the usual assumption that conditionally on Θ the number of claims and the claim sizes are independent. For simplicity, we further assume that the distribution of the claim sizes does not depend on Θ. Then the pure risk premium is given by

$$\mu(\Theta) = \xi_1 \mu_1(\Theta) + \xi_2 \mu_2(\Theta),$$

where ξ_1 resp. ξ_2 are the expected values of the claim severities of "normal" resp. "large" claims. Because of the linearity property of the credibility estimator, we have

$$\widehat{\widehat{\mu(\Theta)}} = \xi_1 \widehat{\widehat{\mu_1(\Theta)}} + \xi_2 \widehat{\widehat{\mu_2(\Theta)}}.$$

In order to find the credibility estimator for $\mu(\Theta)$ for all possible severities, we need to determine the *credibility estimator of the vector* $\boldsymbol{\mu}(\Theta)$, i.e.

$$\widehat{\widehat{\boldsymbol{\mu}(\Theta)}}' = \left(\widehat{\widehat{\mu_1(\Theta)}}, \widehat{\widehat{\mu_2(\Theta)}}\right).$$

Here, the vector $\boldsymbol{\mu}(\Theta)$ is two-dimensional. In the group accident example mentioned above, $\boldsymbol{\mu}(\Theta)$ would be a vector with dimension four. The point is that the quantity to be estimated is *multidimensional* and hence the title *multidimensional credibility*.

We might first ask ourselves if it would make sense to estimate each of the components of the vector separately on the basis of each of the associated observations (= corresponding components of the observation vector) and in this way reduce the problem to two one-dimensional problems. We could do that, if the components of the vector $\boldsymbol{\mu}(\Theta)$ were independent. However, in most practical situations there is no a priori evidence that they are independent. On the contrary, there is mostly some evidence that there are dependences between the components and that we can therefore learn something from the observations of one component with respect to the others. For instance, in the example of group accident insurance, it is known that people who live "dangerously" professionally, tend also to take more risks in their free time. There is therefore a correlation between accidents at work and "not at work". This in turn means that we can learn something about the expected aggregate claim amount for accidents at work from the claim experience of accidents "not at work", and vice versa.

The essential point of *multidimensional credibility* is not only to *estimate a multidimensional vector* $\boldsymbol{\mu}(\Theta)$, but also *to use* for this estimation *all available observations simultaneously.*

7.2 The Abstract Multidimensional Credibility Model

In this section we concentrate on the mathematical structure which is essential for the results of this chapter. This structure is very general. It also covers the credibility regression model which will be the subject of Chapter 8, where we can use the results derived in this chapter. A concrete multidimensional model will be considered in Section 7.3.

7.2.1 Model

As already explained, one essential feature of multidimensional credibility is that the quantity $\boldsymbol{\mu}(\Theta)$ to be estimated is multidimensional, i.e. a vector of length p.

In the abstract multidimensional credibility model, we assume that there is a random vector $\mathbf{X}$ with the same dimension as $\boldsymbol{\mu}(\Theta)$, which is individually unbiased, i.e. for which $E[\mathbf{X}|\Theta] = \boldsymbol{\mu}(\Theta)$. The goal is to find the credibility estimator based on $\mathbf{X}$.

The vector $\mathbf{X}$ considered here must not be the same as the observation vector and, as a rule, this is not the case. As we will see, $\mathbf{X}$ is in general a compression of the original observations. For the moment, we leave open the question of how the appropriate compression of the data is determined. This problem will be treated in Section 7.4.

The vector $\boldsymbol{\mu}(\Theta)$, which we want to estimate, is also not further specified here. In most concrete applications this arises in a very natural way.

In the following, we just want to derive the form of the credibility estimator for $\boldsymbol{\mu}(\Theta)$ based on this simple structure as just defined. For the homogeneous credibility estimator we need a corresponding collective. We will assume that we have a portfolio of contracts having the required structure. This leads us to the following:

Model Assumptions 7.1 (abstract multidimensional credibility)
For each risk i we are given a p-dimensional vector $\mathbf{X}_i$, for which it holds that

i) Conditionally, given Θ_i,

$$E[\mathbf{X}_i'|\Theta_i] = \boldsymbol{\mu}(\Theta_i)' = (\mu_1(\Theta_i), \mu_2(\Theta_i), \ldots, \mu_p(\Theta_i)), \tag{7.1}$$

$$\operatorname{Cov}(\mathbf{X}_i, \mathbf{X}_i'|\Theta_i) = \Sigma_i(\Theta_i). \tag{7.2}$$

ii) The pairs $(\Theta_1, \mathbf{X}_1)$, $(\Theta_2, \mathbf{X}_2)$,... are independent, and Θ_1, Θ_2,... are independent and identically distributed.

Note that no assumptions are made about the conditional probability structure *within* $\mathbf{X}_i$, i.e. $\Sigma_i(\Theta_i)$ in (7.2) can be any covariance matrix.

For each risk i, we are interested in estimating the associated vector $\boldsymbol{\mu}(\Theta_i)$.

As in the earlier models, this estimator will depend on the values of the structural parameters. The *structural parameters of the multidimensional credibility model are*

$$\boldsymbol{\mu} := E[\boldsymbol{\mu}(\Theta_i)], \tag{7.3}$$

$$S_i := E[\Sigma_i(\Theta_i)], \tag{7.4}$$

$$T := \operatorname{Cov}\left(\boldsymbol{\mu}(\Theta_i), \boldsymbol{\mu}(\Theta_i)'\right). \tag{7.5}$$

Remarks:

- $\boldsymbol{\mu}$ is a vector of dimension p, S_i and T are $p \times p$ matrices.
- For the models used in insurance, it is almost always true that S_i can be obtained from a common matrix S and appropriately chosen weights.
- On the basis of the model assumptions, the different contracts or risks are considered a priori equal, i.e. a priori no differences among them are recognized.

7.2.2 The (Inhomogeneous) Multidimensional Credibility Estimator

The multidimensional credibility estimator is defined componentwise. We thus have to find, for each component of the vector $\boldsymbol{\mu}(\Theta_i)$, the associated credibility estimator.

As in the Bühlmann–Straub model, one easily sees that the credibility estimator for $\boldsymbol{\mu}(\Theta_i)$ depends only on the components of the vector $\mathbf{X}_i$ and not on those associated with the other contracts. *In order to simplify the notation in this section we omit the subscript i.*

By definition

$$\widehat{\widehat{\boldsymbol{\mu}(\Theta)}} := \operatorname{Pro}(\boldsymbol{\mu}(\Theta)|\, L(\mathbf{X},1)),$$

where the projection operator should be understood componentwise and $L(\mathbf{X},1)$ is the linear subspace spanned by 1 and the components of the vector $\mathbf{X}$ (cf. Definition 3.8).

Now, for $k = 1, 2, \ldots, p$, let

$$\widehat{\widehat{\mu_k(\Theta)}} = a_{k0} + \sum_{j=1}^{p} a_{kj} X_j$$

be the credibility estimator for the components of $\boldsymbol{\mu}(\Theta)$. Any such estimator must satisfy the normal equations (see Corollary 3.17):

$$\mu_k = a_{k0} + \sum_{j=1}^{p} a_{kj} \mu_j, \tag{7.6}$$

$$\operatorname{Cov}(\mu_k(\Theta), X_m) = \sum_{j=1}^{p} a_{kj} \operatorname{Cov}(X_j, X_m) \quad \text{for } m = 1, 2, \ldots, p. \tag{7.7}$$

More elegantly, using matrix notation we can write

$$\widehat{\widehat{\boldsymbol{\mu}(\Theta)}} = \mathbf{a} + A \cdot \mathbf{X},$$

where

$$\mathbf{a} = \begin{pmatrix} a_{10} \\ \vdots \\ a_{k0} \\ \vdots \\ a_{p0} \end{pmatrix}, \qquad A = \begin{pmatrix} a_{11}, a_{12}, \ldots, a_{1p} \\ \vdots \\ a_{k1}, a_{k2}, \ldots, a_{kp} \\ \vdots \\ a_{p1}, a_{p2}, \ldots, a_{pp} \end{pmatrix}.$$

The normal equations (7.6) and (7.7) in matrix notation are given by

$$\boldsymbol{\mu} = \mathbf{a} + A \cdot \boldsymbol{\mu}, \tag{7.8}$$

$$\operatorname{Cov}(\boldsymbol{\mu}(\Theta), \mathbf{X}') = A \cdot \operatorname{Cov}(\mathbf{X}, \mathbf{X}'). \tag{7.9}$$

Since

$$\begin{aligned} \operatorname{Cov}(\boldsymbol{\mu}(\Theta), \mathbf{X}') &= E\left[\operatorname{Cov}(\boldsymbol{\mu}(\Theta), \mathbf{X}'|\, \Theta)\right] + \operatorname{Cov}(\boldsymbol{\mu}(\Theta), E\left[\mathbf{X}'|\, \Theta\right]) \\ &= \operatorname{Cov}\left[\boldsymbol{\mu}(\Theta), \boldsymbol{\mu}(\Theta)'\right] \\ &= T, \\ \operatorname{Cov}(\mathbf{X}, \mathbf{X}') &= E\left[\operatorname{Cov}(\mathbf{X}, \mathbf{X}'|\, \Theta)\right] + \operatorname{Cov}(E\left[\mathbf{X}|\, \Theta\right], E\left[\mathbf{X}'|\, \Theta\right]) \\ &= S + T, \end{aligned}$$

we get from (7.9)

$$T = A(T+S),$$

and therefore

$$A = T(T+S)^{-1}.$$

Thus we have derived the following multidimensional credibility formula:

Theorem 7.2 (credibility estimator). *The (inhomogeneous) multidimensional credibility estimator is given by*

$$\widehat{\widehat{\boldsymbol{\mu}(\Theta_i)}} = A_i \mathbf{X}_i + (I - A_i)\,\boldsymbol{\mu}, \tag{7.10}$$

where $\boldsymbol{\mu}$, T, S_i were defined in (7.3)–(7.5), and

$$A_i = T(T+S_i)^{-1}.$$

Remarks:

- A_i is called the credibility matrix.
- The matrix $T + S_i$ is a covariance matrix and is therefore regular for linearly independent components of the vector $\mathbf{X}$.
- Since T and S_i are symmetric, we also have that $A_i' = (T+S_i)^{-1} T$. Notice, however, that, in general, $A_i \neq A_i'$, i.e. the credibility matrix A_i is not necessarily symmetric, even though T and S_i are symmetric.
- The estimator $\widehat{\widehat{\boldsymbol{\mu}(\Theta_i)}}$ is a weighted average of $\mathbf{X}_i$ and $\boldsymbol{\mu}$. The weighting results from the credibility matrix A_i and its complement $I - A_i$. It also holds that

$$\begin{aligned} A_i &= \left[T\,(T+S_i)^{-1}\, S_i\right] S_i^{-1} \\ &= \left(T^{-1} + S_i^{-1}\right)^{-1} S_i^{-1}, & (7.11) \\ I - A_i &= \left(T^{-1} + S_i^{-1}\right)^{-1} T^{-1}. & (7.12) \end{aligned}$$

  The weights are again "proportional" to their precision (inverse of the squared error matrix). In particular, we have that

$$\begin{aligned} T &= E\left[(\boldsymbol{\mu} - \boldsymbol{\mu}(\Theta_i))(\boldsymbol{\mu} - \boldsymbol{\mu}(\Theta_i))'\right] & (7.13) \\ &= \text{quadratic loss matrix of } \boldsymbol{\mu} \text{ with respect to } \boldsymbol{\mu}(\Theta_i), \\ S_i &= E\left[(\mathbf{X}_i - \boldsymbol{\mu}(\Theta_i))(\mathbf{X}_i - \boldsymbol{\mu}(\Theta_i))'\right] & (7.14) \\ &= \text{quadratic loss matrix of } \mathbf{X}_i \text{ with respect to } \boldsymbol{\mu}(\Theta_i). \end{aligned}$$

  The proportionality factor $\left(T^{-1} + S_i^{-1}\right)^{-1}$ is to be multiplied from the left.

7.2.3 The Homogeneous Credibility Estimator

In this section we will derive the homogeneous multidimensional credibility estimator. In contrast to the inhomogeneous estimator, this estimator has no constant term $\boldsymbol{\mu}$, i.e. the homogeneous credibility estimator contains an "automatically built-in estimator" for the a priori expected value $\boldsymbol{\mu}$.

By definition

$$\widehat{\widehat{\boldsymbol{\mu}(\Theta)}}^{\text{hom}} := \text{Pro}(\boldsymbol{\mu}(\Theta)|\, L_e(\mathbf{X}_1, \ldots, \mathbf{X}_I)), \tag{7.15}$$

where the projection operator is to be understood componentwise, and $L_e(\mathbf{X}_1, \ldots, \mathbf{X}_I)$ is again the affine subspace, spanned by the components of the vectors $\mathbf{X}_i$ with the corresponding side constraint (componentwise unbiasedness over the collective).

Theorem 7.3 (homogeneous credibility estimator). *The homogeneous multidimensional credibility estimator is given by*

$$\widehat{\widehat{\boldsymbol{\mu}(\Theta_i)}}^{\text{hom}} = A_i\, \mathbf{X}_i + (I - A_i)\, \widehat{\widehat{\boldsymbol{\mu}}}, \tag{7.16}$$

where T, S_i were defined in (7.4)–(7.5), and

$$A_i = T\, (T + S_i)^{-1}, \tag{7.17}$$

$$\widehat{\widehat{\boldsymbol{\mu}}} = \left(\sum_{i=1}^{I} A_i\right)^{-1} \sum_{i=1}^{I} A_i \mathbf{X}_i. \tag{7.18}$$

Remark:

- As an estimator of $\boldsymbol{\mu}$ we again take the credibility-weighted mean.

Proof of Theorem 7.3: Because of the iterative property of projection operators and since $L_e(\mathbf{X}_1, \ldots, \mathbf{X}_I) \subset L(\mathbf{X}_1, \ldots, \mathbf{X}_I, 1)$, we have that

$$\begin{aligned}
\widehat{\widehat{\boldsymbol{\mu}(\Theta_i)}}^{\text{hom}} &= \text{Pro}\left(\boldsymbol{\mu}(\Theta_i)|\, L_e(\mathbf{X}_1, \ldots, \mathbf{X}_I)\right) \\
&= \text{Pro}\left(\text{Pro}\left(\boldsymbol{\mu}(\Theta_i)|\, L(\mathbf{X}_1, \ldots, \mathbf{X}_I, 1)\right)|\, L_e(\mathbf{X}_1, \ldots, \mathbf{X}_I)\right) \\
&= \text{Pro}\left(A_i \mathbf{X}_i + (I - A_i)\, \boldsymbol{\mu}|\, L_e(\mathbf{X}_1, \ldots, \mathbf{X}_I)\right) \\
&= A_i \mathbf{X}_i + (I - A_i)\, \text{Pro}\left(\boldsymbol{\mu}|\, L_e(\mathbf{X}_1, \ldots, \mathbf{X}_I)\right).
\end{aligned}$$

It remains then to show that

$$\widehat{\widehat{\boldsymbol{\mu}}} := \text{Pro}\left(\boldsymbol{\mu}|\, L_e(\mathbf{X}_1, \ldots, \mathbf{X}_I)\right) = \left(\sum_{i=1}^{I} A_i\right)^{-1} \sum_{i=1}^{I} A_i \mathbf{X}_i.$$

We check the normal equations for the homogeneous credibility estimator (see Corollary 3.3), i.e. we must show that for all $\sum_{i=1}^{I} B_i \mathbf{X}_i$ with $\sum_{i=1}^{I} B_i = I$ it holds that

$$\operatorname{Cov}\left(\widehat{\widehat{\boldsymbol{\mu}}}, \left(\sum_{i=1}^{I} B_i \mathbf{X}_i\right)'\right) = \operatorname{Cov}\left(\widehat{\widehat{\boldsymbol{\mu}}}, \widehat{\widehat{\boldsymbol{\mu}}}'\right). \tag{7.19}$$

Because $\operatorname{Cov}(\mathbf{X}_i, \mathbf{X}_i') = T + S_i$ and because of the independence of the $\mathbf{X}_i$ we have that

$$\begin{aligned}
\operatorname{Cov}\left(\widehat{\widehat{\boldsymbol{\mu}}}, \widehat{\widehat{\boldsymbol{\mu}}}'\right) &= \left(\sum_{i=1}^{I} A_i\right)^{-1} \left(\sum_{i=1}^{I} A_i \operatorname{Cov}\left(\mathbf{X}_i, \mathbf{X}_i'\right) A_i'\right) \left(\sum_{i=1}^{I} A_i'\right)^{-1} \\
&= \left(\sum_{i=1}^{I} A_i\right)^{-1} \left(\sum_{i=1}^{I} T\, A_i'\right) \left(\sum_{i=1}^{I} A_i'\right)^{-1} \\
&= \left(\sum_{i=1}^{I} A_i\right)^{-1} T\,.
\end{aligned} \tag{7.20}$$

Since $\operatorname{Cov}\left(\widehat{\widehat{\boldsymbol{\mu}}}, \widehat{\widehat{\boldsymbol{\mu}}}'\right)$ is symmetric, we also have

$$\operatorname{Cov}\left(\widehat{\widehat{\boldsymbol{\mu}}}, \widehat{\widehat{\boldsymbol{\mu}}}'\right) = T \left(\sum_{i=1}^{I} A_i'\right)^{-1}. \tag{7.21}$$

Analogously (because $\sum_{i=1}^{I} B_i = I$)

$$\begin{aligned}
\operatorname{Cov}\left(\widehat{\widehat{\boldsymbol{\mu}}}, \left(\sum_{i=1}^{I} B_i \mathbf{X}_i\right)'\right) &= \left(\sum_{j=1}^{I} A_j\right)^{-1} \left(\sum_{i=1}^{I} A_i \operatorname{Cov}\left(\mathbf{X}_i, \mathbf{X}_i'\right) B_i'\right) \\
&= \left(\sum_{j=1}^{I} A_j\right)^{-1} \sum_{i=1}^{I} (T\, B_i') \\
&= \left(\sum_{j=1}^{I} A_j\right)^{-1} T,
\end{aligned} \tag{7.22}$$

with which we have shown (7.19). This ends the proof of Theorem 7.3. □

7.2.4 The Quadratic Loss of the Multidimensional Credibility Estimators

As a rule, we are not primarily interested in the individual components of the vector $\boldsymbol{\mu}(\Theta)$, but rather in a linear combination

$$\mu(\Theta) := \sum_{k=1}^{p} a_k \mu_k(\Theta) = \mathbf{a}' \boldsymbol{\mu}(\Theta).$$

Let

$$\widehat{\mu(\Theta)} := \sum_{k=1}^{p} a_k \widehat{\mu_k(\Theta)} = \mathbf{a}' \widehat{\boldsymbol{\mu}(\Theta)}$$

be an estimator for $\mu(\Theta)$. This has the quadratic loss

$$\begin{aligned} E\left[\left(\widehat{\mu(\Theta)} - \mu(\Theta)\right)^2\right] &= E\left[\left(\sum a_k \left(\widehat{\mu_k(\Theta)} - \mu_k(\Theta)\right)\right)^2\right] \\ &= \mathbf{a}' E\left[\left(\widehat{\boldsymbol{\mu}(\Theta)} - \boldsymbol{\mu}(\Theta)\right) \cdot \left(\widehat{\boldsymbol{\mu}(\Theta)} - \boldsymbol{\mu}(\Theta)\right)'\right] \mathbf{a} \\ &= \mathbf{a}' V_2 \mathbf{a}. \end{aligned}$$

It makes sense then to define the quadratic loss of the vector $\widehat{\boldsymbol{\mu}(\Theta)}$ by means of the matrix V_2.

Definition 7.4. *The quadratic loss matrix of the estimator $\widehat{\boldsymbol{\mu}(\Theta)}$ is defined as*

$$V_2 := E\left[\left(\widehat{\boldsymbol{\mu}(\Theta)} - \boldsymbol{\mu}(\Theta)\right) \cdot \left(\widehat{\boldsymbol{\mu}(\Theta)} - \boldsymbol{\mu}(\Theta)\right)'\right],$$

where the expected value is to be understood componentwise.

We have the following result for the credibility estimator:

Theorem 7.5. *The quadratic loss matrix of the multidimensional credibility estimator $\widehat{\widehat{\boldsymbol{\mu}(\Theta_i)}}$ (cf. (7.10)) is given by*

$$E\left[\left(\widehat{\widehat{\boldsymbol{\mu}(\Theta_i)}} - \boldsymbol{\mu}(\Theta_i)\right) \cdot \left(\widehat{\widehat{\boldsymbol{\mu}(\Theta_i)}} - \boldsymbol{\mu}(\Theta_i)\right)'\right] = (I - A_i)\, T = A_i S_i. \quad (7.23)$$

Remarks:

- Note the similarity between formula (7.23) (multidimensional case) and formula (4.11) (one dimensional case).
- Note also, that T and S_i are respectively the quadratic loss matrices of $\boldsymbol{\mu}$ and $\mathbf{X}_i$ with respect to $\boldsymbol{\mu}(\Theta_i)$ (cf. (7.13) and (7.14)).

Proof of Theorem 7.5: We calculate the quadratic loss matrix:

$$\begin{aligned} &E\left[\left(\widehat{\widehat{\boldsymbol{\mu}(\Theta_i)}} - \boldsymbol{\mu}(\Theta_i)\right) \cdot \left(\widehat{\widehat{\boldsymbol{\mu}(\Theta_i)}} - \boldsymbol{\mu}(\Theta_i)\right)'\right] \\ &= A_i\, E\left[(\mathbf{X}_i - \boldsymbol{\mu}(\Theta_i)) \cdot (\mathbf{X}_i - \boldsymbol{\mu}(\Theta_i))'\right] A_i' \\ &\quad + (I - A_i)\, E\left[(\boldsymbol{\mu} - \boldsymbol{\mu}(\Theta_i)) \cdot (\boldsymbol{\mu} - \boldsymbol{\mu}(\Theta_i))'\right] (I - A_i)' \qquad (7.24) \\ &= T\,(T + S_i)^{-1} S_i\,(T + S_i)^{-1} T + S_i\,(T + S_i)^{-1} T\,(T + S_i)^{-1} S_i \\ &= S_i\,(T + S_i)^{-1} T = (I - A_i)\, T, \end{aligned}$$

where we have used in the last step the fact that

$$S_i\,(T+S_i)^{-1}\,T = T\,(T+S_i)^{-1}\,S_i. \tag{7.25}$$

One sees this directly by taking S_i and T inside the brackets on both sides, giving $\left(T^{-1}+S_i^{-1}\right)^{-1}$ on both sides. Note also, that the right-hand side of (7.25) is equal to A_iS_i. This completes the proof of Theorem 7.5. □

Theorem 7.6. *The quadratic loss matrix of the homogeneous multidimensional credibility estimator is given by*

$$E\left[\left(\widehat{\widehat{\boldsymbol{\mu}(\Theta_i)}}^{\text{hom}} - \boldsymbol{\mu}(\Theta_i)\right)\cdot\left(\widehat{\widehat{\boldsymbol{\mu}(\Theta_i)}}^{\text{hom}} - \boldsymbol{\mu}(\Theta_i)\right)'\right]$$
$$= (I-A_i)\,T\left[I+\left(\sum_{i=1}^{I}A_i'\right)^{-1}(I-A_i)'\right]. \tag{7.26}$$

Remark:

- Note again the similarity between (7.26) (multidimensional quadratic loss) and (4.21) (one-dimensional quadratic loss).

Proof of Theorem 7.6: Take $\mu(\Theta_i) = \mathbf{a}'\boldsymbol{\mu}(\Theta_i)$. From Theorem 3.14 we have

$$E\left[\left(\widehat{\widehat{\mu(\Theta_i)}}^{\text{hom}} - \mu(\Theta_i)\right)^2\right]$$
$$= E\left[\left(\widehat{\widehat{\mu(\Theta_i)}} - \mu(\Theta_i)\right)^2\right] + E\left[\left(\widehat{\widehat{\mu(\Theta_i)}} - \widehat{\widehat{\mu(\Theta_i)}}^{\text{hom}}\right)^2\right]. \tag{7.27}$$

Since (7.27) holds for all vectors $\mathbf{a}$, we must have for the loss matrices

$$E\left[\left(\widehat{\widehat{\boldsymbol{\mu}(\Theta_i)}}^{\text{hom}} - \boldsymbol{\mu}(\Theta_i)\right)\cdot\left(\widehat{\widehat{\boldsymbol{\mu}(\Theta_i)}}^{\text{hom}} - \boldsymbol{\mu}(\Theta_i)\right)'\right]$$
$$= E\left[\left(\widehat{\widehat{\boldsymbol{\mu}(\Theta_i)}} - \boldsymbol{\mu}(\Theta_i)\right)\cdot\left(\widehat{\widehat{\boldsymbol{\mu}(\Theta_i)}} - \boldsymbol{\mu}(\Theta_i)\right)'\right]$$
$$+\,E\left[\left(\widehat{\widehat{\boldsymbol{\mu}(\Theta_i)}}^{\text{hom}} - \widehat{\widehat{\boldsymbol{\mu}(\Theta_i)}}\right)\cdot\left(\widehat{\widehat{\boldsymbol{\mu}(\Theta_i)}}^{\text{hom}} - \widehat{\widehat{\boldsymbol{\mu}(\Theta_i)}}\right)'\right].$$

For the first summand we have from Theorem 7.5

$$E\left[\left(\widehat{\widehat{\mu(\Theta_i)}} - \mu(\Theta_i)\right) \cdot \left(\widehat{\widehat{\mu(\Theta_i)}} - \mu(\Theta_i)\right)'\right] = (I - A_i)\,T. \tag{7.28}$$

For the second summand, applying (7.16) and (7.20) we get

$$\begin{aligned}
&E\left(\left(\widehat{\widehat{\boldsymbol{\mu}(\Theta_i)}}^{\mathrm{hom}} - \widehat{\widehat{\boldsymbol{\mu}(\Theta_i)}}\right) \cdot \left(\widehat{\widehat{\boldsymbol{\mu}(\Theta_i)}}^{\mathrm{hom}} - \widehat{\widehat{\boldsymbol{\mu}(\Theta_i)}}\right)'\right) \\
&= (I - A_i)\,\mathrm{Cov}\left(\widehat{\widehat{\boldsymbol{\mu}}}, \widehat{\widehat{\boldsymbol{\mu}}}'\right)(I - A_i)' \\
&= (I - A_i)\,T\left(\sum_{j=1}^{I} A_j'\right)^{-1}(I - A_i)'.
\end{aligned}$$

Theorem 7.6 is thus proved. □

7.3 The Multidimensional Bühlmann–Straub Model

7.3.1 Motivation and Interpretation

In the Bühlmann–Straub model, we have associated with each risk i one-dimensional observations X_{ij} $(j = 1, 2, \ldots, n)$ satisfying certain model assumptions (see Model Assumptions 4.1). In the *multidimensional Bühlmann–Straub model,* these *observations* are *multidimensional,* i.e. the observation associated with risk i in year j is itself a vector

$$\mathbf{X}'_{ij} = (X_{ij}^{(1)}, X_{ij}^{(2)}, \ldots, X_{ij}^{(p)}).$$

Example 7.1 (layering)

$$\mathbf{X}'_{ij} = (X_{ij}^{(1)}, X_{ij}^{(2)}, \ldots, X_{ij}^{(p)}),$$

where
$X_{ij}^{(1)}$= the claims ratio (or the claim frequency or the claims average) associated with the claims in layer 1 (e.g. claims less than CHF 1 000),
$X_{ij}^{(2)}$= the claims ratio (or the claim frequency or the claims average) associated with the claims in layer 2 (e.g. claims between CHF 1 000 and CHF 50 000),
and so on.

A special case of this example would be the simple differentiation between normal claims and large claims, which is equivalent to a two-layer system (claims under and over a specific, fixed amount).
An example of claim types are the distinction between pure physical damage and bodily injury claims in motor liability. Then one might consider

$$\mathbf{X}'_{ij} = \left(X_{ij}^{(1)}, X_{ij}^{(2)} \right),$$

where
$X_{ij}^{(1)}$= the claims ratio (or the claim frequency) associated with pure physical damage and
$X_{ij}^{(2)}$= the claims ratio (or the claim frequency) associated with bodily injury claims.

Example 7.2 (differentiation according to classes)

$$\mathbf{X}'_{ij} = (X_{ij}^{(1)}, X_{ij}^{(2)}, \dots, X_{ij}^{(p)}),$$

where
$X_{ij}^{(1)}$= claims ratio in the first class,
$X_{ij}^{(2)}$= claims ratio in the second class,
and so on.

7.3.2 Definition of the Model

We are given a portfolio of I risks or "risk categories", and let

$\mathbf{X}'_{ij} = (X_{ij}^{(1)}, X_{ij}^{(2)}, \dots, X_{ij}^{(p)})$
= observation vector (of length p) of the ith risk in year j,
$\mathbf{w}'_{ij} = (w_{ij}^{(1)}, w_{ij}^{(2)}, \dots, w_{ij}^{(p)})$
= associated vector of known weights.

Remark:

- Here, $w_{ij}^{(k)}$ denotes the weight associated with the kth component of the vector $\mathbf{X}_{ij}$. Possibly, $w_{ij}^{(k)} = w_{ij}$ for $k = 1, 2, \dots, p$, i.e. the weights for the different components can be all the same. This will typically be the case if we consider claim frequencies, burning costs or claims ratios for different layers numbered $1, 2, \dots, p$.

Model Assumptions 7.7 (multidim. Bühlmann-Straub model)
The risk i is characterized by its individual risk profile ϑ_i, which is itself the realization of a random variable Θ_i, and we have that:

A1:Conditionally, given Θ_i, the $\mathbf{X}_{ij}$, $j = 1, 2, \dots, n$, are independent with

$$E[\mathbf{X}_{ij} \mid \Theta_i] = \boldsymbol{\mu}(\Theta_i) \quad \textit{(vector of length } p\textit{)}, \tag{7.29}$$

$$\operatorname{Cov}\left(\mathbf{X}_{ij}, \mathbf{X}'_{ij} \mid \Theta_i \right) = \begin{pmatrix} \frac{\sigma_1^2(\Theta_i)}{w_{ij}^{(1)}} & 0 & \cdots & 0 \\ 0 & \ddots & & \vdots \\ \vdots & & \ddots & 0 \\ 0 & \cdots & 0 & \frac{\sigma_p^2(\Theta_i)}{w_{ij}^{(p)}} \end{pmatrix}. \tag{7.30}$$

A2: The pairs $\{(\Theta_i, \mathbf{X}_i) : i = 1, 2, \ldots, I\}$, *where* $\mathbf{X}_i' = (\mathbf{X}_{i1}', \mathbf{X}_{i2}', \ldots, \mathbf{X}_{in}')$, *are independent, and* $\Theta_1, \Theta_2, \ldots$ *are independent and identically distributed.*

Remarks:

- According to (7.30), the components of $\mathbf{X}_{ij}$ are assumed to be conditionally, given Θ_i, uncorrelated. This is the *standard case* most often encountered in practice. In the remainder of Section 7.3 we will refer to (7.30) as the *standard assumption.*
- However, if one considers for instance claims ratios of different layers, then this standard assumption is not fulfilled any more. In this case, however, the weights of the different components will be identical. Fortunately, all results of Section 7.3 (multidimensional Bühlmann–Straub) also hold true under the following:

  Alternative assumption to (7.30):
  the weights for the different components are all the same ($w_{ij}^{(k)} = w_{ij}$ for $k = 1, 2, \ldots, p$) and

$$\operatorname{Cov}\left(\mathbf{X}_{ij}, \mathbf{X}_{ij}' \mid \Theta_i\right) = \frac{1}{w_{ij}} \sum (\Theta_i), \tag{7.31}$$

  where $\sum(\Theta_i)$ is any covariance matrix and not necessarily diagonal. In the remainder of Section 7.3 we will refer to (7.31) as the *alternative standard assumption.*
- The standard assumption and the alternative standard assumption cover most cases encountered in practice.
- A more general assumption on the conditional covariance structure would be

$$\operatorname{Cov}\left(X_{ij}^{(k)}, X_{ij}^{(l)} \mid \Theta_i\right) = \left(\frac{\sigma_{kl}(\Theta_i)}{(w_{ij}^{(k)} w_{ij}^{(l)})^{1/2}}\right). \tag{7.32}$$

  Under this more general assumption, however, the following results are not generally valid.

Notation:
In the following we will denote by W_{ij}, $W_{ij}^{-1/2}$, $W_{i\bullet}^{-1}$, etc. diagonal $p \times p$ matrices with corresponding elements in the diagonal, e.g.

$$W_{ij} = \begin{pmatrix} w_{ij}^{(1)} & & 0 \\ & \ddots & \\ 0 & & w_{ij}^{(p)} \end{pmatrix}, \quad W_{i\bullet}^{-1} = \begin{pmatrix} \left(w_{i\bullet}^{(1)}\right)^{-1} & & 0 \\ & \ddots & \\ 0 & & \left(w_{i\bullet}^{(p)}\right)^{-1} \end{pmatrix}. \tag{7.33}$$

Remarks:

- All of the above assumptions on the conditional covariance structure ((7.30), (7.31), (7.32)) can also be written in the following way:
$$\mathrm{Cov}\left(\mathbf{X}_{ij}, \mathbf{X}'_{ij} \mid \Theta_i\right) = W_{ij}^{-1/2}\, \Sigma(\Theta_i)\, W_{ij}^{-1/2} \quad (p \times p \text{ matrix}).$$
- The functions $\boldsymbol{\mu}(\cdot)$ and $\Sigma(\cdot)$ are independent of i and j. The unconditional covariance matrix of $\mathbf{X}_{ij}$ depends on the individual risk only via the weights $w_{ij}^{(k)}$.
- It is straightforward to check that in the case where the observations are one-dimensional, the assumptions of the one-dimensional Bühlmann–Straub model are satisfied.
- The multidimensional Bühlmann–Straub model does not fully fit into the framework of the abstract multidimensional credibility model described in Section 7.2 in that not just one, but many vectors of observations are available for the ith risk. In order that we can use the results derived in Section 7.2, we must first compress the data in an appropriate manner so that we have one single observation vector for each risk i.

The following quantities are of interest:

Risk i	**Collective/Portfolio**
$\boldsymbol{\mu}(\Theta_i) = E\left[\mathbf{X}_{ij} \middle\vert \Theta_i\right]$,	$\boldsymbol{\mu} := E\left[\boldsymbol{\mu}(\Theta_i)\right]$,
$\Sigma(\Theta_i) = \left(W_{ij} E\left[\mathrm{Cov}\left(\mathbf{X}_{ij}, \mathbf{X}'_{ij} \middle\vert \Theta_i\right)\right] W_{ij}\right)$,	$S := E\left[\Sigma(\Theta_i)\right]$,
	$T := \mathrm{Cov}\left(\boldsymbol{\mu}(\Theta_i), \boldsymbol{\mu}(\Theta_i)'\right)$.

Note that $\Sigma(\Theta_i)$ is the conditional covariance normalized for unit weights.

7.3.3 Credibility Formulae in the Multidimensional Bühlmann–Straub Model

In order to be able to apply Theorem 7.2, for each risk i, we use instead of $\mathbf{X}_{i1}, \ldots, \mathbf{X}_{in}$ (sequence of multidimensional one-period claims ratios) a compressed observation vector $\mathbf{B}_i$ (single, multidimensional n-period claims ratio). We explicitly choose

$$\mathbf{B}_i = \left(\sum_j \frac{w_{ij}^{(1)}}{w_{i\bullet}^{(1)}} X_{ij}^{(1)}, \ldots, \sum_j \frac{w_{ij}^{(p)}}{w_{i\bullet}^{(p)}} X_{ij}^{(p)}\right)',$$

the average over time of the observations. This choice is intuitive. Also, it is easy to check (from Theorem A.3 of Appendix A), that

$$\mathbf{B}_i = \mathrm{Pro}\left(\boldsymbol{\mu}(\Theta_i) \middle\vert L_e^{ind}(\mathbf{X}_i)\right),$$

where

$$L_e^{ind}(\mathbf{X}_i) = \left\{ \widehat{\boldsymbol{\mu}(\Theta_i)} : \ \mu_k(\Theta_i) = \sum_j a_{ij}^{(k)} X_{ij} \text{ for } k = 1, \ldots, p; \right.$$
$$\left. E\left[\widehat{\boldsymbol{\mu}(\Theta_i)} \middle| \Theta_i \right] = \boldsymbol{\mu}(\Theta_i). \right\}$$

Remark:

- As in Chapter 4 (see remarks on page 150) we denote by $L_e^{ind}(.)$ the space of *individually unbiased* estimators. What that means exactly becomes always clear from the context. For instance, $L_e^{ind}(\mathbf{X}_i)$ in $\text{Pro}(\boldsymbol{\mu}(\Theta_i)|\, L_e^{ind}(\mathbf{X}_i))$ is the space of estimators which are linear in the components of $\mathbf{X}_i$ and whose conditional expected value given Θ_i is equal to $\boldsymbol{\mu}(\Theta_i)$.

We will see later, that $\mathbf{B}_i$ is indeed an allowable compression, i.e. that the inhomogeneous as well as the homogeneous credibility estimator depend on the data only via $\mathbf{B}_i$.

In matrix notation

$$\mathbf{B}_i = W_{i\bullet}^{-1} \sum_{j=1}^{n} W_{ij}\, \mathbf{X}_{ij}.$$

For the compressed data $\mathbf{B}_i$ we have that

$$E\left[\mathbf{B}_i \middle| \Theta_i\right] = \boldsymbol{\mu}(\Theta_i), \tag{7.34}$$

$$\begin{aligned} \text{Cov}\left(\mathbf{B}_i, \mathbf{B}_i' \middle| \Theta_i\right) &= W_{i\bullet}^{-1} \left(\sum_{j=1}^{n} W_{ij}\, E\left[\text{Cov}\left(\mathbf{X}_{ij}, \mathbf{X}_{ij}' \middle| \Theta_i\right)\right] W_{ij} \right) W_{i\bullet}^{-1} \\ &= W_{i\bullet}^{-1} \cdot W_{i\bullet}^{1/2} \cdot \Sigma(\Theta_i) \cdot W_{i\bullet}^{1/2} \cdot W_{i\bullet}^{-1} \qquad (7.35) \\ &= W_{i\bullet}^{-1/2}\, \Sigma(\Theta_i)\, W_{i\bullet}^{-1/2}. \end{aligned}$$

Remark:

- Equation (7.35) would not be generally valid under the general conditional covariance structure (7.32), but holds true under the standard assumptions (7.30) or the alternative standard assumption (7.31).

We can now apply Theorem 7.2 to $\mathbf{B}_i$.

Theorem 7.8 (credibility estimator). *The (inhomogeneous) credibility estimator in the multidimensional Bühlmann–Straub model (Model Assumptions 7.7 or the alternative standard assumption (7.31)) is given by*

$$\widehat{\widehat{\boldsymbol{\mu}(\Theta_i)}} = A_i\, \mathbf{B}_i + (I - A_i)\, \boldsymbol{\mu}\ , \tag{7.36}$$

$$\text{where} \quad A_i = T\left(T + W_{i\bullet}^{-1/2} S\, W_{i\bullet}^{-1/2}\right)^{-1}. \tag{7.37}$$

Remarks:

- Note that under the standard assumption (7.30)

$$W_{i\bullet}^{-1/2} S W_{i\bullet}^{-1/2} = \begin{pmatrix} \frac{\sigma_1^2}{w_{i\bullet}^{(1)}} & & 0 \\ & \ddots & \\ 0 & & \frac{\sigma_p^2}{w_{i\bullet}^{(p)}} \end{pmatrix}, \tag{7.38}$$

  and that under the alternative standard assumption (7.31)

$$W_{i\bullet}^{-1/2} S W_{i\bullet}^{-1/2} = \frac{1}{w_{i\bullet}} S, \tag{7.39}$$
$$\text{where} \quad S = E\left[\Sigma(\Theta_i)\right].$$

- The credibility matrix is not so easy to interpret. As an example consider the case where the observation vector is of dimension 2 and where the first component is the frequency of "normal" claims and the second component is the frequency of "large" claims. Then a small number in the left lower corner of the credibility matrix does not necessarily mean that the influence of the observed normal claim frequency on the estimated large claim frequency is small. The reason is that the "normal" claim frequency might be 100 times bigger than the large claim frequency. In order that the credibility matrix gives a better picture of the influence of the different observation components on the credibility estimates, it is advisable to first bring the different components of the observation vector on the same scale, i.e. to consider instead of the original observations the normalized or relative observation vectors

$$\mathbf{Y}'_{ij} = (Y_{ij}^{(1)}, Y_{ij}^{(2)}, \ldots, Y_{ij}^{(p)}),$$

  where

$$Y_{ij}^{(k)} := \frac{X_{ij}^{(k)}}{\mu_k}.$$

  Then the entries of the credibility matrix reflect much better the influence of the corresponding observation components and the credibility estimator becomes

$$\widehat{\widehat{\boldsymbol{\mu}(\Theta_i)}} = A_i \, \mathbf{B}_i + (I - A_i)\, \mathbf{1}.$$

- Explicit formulae of the credibility matrix in the case of two dimensions can be found in Exercise 7.3 and in Exercise 7.4.

Proof of Theorem 7.8:

From Theorem 7.2 we know that $\widehat{\widehat{\boldsymbol{\mu}(\Theta_i)}}$ is the credibility estimator based on $\mathbf{B}_i$. It remains to show that $\widehat{\widehat{\boldsymbol{\mu}(\Theta_i)}}$ is also the credibility estimator based on all data (admissibility of the compression $\mathbf{B}_i$). To do this we must check the orthogonality conditions

$$\boldsymbol{\mu}(\Theta_i) - \widehat{\boldsymbol{\mu}(\Theta_i)} \perp \mathbf{X}_{kj} \quad \text{for all } k,\ j. \tag{7.40}$$

For $k \neq i$ we have that (7.40) holds because of the independence of the risks. It then remains to show (7.40) for $k = i$,

$$\begin{aligned} &E\left[(\boldsymbol{\mu}(\Theta_i) - A_i\mathbf{B}_i - (I - A_i)\,\boldsymbol{\mu}) \cdot \mathbf{X}'_{ij}\right] \\ &\quad = A_i E\left[(\boldsymbol{\mu}(\Theta_i) - \mathbf{B}_i)\ \mathbf{X}'_{ij}\right] + (I - A_i) E\left[(\boldsymbol{\mu}(\Theta_i) - \boldsymbol{\mu})\ \mathbf{X}'_{ij}\right]. \end{aligned} \tag{7.41}$$

From

$$\mathbf{B}_i = \mathrm{Pro}(\,\boldsymbol{\mu}(\Theta_i)|\, L_e^{ind}\,(\mathbf{X}_i))$$

it follows that

$$\begin{aligned} E\left[(\boldsymbol{\mu}(\Theta_i) - \mathbf{B}_i)\ \mathbf{X}'_{ij}\right] &= E\left[(\boldsymbol{\mu}(\Theta_i) - \mathbf{B}_i)\ \mathbf{B}'_i\right] \\ &= E\left[(\boldsymbol{\mu}(\Theta_i) - \mathbf{B}_i)\,(\mathbf{B}_i - \boldsymbol{\mu}(\Theta_i))'\right] \\ &= -E\left[\mathrm{Cov}\,(\mathbf{B}_i, \mathbf{B}'_i|\,\Theta_i)\right] \\ &= -W_{i\bullet}^{-1/2} S\, W_{i\bullet}^{-1/2}. \end{aligned} \tag{7.42}$$

For the second summand in (7.41), we get by conditioning on Θ_i

$$\begin{aligned} E\left[(\boldsymbol{\mu}(\Theta_i) - \boldsymbol{\mu})\,\mathbf{X}'_{ij}\right] &= E\left[(\boldsymbol{\mu}(\Theta_i) - \boldsymbol{\mu})\,\boldsymbol{\mu}(\Theta_i)'\right] \\ &= E\left[(\boldsymbol{\mu}(\Theta_i) - \boldsymbol{\mu})\,(\boldsymbol{\mu}(\Theta_i) - \boldsymbol{\mu})'\right] \\ &= T. \end{aligned} \tag{7.43}$$

Plugging (7.42) and (7.43) into (7.41) we see that

$$\begin{aligned} E\left[(\boldsymbol{\mu}(\Theta_i) - A_i\mathbf{B}_i - (I - A_i)\,\boldsymbol{\mu}) \cdot \mathbf{X}'_{ij}\right] &= -A_i\left(W_{i\bullet}^{-1/2} S\, W_{i\bullet}^{-1/2} + T\right) + T \\ &= -T + T \\ &= \mathbf{0}. \end{aligned}$$

This completes the proof of Theorem 7.8. □

Theorem 7.9 (homogeneous credibility estimator). *The homogenous credibility estimator in the multidimensional Bühlmann–Straub model (Model Assumptions 7.7 or the alternative assumption (7.31)) is given by*

$$\widehat{\widehat{\boldsymbol{\mu}\,(\Theta_i)}}^{\text{hom}} = A_i\,\mathbf{B}_i + (I - A_i)\,\widehat{\widehat{\boldsymbol{\mu}}}, \tag{7.44}$$

$$\textit{where}\quad \widehat{\widehat{\boldsymbol{\mu}}} = \left(\sum_{i=1}^{I} A_i\right)^{-1} \sum_{i=1}^{I} A_i\mathbf{B}_i. \tag{7.45}$$

Proof:
It follows from Theorem 7.3 that (7.44) is the homogeneous credibility estimator based on the statistic of the $\mathbf{B}_i$. We will show later, in a more general framework (cf. Section 7.4, Theorem 7.14), that the $\mathbf{B}_i$ are a linear sufficient statistic, which means that the credibility estimators depend on the data only through this statistic. This concludes the proof of Theorem 7.9. □

The next theorem shows that the balance property of the homogeneous credibility estimator formulated in Theorem 4.5 holds also componentwise in the multidimensional model.

Theorem 7.10 (balance property). *Given Model Assumptions 7.7 or the alternative standard assumption (7.31) we have for each component*

$$\sum_{ij} w_{ij}^{(k)} \widehat{\widehat{\mu_k(\Theta_i)}}^{\text{hom}} = \sum_{ij} w_{ij}^{(k)} X_{ij}^{(k)} \qquad \textit{for } k = 1, 2, \ldots, p. \tag{7.46}$$

Remark:

- As an example, assume that we want to estimate the burning cost (or the claim frequency or the claim average) for "normal" and "large" claims. Theorem 7.10 then guarantees that the homogeneous credibility estimator applied to the past observation period and weighted over the whole portfolio coincides with the observed average burning cost (claim frequency, claim average) for each component (e.g. normal and large claims).

Proof of Theorem 7.10:
(7.46) is equivalent to

$$\sum_{i=1}^{I} W_{i\bullet}^{1/2} \left(\widehat{\widehat{\boldsymbol{\mu}(\Theta_i)}}^{\text{hom}} - \mathbf{B}_i \right) W_{i\bullet}^{1/2} = \mathbf{0}.$$

$$\begin{aligned}
&\sum_{i=1}^{I} W_{i\bullet}^{1/2} \left(\widehat{\widehat{\boldsymbol{\mu}(\Theta_i)}}^{\text{hom}} - \mathbf{B}_i \right) W_{i\bullet}^{1/2} \\
&= \sum_{i=1}^{I} W_{i\bullet}^{1/2} \left(W_{i\bullet}^{-1/2} S\, W_{i\bullet}^{-1/2} \right) \left(T + W_{i\bullet}^{-1/2} S\, W_{i\bullet}^{-1/2} \right)^{-1} \left(\widehat{\widehat{\boldsymbol{\mu}}} - \mathbf{B}_i \right) W_{i\bullet}^{1/2} \\
&= \sum_{i=1}^{I} W_{i\bullet}^{1/2} \left(W_{i\bullet}^{-1/2} S\, W_{i\bullet}^{-1/2} \right) W_{i\bullet}^{1/2} \left(T + W_{i\bullet}^{-1/2} S\, W_{i\bullet}^{-1/2} \right)^{-1} \left(\widehat{\widehat{\boldsymbol{\mu}}} - \mathbf{B}_i \right) \\
&= S\, T^{-1} \sum_{i=1}^{I} T \left(T + W_{i\bullet}^{-1/2} S\, W_{i\bullet}^{-1/2} \right)^{-1} \left(\widehat{\widehat{\boldsymbol{\mu}}} - \mathbf{B}_i \right)
\end{aligned}$$

$$= S T^{-1} \sum_{i=1}^{I} A_i \left(\widehat{\widehat{\boldsymbol{\mu}}} - \mathbf{B}_i \right)$$
$$= \mathbf{0}.$$

□

7.3.4 Quadratic Loss

The *quadratic loss of the multidimensional credibility estimators* is directly given by Theorem 7.5 and Theorem 7.6, where the credibility matrix A_i is given by (7.37) and where $S_i = W_{i\bullet}^{-1/2} S W_{i\bullet}^{-1/2}$ is given by (7.38) (standard assumption) or by (7.39) (alternative standard assumption).

7.3.5 Estimation of Structural Parameters

Already in the remarks to Model Assumptions 7.7 several forms of the covariance matrix $\Sigma(\Theta_i)$ and, as a consequence, of the matrix S were discussed. In the *standard situation,* these matrices are assumed to be diagonal. Below we give estimators for this standard situation.

The diagonal elements of S and T can be estimated in the same way as the structural parameters in the one-dimensional Bühlmann–Straub model. For the non-diagonal elements of T we can use analogous statistics. This leads to the following estimators:

Diagonal elements of S:

$$\widehat{\sigma_k^2} = \frac{1}{I} \sum_{i=1}^{I} \frac{1}{n-1} \sum_{j=1}^{n} w_{ij}^{(k)} \left(X_{ij}^{(k)} - B_i^{(k)} \right)^2 .$$

Diagonal elements of T:

$$\widehat{\tau_k^2} = \min \left(\widetilde{\tau_k^2}, 0 \right),$$

where

$$\widetilde{\tau_k^2} = c \cdot \left\{ \frac{I}{I-1} \sum_{i=1}^{I} \frac{w_{i\bullet}^{(k)}}{w_{\bullet\bullet}^{(k)}} \left(B_i^{(k)} - \overline{B}^{(k)} \right)^2 - \frac{I \widehat{\sigma_k^2}}{w_{\bullet\bullet}^{(k)}} \right\},$$

$$\text{where } c = \frac{I-1}{I} \left\{ \sum_{i=1}^{I} \frac{w_{i\bullet}^{(k)}}{w_{\bullet\bullet}^{(k)}} \left(1 - \frac{w_{i\bullet}^{(k)}}{w_{\bullet\bullet}^{(k)}} \right) \right\}^{-1},$$

$$\overline{B}^{(k)} = \sum_{i=1}^{I} \frac{w_{i\bullet}^{(k)}}{w_{\bullet\bullet}^{(k)}} B_i^{(k)}.$$

non-diagonal elements of T:
a straightforward calculation shows that the following two estimators are unbiased estimators of τ_{kl}:

$$\widetilde{\tau_{kl}}^{*} = c_1 \cdot \left\{ \frac{I}{I-1} \sum_{i=1}^{I} \frac{w_{i\bullet}^{(k)}}{w_{\bullet\bullet}^{(k)}} \left(B_i^{(k)} - \overline{B}^{(k)}\right)\left(B_i^{(l)} - \overline{B}^{(l)}\right) \right\},$$

$$\text{where } c_1 = \frac{I-1}{I} \left\{ \sum_{i=1}^{I} \frac{w_{i\bullet}^{(k)}}{w_{\bullet\bullet}^{(k)}} \left(1 - \frac{w_{i\bullet}^{(k)}}{w_{\bullet\bullet}^{(k)}}\right) \right\}^{-1},$$

$$\widetilde{\tau_{kl}}^{**} = c_2 \cdot \left\{ \frac{I}{I-1} \sum_{i=1}^{I} \frac{w_{i\bullet}^{(l)}}{w_{\bullet\bullet}^{(l)}} \left(B_i^{(k)} - \overline{B}^{(k)}\right)\left(B_i^{(l)} - \overline{B}^{(l)}\right) \right\},$$

$$\text{where } c_2 = \frac{I-1}{I} \left\{ \sum_{i=1}^{I} \frac{w_{i\bullet}^{(l)}}{w_{\bullet\bullet}^{(l)}} \left(1 - \frac{w_{i\bullet}^{(l)}}{w_{\bullet\bullet}^{(l)}}\right) \right\}^{-1}.$$

Since $\tau_{kl}^2 \leq \tau_k^2 \tau_l^2$, we suggest the following estimators:

$$\widehat{\tau_{kl}} = \text{signum}\left(\frac{\widetilde{\tau_{kl}}^{*} + \widetilde{\tau_{kl}}^{**}}{2}\right) \cdot \min\left(\frac{\left|\widetilde{\tau_{kl}}^{*} + \widetilde{\tau_{kl}}^{**}\right|}{2}, \sqrt{\widehat{\tau_k^2} \cdot \widehat{\tau_l^2}}\right). \tag{7.47}$$

Remarks:

- If there is more structure in the model, then this structure should be taken into account. For instance, if we apply multidimensional credibility to estimate claim frequencies and if we assume the claim numbers to be Poisson distributed, then we can use this structure to use another estimator for σ_i^2 analogously to those discussed in Section 4.10.
- Since T is a covariance matrix, it is positive definite. However, for dimension $p \geq 3$, the matrix $\widehat{T}$ resulting from (7.47) is not necessarily positive definite. To achieve positive definiteness the following procedure is often applied in statistics: find an orthogonal transformation O, such that

  $$\widetilde{T} := O \cdot \widehat{T}$$

  is diagonal. If all diagonal elements of $\widetilde{T}$ are ≥ 0, then $\widehat{T}$ is positive definite. If not, then replace the negative diagonal elements of $\widetilde{T}$ by zero and transform back to get a modified, positive (more precise: nonnegative definite) definite estimator for T. The orthogonal transformation matrix O is obtained by taking the eigenvectors of $\widehat{T}$ as columns of O.

  An alternative to (7.47) is to take as "starting point" the matrix $\widehat{T}^{*}$ with the entries

  $$\widehat{\tau_{kl}}^{*} = \left(\frac{\widetilde{\tau_{kl}}^{*} + \widetilde{\tau_{kl}}^{**}}{2}\right) \tag{7.48}$$

  and then to apply the above orthogonal transformation procedure on $\widehat{T}^{*}$.

7.4 General Remarks About Data Compression and Its Optimality

7.4.1 Introduction

The basic idea behind multidimensional credibility is that the quantity $\boldsymbol{\mu}(\Theta)$ to be estimated is multidimensional, i.e. a vector of length p. In Section 7.2 we assumed that in the "abstract" situation, there is exactly one data vector $\mathbf{X}$ having the same dimension as $\boldsymbol{\mu}(\Theta)$, for which we have $E[\mathbf{X}|\,\Theta] = \boldsymbol{\mu}(\Theta)$. We have pointed out that in this "abstract" framework, the data vector $\mathbf{X}$ is, as a rule, not the same as the observation vector, but is in most cases an appropriate compression of the original data. We have, until now, left open the question as to how one accomplishes this compression without losing relevant information. In Section 7.3 we have shown what the data compression looks like in the multidimensional Bühlmann–Straub model.

The reader who wants to understand what to do if $\mathbf{X}$ has a lower dimension than the estimand $\boldsymbol{\mu}(\Theta)$ is referred to Chapter 10.

In this section we consider, quite generally, what the appropriate compression is in the multidimensional case. We will see the validity of the intuitive general principle. We will also see that the compressed data are a *linear sufficient statistics,* i.e. that the inhomogeneous as well as the homogeneous credibility estimators depend only on the compressed data.

First, however, we want to investigate if we could proceed in another way, in that we summarize all the data associated with one contract into one vector $\mathbf{X}$ and then define the vector to be estimated as

$$\boldsymbol{\mu}(\Theta) := E[\mathbf{X}|\,\Theta].$$

As mentioned in Section 7.2, we are in principle free in the choice of the vector $\boldsymbol{\mu}(\Theta)$. With this definition, we would be in the set-up of the defined abstract multidimensional credibility model described in Section 7.2, and we could apply the theory that we derived there. This procedure is theoretically correct, but is not to be recommended. It leads to unnecessarily complicated inversion of high-dimensional matrices. The problem is that we would be working in too high a dimension. For example, in the simple (not multidimensional!) Bühlmann–Straub model, the vector $\boldsymbol{\mu}(\Theta)$ would be composed of n components all of which are equal to $\mu(\Theta)$. In this way we make a one-dimensional problem into a complicated multidimensional problem. In general, we should try to keep the dimension of $\boldsymbol{\mu}(\Theta)$ minimal: in particular, the elements of $\boldsymbol{\mu}(\Theta)$ should not be linearly dependent.

7.4.2 General Multidimensional Data Structure

We assume the following, very general data structure. For each risk $i = 1, \dots, I$ let

$$\boldsymbol{\mu}(\Theta_i)' = (\mu_1(\Theta_i), \ldots, \mu_p(\Theta_i))$$

be the vector to be estimated and let

$$\mathbf{X}_i' = (X_{i1}', X_{i2}', \ldots, X_{in_i}')$$

be the vector of observations.

In this general structure, multidimensional means that the vector to be estimated, $\boldsymbol{\mu}(\Theta_i)$, is multidimensional, i.e. the length p is greater than one. On the other hand, we summarize all "raw" observations associated with the ith risk, whether one-dimensional or multidimensional, in one large observation vector $\mathbf{X}_i$.

Notice that we do not make any assumptions about the internal probability structure of the vector $\mathbf{X}_i$. In particular, some components can be (non-linear) functions of the same observed random variable, e.g.

$$X_{i1} = Y, \quad X_{i2} = Y^2, \quad X_{i3} = \log Y, \quad \text{etc.}$$

The variables we choose to use as components depend on the vector $\boldsymbol{\mu}(\Theta_i)$ which we want to estimate. For example, in the multidimensional Bühlmann–Straub model, if we wanted to estimate the individual variances as well, then we would also need beside

$$\mathbf{X}_{ij}' = (X_{ij}^{(1)}, \ldots, X_{ij}^{(p)})$$

the vector of squared observations

$$\mathbf{Z}_{ij}' = ((X_{ij}^{(1)})^2, \ldots, (X_{ij}^{(p)})^2).$$

In this case the vector $\mathbf{X}_i$ would be the listing of all the relevant quantities for the ith risk, i.e.

$$\mathbf{X}_i' = (\mathbf{X}_{i1}', \mathbf{Z}_{i1}', \mathbf{X}_{i2}', \mathbf{Z}_{i2}', \ldots, \mathbf{X}_{in}', \mathbf{Z}_{in}').$$

With this interpretation of the vector $\mathbf{X}_i$ of all observed variables, we can assume that there exists a matrix C_i with the property that

$$E\left[C_i \mathbf{X}_i \middle| \Theta_i\right] = \boldsymbol{\mu}(\Theta_i),$$

i.e. $\boldsymbol{\mu}(\Theta_i)$ is linearly estimable from the data $\mathbf{X}_i$.

On the other hand, the vectors $\mathbf{X}_i$ should not be too rich. For this reason we make the additional assumption, that the conditional expectation of each component of the vector $\mathbf{X}_i$ can be written as a linear combination of the components of $\boldsymbol{\mu}(\Theta_i)$. In other words, the linear space spanned by the vectors $E\left[\mathbf{X}_i \middle| \Theta_i\right]$ should be the same as the linear space spanned by the components of $\boldsymbol{\mu}(\Theta_i)$. These considerations lead to the following model assumptions:

Model Assumptions 7.11

We are given a portfolio of I risks. Each risk i ($i = 1, 2, \ldots, I$) is characterized by its individual (unknown) risk profile ϑ_i, which is itself the realization of a random variable Θ_i. We assume that

MD1: For each i there exists a matrix Y_i for which conditionally, given Θ_i,

$$E[\mathbf{X}_i|\Theta_i] = Y_i\, \boldsymbol{\mu}(\Theta_i), \tag{7.49}$$

where Y_i is a $n_i \times p$ matrix of rank p.

MD2: The pairs $(\Theta_1, \mathbf{X}_1)$, $(\Theta_2, \mathbf{X}_2), \ldots$ are independent, and $\Theta_1, \Theta_2, \ldots$ are independent and identically distributed.

Remarks:

- On MD1: In the typical multidimensional credibility model, e.g. in the multidimensional Bühlmann–Straub model, the matrix Y_i is composed entirely of the elements 1 and 0. In the regression case, which we will discuss in the next chapter, the elements of Y_i can be any real numbers for continuous covariates and 0 or 1 for categorical covariates.
- By pre-multiplying (7.49) with $C_i = (Y_i', Y_i)^{-1} Y_i'$ we get

$$\boldsymbol{\mu}(\Theta_i) = C_i\, E[\mathbf{X}_i|\Theta_i] \tag{7.50}$$

  where C_i is a $p \times n_i$ matrix of rank p.

7.4.3 Optimal Data Compression

By optimal data compression we mean the reduction of the vector $\mathbf{X}_i$ to a vector $\mathbf{B}_i$, having the same dimension as the vector $\boldsymbol{\mu}(\Theta_i)$ which is to be estimated, without being penalized by losing precision, i.e. the credibility estimator depends on the data only through $\mathbf{B}_i$. The following theorem shows how the optimal compression looks under Model Assumptions 7.11.

Theorem 7.12 (optimal data compression in the multidimensional case). *The optimal data compression is given by*

$$\mathbf{B}_i := \mathrm{Pro}\,(\boldsymbol{\mu}(\Theta_i)|\, L_e^{ind}(\mathbf{X}_i)), \tag{7.51}$$

where

$$L_e^{ind}(\mathbf{X}_i) = \Big\{\widehat{\boldsymbol{\mu}(\Theta_i)} \,:\, \mu_k(\Theta_i) = \sum\nolimits_j a_{ij}^{(k)} X_{ij} \text{ for } k = 1, \ldots, p; \; E\left[\widehat{\boldsymbol{\mu}(\Theta_i)}\Big|\,\Theta_i\right] = \boldsymbol{\mu}(\Theta_i).\Big\}$$

Remarks:

- Note that, because of MD1, the subspace $L_e^{ind}(\mathbf{X}_i)$ is not empty.
- As in Chapter 6 (see remarks on page 150) we denote by $L_e^{ind}(.)$ the space of individually unbiased estimators. As in Chapter 6, the notation of $L_e^{ind}(.)$ does not reflect what quantity we want to estimate and what the individual unbiasedness condition exactly means. However, this always becomes clear from the context. For instance, it is clear how $L_e^{ind}(\mathbf{X}_i)$ in (7.51) has to be interpreted: it is the space of estimators of $\boldsymbol{\mu}(\Theta_i)$ which are linear in the components of $\mathbf{X}_i$ and whose conditional expected value given Θ_i is equal to $\boldsymbol{\mu}(\Theta_i)$.

Proof of Theorem 7.12:
We have to show that under Model Assumptions 7.11 the credibility estimator depends on the observations $\mathbf{X}_i$ only via $\mathbf{B}_i$. This is a consequence of the following theorem. □

Theorem 7.13 (credibility estimator). *The (inhomogeneous) credibility estimator under Model Assumptions 7.11 is given by*

$$\widehat{\widehat{\boldsymbol{\mu}(\Theta_i)}} = A_i \mathbf{B}_i + (I - A_i)\,\boldsymbol{\mu}, \tag{7.52}$$

$$\text{where } A_i = T\,(T + S_i)^{-1}, \tag{7.53}$$

$$\boldsymbol{\mu} = E\left[\boldsymbol{\mu}(\Theta_i)\right],$$

$$S_i = E\left[\operatorname{Cov}\left(\mathbf{B}_i, \mathbf{B}_i' \middle| \Theta_i\right)\right],$$

$$T = \operatorname{Cov}\left(\boldsymbol{\mu}\left(\Theta_i\right), \boldsymbol{\mu}\left(\Theta_i\right)'\right).$$

Remarks:

- Because of the independence of the risks, the inhomogeneous credibility estimator $\widehat{\widehat{\boldsymbol{\mu}(\Theta_i)}}$ depends only on the data from the ith risk.
- Note that

$$\begin{aligned} S_i &= E\left[\left(\mathbf{B}_i - \boldsymbol{\mu}\left(\Theta_i\right)\right)\left(\mathbf{B}_i - \boldsymbol{\mu}\left(\Theta_i\right)\right)'\right] \\ &= \text{quadratic loss matrix of } \mathbf{B}_i \text{ (with respect to } \boldsymbol{\mu}\left(\Theta_i\right)), \\ T &= E\left(\left(\boldsymbol{\mu} - \boldsymbol{\mu}\left(\Theta_i\right)\right)\left(\boldsymbol{\mu} - \boldsymbol{\mu}\left(\Theta_i\right)\right)'\right) \\ &= \text{squared loss matrix of } \boldsymbol{\mu} \text{ (with respect to } \boldsymbol{\mu}\left(\Theta_i\right)). \end{aligned}$$

  From this we see that (7.52) corresponds again to the general intuitive principle that the credibility estimator is a weighted average of $\mathbf{B}_i$ (best individual, unbiased estimator based on the data) and $\boldsymbol{\mu}$ (best estimator based on a priori knowledge) with weights being proportional to the corresponding precisions (inverse of the quadratic loss).
- The quadratic loss of the credibility estimator (7.52) is given by Theorem 7.5.

Proof of Theorem 7.13:
According to Theorem 7.2, (7.52) is the credibility estimator based on $\mathbf{B}_i$ and therefore

$$\boldsymbol{\mu}(\Theta_i) - \widehat{\widehat{\boldsymbol{\mu}(\Theta_i)}} \perp \mathbf{B}_i.$$

We must also show that this is the credibility estimator based on all the data, i.e. that

$$\boldsymbol{\mu}(\Theta_i) - \widehat{\widehat{\boldsymbol{\mu}(\Theta_i)}} \perp \mathbf{X}_i. \tag{7.54}$$

Hence (7.54) is equivalent to

$$\boldsymbol{\mu}(\Theta_i) - \widehat{\widehat{\boldsymbol{\mu}(\Theta_i)}} \perp \mathbf{X}_i - Y_i \mathbf{B}_i, \tag{7.55}$$

where Y_i is the matrix given by (7.49). We can rewrite the left-hand side of the above equation:

$$\boldsymbol{\mu}(\Theta_i) - \widehat{\widehat{\boldsymbol{\mu}(\Theta_i)}} = A_i\left(\boldsymbol{\mu}(\Theta_i) - \mathbf{B}_i\right) + \left(I - A_i\right)\left(\boldsymbol{\mu}(\Theta_i) - \boldsymbol{\mu}\right).$$

We first show that the first summand $\boldsymbol{\mu}(\Theta_i) - \mathbf{B}_i \perp \mathbf{X}_i - Y_i\mathbf{B}_i$. Since $\mathbf{B}_i = \operatorname{Pro}(\boldsymbol{\mu}(\Theta_i)|\, L_e^{ind}(\mathbf{X}_i))$ and $Y_i\mathbf{B}_i = \operatorname{Pro}(Y_i\boldsymbol{\mu}(\Theta_i)|\, L_e^{ind}(\mathbf{X}_i))$, we have

$$Y_i\left(\boldsymbol{\mu}(\Theta_i) - \mathbf{B}_i\right) \perp \mathbf{X}_i - Y_i\mathbf{B}_i. \tag{7.56}$$

Since Y_i has rank p, the left-hand side is composed of p linearly independent, linear combinations of $\boldsymbol{\mu}(\Theta_i) - \mathbf{B}_i$, all of which are orthogonal to $\mathbf{X}_i - Y_i\mathbf{B}_i$. From this it follows that all of the components of $\boldsymbol{\mu}(\Theta_i) - \mathbf{B}_i$ are orthogonal to $\mathbf{X}_i - Y_i\mathbf{B}_i$. (Formally, we can also calculate this by multiplying the left-hand side of (7.56) by the matrix C_i given by (7.50), and then using the identity $C_iY_i = I$ (all matrices are of rank p).)

The orthogonality of the second summand $\boldsymbol{\mu}(\Theta_i) - \boldsymbol{\mu}$ is easy to see. If we condition on Θ_i, we get that

$$E\left[\left(\boldsymbol{\mu}(\Theta_i) - \boldsymbol{\mu}\right)\cdot\left(\mathbf{X}_i - Y_i\mathbf{B}_i\right)'\right] = 0.$$

This completes the proof of Theorem 7.13. □

The compression $\mathbf{B}_i$ contains, as we saw above, the full information of the data in the inhomogeneous credibility formula. We would be surprised if $\mathbf{B}_i$ did not also contain the full information for the homogeneous credibility estimator. We formulate this fact in the next theorem.

Theorem 7.14 (linear sufficient statistic). $\{\mathbf{B}_i : i = 1, 2, \ldots, I\}$ *is a linear sufficient statistic, i.e. both the inhomogeneous and the homogeneous credibility estimator depend on the data only through this statistic.*

Remark on the notation:

- In the following proof and in the remaining part of this subsection, capital I is used for the number of risks in the portfolio as well as for the identity matrix. However the meaning of I is always clear from the context.

Proof of Theorem 7.14: We have already shown in Theorem 7.13 that the inhomogeneous credibility estimator depends only on $\mathbf{B}_i$. It remains only to show that this is also true for the homogeneous credibility estimator. According to Theorem 7.13 we have that

$$\widehat{\widehat{\boldsymbol{\mu}(\Theta_i)}} = \operatorname{Pro}\left(\boldsymbol{\mu}(\Theta_i)|\, L(1, \mathbf{X}_1, \ldots, \mathbf{X}_I)\right) = A_i\mathbf{B}_i + \left(I - A_i\right)\boldsymbol{\mu}.$$

Therefore, because of the normalized linearity property of projections on affine spaces we have that

$$\begin{aligned}\widehat{\widehat{\boldsymbol{\mu}(\Theta_i)}}^{\text{hom}} &:= \text{Pro}\left(\boldsymbol{\mu}(\Theta_i)\middle| L_e(\mathbf{X}_1,\ldots,\mathbf{X}_I)\right)\\ &= A_i\,\mathbf{B}_i + (I - A_i)\,\widehat{\widehat{\boldsymbol{\mu}}}\,, \\ \text{where } \widehat{\widehat{\boldsymbol{\mu}}} &= \text{Pro}\left(\boldsymbol{\mu}\middle| L_e(\mathbf{X}_1,\ldots,\mathbf{X}_I)\right).\end{aligned} \tag{7.57}$$

By first projecting on a larger space and using the iterativity property we get

$$\widehat{\widehat{\boldsymbol{\mu}}} = \text{Pro}\left(\text{Pro}\left(\boldsymbol{\mu}\middle| L_e(\mathbf{X}_1,\ldots,\mathbf{X}_I,\boldsymbol{\mu}(\Theta_1),\ldots,\boldsymbol{\mu}(\Theta_I))\right)\middle| L_e(\mathbf{X}_1,\ldots,\mathbf{X}_I)\right).$$

It is intuitively obvious and easy to prove that

$$\text{Pro}\left(\boldsymbol{\mu}\middle| L_e(\mathbf{X}_1,\ldots,\mathbf{X}_I,\boldsymbol{\mu}(\Theta_1),\ldots,\boldsymbol{\mu}(\Theta_I))\right) = \frac{1}{I}\sum_{i=1}^{I}\boldsymbol{\mu}(\Theta_i).$$

From the above equations we obtain

$$\begin{aligned}\widehat{\widehat{\boldsymbol{\mu}}} &= \text{Pro}\left(\frac{1}{I}\sum_{i=1}^{I}\boldsymbol{\mu}(\Theta_i)\middle| L_e(\mathbf{X}_1,\ldots,\mathbf{X}_I)\right)\\ &= \frac{1}{I}\sum_{i=1}^{I}\widehat{\widehat{\boldsymbol{\mu}(\Theta_i)}}^{\text{hom}}\\ &= \frac{1}{I}\sum_{i=1}^{I}\left\{A_i\mathbf{B}_i + (I - A_i)\,\widehat{\widehat{\boldsymbol{\mu}}}\right\}.\end{aligned} \tag{7.58}$$

In (7.58) only the compressed data vectors $\mathbf{B}_i$ appear, which completes the proof of Theorem 7.14. □

Corollary 7.15 (homogeneous credibility estimator). *The homogeneous credibility estimator under Model Assumptions 7.11 is*

$$\widehat{\widehat{\boldsymbol{\mu}(\Theta_i)}}^{\text{hom}} = A_i\,\mathbf{B}_i + (I - A_i)\,\widehat{\widehat{\boldsymbol{\mu}}}\,, \tag{7.59}$$

where A_i was defined in (7.52) and where

$$\widehat{\widehat{\boldsymbol{\mu}}} = \left(\sum_{i=1}^{I}A_i\right)^{-1}\sum_{i=1}^{I}A_i\mathbf{B}_i. \tag{7.60}$$

Proof:
From (7.58) we get

$$\sum_{i=1}^{I}A_i\mathbf{B}_i - \left(\sum_{i=1}^{I}A_i\right)\widehat{\widehat{\boldsymbol{\mu}}} = \mathbf{0}.$$

The corollary then follows from (7.57). □

Remark

- The quadratic loss of the homogeneous credibility estimator is given by Theorem 7.6.

7.5 Exercises

Exercise 7.1

Consider the situation of Exercise 6.1. Calculate the values of the (empirical) homogeneous credibility estimators for each risk group i by considering the observed claims ratios $B_i^{(1)}$ (own company data) and $B_i^{(2)}$ (other companies' data) simultaneously. Assume that Model Assumptions 7.7 are fulfilled for the tuples $(B_i^{(1)}, B_i^{(2)})$.

Beside the point estimates, calculate the same measure of precision as in Exercise 6.1.

Exercise 7.2

Consider the same data set as in Exercise 4.6 and Exercise 6.2. We are in particular interested in estimating the large claim frequency by making best use of the data.

Calculate the homogeneous credibility estimator for the large claim frequency in relative terms by looking for each risk group A1,..., B5 at the observed standardized frequencies of both normal and large claims.

Remarks:

- In order that the credibility matrix becomes easier to interpret, it is advisable to bring both components onto the same scale and to look at the standardized rather than at the absolute claim frequencies, where the standardized claim frequency for risk group i and component k is defined as $\widetilde{F}_i^{(k)} = F_i^{(k)}/F_\bullet^{(k)}$, where $F_\bullet^{(k)}$ is the observed claim frequency over the whole collective. Note that $\widetilde{F}_i^{(k)} = N_{i\bullet}^{(k)}/\nu_{i\bullet}^{(k)}$, where $\nu_{i\bullet}^{(k)} = w_{i\bullet}^{(k)} F_\bullet^{(k)}$, i.e. $\widetilde{F}_i^{(k)}$ is the frequency with respect to the weight $\nu_{i\bullet}^{(k)}$.
- Assume that the claim numbers are Poisson distributed. Estimate the structural parameters σ_k^2 and μ_k with an analogous iterative procedure as described in Section 4.10 following Corollary 4.8, but now adapted to the multidimensional case.

Exercise 7.3

Work with Model Assumptions 7.7. Assume that the observation vectors $\mathbf{X}_{ij}$ have dimension 2 and that the weights are equal for both components, i.e. $w_{ij}^{(1)} = w_{ij}^{(2)} = w_{ij}$.

Show that the credibility estimators are then given by

$$\widehat{\widehat{\mu_l\,(\Theta_i)}} = \mu_l + a_{l1}^{(i)}(B_1^{(i)} - \mu_1) + a_{l2}^{(i)}(B_2^{(i)} - \mu_1), \quad l = 1, 2,$$

with credibility weights

$$
\begin{aligned}
a_{11}^{(i)} &= \frac{w_{i\bullet}}{w_{i\bullet} + \kappa_1 \left(1 + \frac{\rho^2}{1-\rho^2+\frac{\kappa_2}{w_{i\bullet}}}\right)}, \\
a_{22}^{(i)} &= \frac{w_{i\bullet}}{w_{i\bullet} + \kappa_2 \left(1 + \frac{\rho^2}{1-\rho^2+\frac{\kappa_1}{w_{i\bullet}}}\right)}, \\
a_{12}^{(i)} &= \frac{\rho}{w_{i\bullet}} \frac{\sigma_1}{\sigma_2} \frac{\sqrt{\kappa_1 \kappa_2}}{\left(1 + \frac{\kappa_1}{w_{i\bullet}}\right)\left(1 + \frac{\kappa_2}{w_{i\bullet}}\right) - \rho^2}, \\
a_{21}^{(i)} &= \frac{\rho}{w_{i\bullet}} \frac{\sigma_2}{\sigma_1} \frac{\sqrt{\kappa_1 \kappa_2}}{\left(1 + \frac{\kappa_1}{w_{i\bullet}}\right)\left(1 + \frac{\kappa_2}{w_{i\bullet}}\right) - \rho^2},
\end{aligned}
$$

where

$$
\begin{aligned}
\kappa_l &= \frac{\sigma_l^2}{\tau_l^2}, \quad l = 1, 2, \\
\rho &= \frac{\tau_{12}}{\tau_1 \tau_2} = \frac{\operatorname{Cov}\left(\mu_1\left(\Theta_i\right), \mu_2\left(\Theta_i\right)\right)}{\sqrt{\operatorname{Var}\left[\mu_1\left(\Theta_i\right)\right] \cdot \operatorname{Var}\left[\mu_2\left(\Theta_i\right)\right]}}.
\end{aligned}
$$

Exercise 7.4

Use the same model as in Exercise 7.3 (two-dimensional observations with equal weights for the components), but now for the claim frequencies and under the assumption that the claim numbers are conditionally, given Θ_i, Poisson distributed.

a) Show that the credibility estimators for the standardized claim frequencies $\widetilde{\lambda}_k\left(\Theta_i\right) = \lambda_k\left(\Theta_i\right)/\lambda_k$ are then given by

$$
\widehat{\widehat{\widetilde{\lambda}_k\left(\Theta_i\right)}} = 1 + a_{k1}^{(i)} \cdot (\widetilde{F}_1^{(i)} - 1) + a_{k2}^{(i)} \cdot (\widetilde{F}_2^{(i)} - 1),
$$

with credibility weights

$$a_{11}^{(i)} = \frac{\nu_1^{(i)}}{\nu_1^{(i)} + \widetilde{\kappa}_1 \left(1 + \frac{\rho^2}{1-\rho^2+\frac{\widetilde{\kappa}_2}{\nu_2^{(i)}}}\right)},$$

$$a_{12}^{(i)} = \rho \cdot \sqrt{\frac{\mu_2}{\mu_1}} \cdot \frac{\sqrt{\frac{\widetilde{\kappa}_1}{\nu_1^{(i)}} \frac{\widetilde{\kappa}_2}{\nu_2^{(i)}}}}{\left(1 + \frac{\widetilde{\kappa}_1}{\nu_1^{(i)}}\right)\left(1 + \frac{\widetilde{\kappa}_2}{\nu_2^{(i)}}\right) - \rho^2},$$

$$a_{21}^{(i)} = \rho \cdot \sqrt{\frac{\mu_1}{\mu_2}} \cdot \frac{\sqrt{\frac{\widetilde{\kappa}_1}{\nu_1^{(i)}} \frac{\widetilde{\kappa}_2}{\nu_2^{(i)}}}}{\left(1 + \frac{\widetilde{\kappa}_1}{\nu_1^{(i)}}\right)\left(1 + \frac{\widetilde{\kappa}_2}{\nu_2^{(i)}}\right) - \rho^2},$$

$$a_{22}^{(i)} = \frac{\nu_2^{(i)}}{\nu_2^{(i)} + \widetilde{\kappa}_2 \left(1 + \frac{\rho^2}{1-\rho^2+\frac{\widetilde{\kappa}_1}{\nu_1^{(i)}}}\right)},$$

where

$$\widetilde{F}_l^{(i)} = \text{observed relative frequencies, } l = 1, 2,$$

$$\widetilde{\kappa}_l = \operatorname{Var}\left(\widetilde{\lambda}_l\left(\Theta_i\right)\right)^{-1} = \operatorname{CoVa}\left(\lambda_l\left(\Theta_i\right)\right)^{-2}, \quad l = 1, 2,$$

$$\nu_l^{(i)} = w_{i\bullet}\lambda_l = \text{a priori expected claim number.}$$

Remark:
Note that

$$\widetilde{\kappa}_l = \lambda_l\, \kappa_l.$$

b) Discuss the special cases $\rho = 0$ and $\rho = 1$.

Exercise 7.5

An actuary wanting to estimate the large claim frequency for different regions is often faced with the following problem: the claims experience in some of the small regions is very small and such regions get very small credibility weights when applying the Bühlmann–Straub model. Often, neighbouring small regions are then put together and treated as one and the same region. Indeed, the risk profiles of neighbouring regions might often be similar, one reason being that people living in one region are often driving to the neighbouring region for work.

The conclusion is that there might be correlations between the risk profiles of different regions and that this correlation might depend on the distance between the regions.

This situation can again be modelled with the multidimensional model. For illustrative purposes and in order to better see the mechanics, we consider the following simple situation with four regions.

Let N_k be the number of large claims in region k and let us assume that the number of a priori expected claims ν_k is known and is the same for all regions, i.e. $\nu_k = \nu$ for $k = 1, 2, \ldots, 4$. The claim numbers N_k are assumed to be conditionally Poisson distributed with Poisson parameter $\lambda_k = \Theta_k \nu$ and that

$$E[\Theta_k] = 1,$$
$$\operatorname{Var}[\Theta_k] = 0.2^2,$$
$$\operatorname{Corr}\left(\boldsymbol{\Theta}, \boldsymbol{\Theta}'\right) = \begin{pmatrix} 1 & 0.2 & 0.2 & 0 \\ 0.2 & 1 & 0.2 & 0.7 \\ 0.2 & 0.2 & 1 & 0.7 \\ 0 & 0.7 & 0.7 & 1 \end{pmatrix}.$$

We want to estimate Θ_k for each region $k < 1, 2, \ldots, 4$.

a) Calculate the multidimensional credibility matrix for $\nu = 5,\ 10,\ 20$.
b) Do you have an explanation as to why the credibility weight $a_{14} \neq 0$, even though the regions 1 and 4 are uncorrelated?
c) Calculate the values of the multidimensional credibility estimators if $\nu = 5$ and if the following observations have been made:
$N_1 = 1,\ N_2 = 3,\ N_3 = 6,\ N_4 = 10.$
d) The same as in c), but now by applying the Bühlmann–Straub model and hence neglecting the correlations between the regions.
e) Compare the results of c) and d).

Exercise 7.6

In group life insurance, a company would like to estimate "individual disability" curves in order to implement experience rating. Denote by i_x the basis disability curve showing the a priori disablement probabilities for the ages $x = 16, 17, \ldots, 65$. The aim is to estimate for the contracts $i = 1, 2, \ldots, I$ individual disability curves i_{ix}.

For simplicity the company considers instead the ages $x = 16, 17, \ldots, 65$ the age groups $A_1 = [16, 20]$, $A_2 = [21, 25]$, ..., $A_{10} = [60, 65]$. For a particular contract the following observations have been made, where ν_k denotes the number of a priori expected disability cases according to the basis disability curve i_x and N_k the number of observed disability cases in the age groups A_k, $k = 1, 2, \ldots, 10$.

age group	ν_k	N_k
16-20	1.60	0
21-25	1.71	0
26-30	1.46	0
31-35	2.45	1
36-40	3.06	2
41-45	4.15	3
46-50	5.28	6
51-55	6.92	8
56-60	9.71	11
61-65	13.16	16
Total	49.50	47

As a basic model assumption we assume that the observed claim numbers N_k are conditionally Poisson distributed with Poisson parameter $\lambda_k = \Theta_k \nu_k$ and $E[\Theta_k] = 1$. From portfolio knowledge the actuaries estimate the differences between contracts to be about $\pm 25\%$, i.e. it is assumed that $\sqrt{\operatorname{Var}(\Theta_k)} = 25\%$. The task is again to estimate Θ_k for $k = 1, 2, \ldots, 10$.

a) Assume that the shape of the individual disability curve is the same as the one of the basis disability curve, i.e. $i_{ix} = \Theta \cdot i_x$ respectively $\Theta_k = \Theta$ for $k = 1, 2, \ldots, 10$. Estimate Θ by looking only at the ratio $N_\bullet / \nu_\bullet$ (total observed number of disabilities divided by the total number of expected number of disabilities) and by making use of the result of Corollary 4.10 (compare also with the methodology of Exercise 2.3).
b) As in a), but estimate now Θ_k by looking only at the ratio N_k / ν_k of each age group $k = 1, 2, \ldots, 10$ separately.
c) Assume now that $\operatorname{Corr}(\Theta_k, \Theta_l) = \rho^{|k-l|}$. Calculate the values of the multidimensional credibility estimator for Θ_k, $k = 1, 2, \ldots, 10$. Do the calculations for $\rho = 0$, 0.25, 0.50, 0.75, 1.00.
d) Compare the results found in c) together with the observed standardized frequencies $\widetilde{F}_k = N_k / \nu_k$ for the different values of ρ.
   Compare the results for $\rho = 0$ and for $\rho = 1$ with the ones found in a) and b).

8

Credibility in the Regression Case

8.1 Motivation

A real problem from insurance practice was the motivation for the well-known paper by Hachemeister [Hac75]. Hachemeister wanted to forecast average claim amounts for bodily injury claims in third party auto liability for various states in the USA. Such data is affected by time trends due to inflation, so that the homogeneity assumption of the Bühlmann–Straub model ($E[X_{ij} | \Theta_i]$ being independent of j) is not appropriate.

On fitting regression lines to the data "observed average claim amounts versus time", Hachemeister noticed considerable differences between the states both in the slope of the fitted regression line and in its intercept. He also noticed that in particular the estimation of the slope was not reliable for the "smaller" states due to the small sample sizes. It was natural to fit also a regression line to the country-wide data (data of all states together). Then the question arises as to how much weight one should assign to the individual regression line of a particular state based on the data of that state alone and the regression line based on the collective experience for all the states.

Hachemeister's work was, therefore, motivated by a practical problem based on simple linear regression. However, he formulated the problem and its solution, for the more general case of a general multiple regression model.

Before introducing the credibility regression model, in the next section we summarize some results from classical regression theory. These results are used to fit a linear model (e.g. a regression line) to the individual data of a particular risk (e.g. the claim sizes of a particular state in Hachemeister's problem).

8.2 The Classical Statistics Point of View

The general linear model in classical statistics is given by

$$\mathbf{X} = Y \cdot \boldsymbol{\beta} + \boldsymbol{\varepsilon}, \tag{8.1}$$

where $\mathbf{X} =$ obs. vector (elements are random variables) of dim. $n \times 1$,
$Y =$ known design matrix of dimension $n \times p$ $(p \leq n)$,
$\boldsymbol{\beta} =$ unknown parameter vector of dimension $p \times 1$,
$\boldsymbol{\varepsilon} =$ vector of the random deviations of dimension $n \times 1$,
with $E[\boldsymbol{\varepsilon}] = \mathbf{0}$ and $\mathrm{Cov}(\boldsymbol{\varepsilon}, \boldsymbol{\varepsilon}') = \Sigma$,

i.e. the vector of random deviations has expected value zero and its covariance matrix is denoted by Σ. In the following we always assume that Y is of full rank p.

Remark:

- Important special cases are:
  - ordinary least squares
    Then
    $$\Sigma = \sigma^2 \cdot I. \tag{8.2}$$
  - weighted least squares
    Then
    $$\Sigma = \sigma^2 \cdot W^{-1}, \tag{8.3}$$
    $$\text{where } W = \begin{pmatrix} w_1 & & 0 \\ & \ddots & \\ 0 & & w_n \end{pmatrix},$$
    $w_1, \ldots, w_n$ are known weights.

The following theorem is given without proof and can be found in most books on classical statistics (e.g. [Wei05], [SL03]).

Theorem 8.1. *The best linear unbiased estimator (BLUE) of $\boldsymbol{\beta}$ is given by*

$$\widehat{\boldsymbol{\beta}} = \left(Y'\Sigma^{-1}Y\right)^{-1} Y'\Sigma^{-1}\mathbf{X}. \tag{8.4}$$

The covariance matrix of $\widehat{\boldsymbol{\beta}}$ is

$$\Sigma_{\widehat{\boldsymbol{\beta}}} = \left(Y'\Sigma^{-1}Y\right)^{-1}. \tag{8.5}$$

Remarks:

- In the case of ordinary least squares $\left(\Sigma = \sigma^2 \cdot I\right)$ we have
  $$\widehat{\boldsymbol{\beta}} = \left(Y'Y\right)^{-1} Y'\mathbf{X}$$
  and
  $$\Sigma_{\widehat{\boldsymbol{\beta}}} = \sigma^2 \cdot \left(Y'Y\right)^{-1}.$$

- In the case of weighted least squares $\left(\Sigma = \sigma^2 \cdot W^{-1}\right)$ we have

$$\widehat{\boldsymbol{\beta}} = \left(Y'WY\right)^{-1} Y'W\mathbf{X}$$

and

$$\Sigma_{\widehat{\boldsymbol{\beta}}} = \sigma^2 \cdot \left(Y'WY\right)^{-1}. \tag{8.6}$$

- "Best" estimator means that

$$E\left[\left(\widehat{\beta}_i - \beta_i\right)^2\right] \stackrel{!}{=} \text{minimum} \qquad \text{for } i = 1, 2, \ldots, p.$$

It is remarkable that this best estimator can be derived from the data as the least squares estimator, i.e.

$$\widehat{\boldsymbol{\beta}} = \begin{cases} \underset{\beta_1,\ldots,\beta_p}{\arg\min} \sum_{j=1}^{n} \left(X_j - (Y\,\boldsymbol{\beta})_j\right)^2 & \text{for } \sum = \sigma^2 \cdot I, \\ \underset{\beta_1,\ldots,\beta_p}{\arg\min} \sum_{j=1}^{n} w_j \left(X_j - (Y\,\boldsymbol{\beta})_j\right)^2 & \text{for } \sum = \sigma^2 \cdot W^{-1}, \\ \underset{\beta_1,\ldots,\beta_p}{\arg\min} \sum_{i,j=1}^{n} \left(X_i - (Y\,\boldsymbol{\beta})_i\right) \Sigma_{ij}^{-1} \left(X_j - (Y\,\boldsymbol{\beta})_j\right) & \text{in the general case.} \end{cases}$$

- In classical statistics it is often further assumed that the ε_i's are normally distributed. If the ε_i's are normally distributed, then the least squares estimators are the best unbiased estimators of all unbiased estimators (not just of all *linear* unbiased estimators). We usually do not make this further assumption.

8.3 The Regression Credibility Model

We are given a portfolio of I risks and we let

$$\mathbf{X}_i' = (X_{i1}, X_{i2}, \ldots, X_{in})$$

be the observation vector of the ith risk and $\mathbf{w}_i' = (w_{i1}, \ldots, w_{in})$ the vector of the associated known weights. For instance, in the Hachemeister problem the entries X_{ij} are the claims averages of state i in the quarter j and w_{ij} are the corresponding number of claims.

We then assume that conditionally, given Θ_i, $\mathbf{X}_i$ fulfils the regression equation (8.1). This assumption leads to an interpretation of the regression vector $\boldsymbol{\beta}$ which is somewhat different to that used in classical statistics: we do not consider the components of $\boldsymbol{\beta}$ as fixed quantities (which are unknown and therefore must be estimated), but rather as random variables, having distributions which are determined by the structure of the collective. These considerations lead us to the credibility regression models of Subsections 8.3.1 and 8.3.2.

8.3.1 The Standard Regression Model

Model Assumptions 8.2 (standard regression case)
The risk i is characterized by an individual risk profile ϑ_i, which is itself the realization of a random variable Θ_i. We make the following assumptions:

R1: Conditionally, given Θ_i, the entries X_{ij}, $j = 1, 2, \ldots, n$ are independent and we have

$$\underbrace{E[\mathbf{X}_i \mid \Theta_i]}_{n \times 1} = \underbrace{Y_i}_{n \times p} \cdot \underbrace{\boldsymbol{\beta}(\Theta_i)}_{p \times 1}, \tag{8.7}$$

$$\begin{aligned} \textit{where } \boldsymbol{\beta}(\Theta_i) &= \textit{regression vector of length } p \leq n, \\ &\quad \textit{with linearly independent components,} \\ Y_i &= \textit{known “design” matrix of rank } p, \\ \operatorname{Var}[X_{ij} \mid \Theta_i] &= \frac{\sigma^2(\Theta_i)}{w_{ij}}. \end{aligned} \tag{8.8}$$

R2: The pairs $(\Theta_1, \mathbf{X}_1)$, $(\Theta_2, \mathbf{X}_2)$,... are independent, and Θ_1, Θ_2,... are independent and identically distributed.

Remarks:

- Assumption *R1* means that conditionally, given Θ_i, the observations X_{ij} ($j = 1, 2, \ldots, n$) fulfil the classical weighted regression assumptions (8.1) and (8.3). The columns of the design matrix Y_i correspond to the observed design variables. Assumption *R2* is the usual assumption that the risks are independent of each other.
- From the conditional independence and condition (8.8) follows that

$$\operatorname{Cov}(\mathbf{X}_i, \mathbf{X}_i' \mid \Theta_i) = \sigma^2(\Theta_i) \cdot W_i^{-1}, \tag{8.9}$$

$$\text{where } W_i = \begin{pmatrix} w_{i1} & & 0 \\ & \ddots & \\ 0 & & w_{in} \end{pmatrix}.$$

- All results in this subsection also hold true if the somewhat weaker condition

$$S_i = E[\operatorname{Cov}(\mathbf{X}_i, \mathbf{X}_i' \mid \Theta_i)] = \sigma^2 \cdot W_i^{-1} \tag{8.10}$$

  is fulfilled. This will become clear from the proofs of Subsection 8.3.2.
- For the case where $n < p$ and $\mathbf{Y}_i$ is not of full rank p, we refer to the remarks on page 242 relating to Corollary 9.9 in Section 9.7.

The goal is to determine the individual premium $\mu_{\mathbf{a}}(\Theta_i) = E[X_{\mathbf{a}} \mid \Theta_i]$ for any given design vector $\mathbf{a}' = (a_1, a_2, \ldots, a_p)$. Given the regression model, the quantity of interest is therefore of the form

$$\mu_{\mathbf{a}}\left(\Theta_i\right) = \mathbf{a}'\,\boldsymbol{\beta}\left(\Theta_i\right). \tag{8.11}$$

In order to determine the credibility estimator for $\mu_{\mathbf{a}}\left(\Theta_i\right)$ for any $\mathbf{a}$, we must determine the credibility estimator for the vector $\boldsymbol{\beta}\left(\Theta_i\right)$.

Notice that *R1* and *R2* are fully compatible with the assumptions *MD1* and *MD2* (Model Assumptions 7.11) for multidimensional credibility. We can therefore directly apply the results from Chapter 7 (Theorem 7.13 and Corollary 7.15). According to Theorem 7.14, in order to construct the credibility estimator, we need to find a formula for the optimal data compression $\mathbf{B}_i = \text{Pro}(\boldsymbol{\beta}(\Theta_i)|\, L_e^{ind}\left(\mathbf{X}_i\right))$.

Before we do that, we define the *structural parameters* in the standard regression model and we look at a few examples. These structural parameters are given by

$$\begin{aligned}
\underbrace{\boldsymbol{\beta}}_{p\times 1} &:= E\left[\boldsymbol{\beta}\left(\Theta_i\right)\right], \\
\sigma^2 &:= E\left[\sigma^2\left(\Theta_i\right)\right], \\
\underbrace{T}_{p\times p} &:= \operatorname{Cov}\left(\boldsymbol{\beta}\left(\Theta_i\right), \boldsymbol{\beta}\left(\Theta_i\right)'\right).
\end{aligned}$$

Remark:

- In the following we always assume that T is of full rank.

The first moment is given by

$$E\left[\mathbf{X}_i|\, \Theta_i\right] = \begin{pmatrix} 1 \\ \vdots \\ 1 \end{pmatrix} \mu\left(\Theta_i\right).$$

Here $\boldsymbol{\beta}\left(\Theta_i\right)$ is one-dimensional and we have

$$\boldsymbol{\beta}\left(\Theta_i\right) = \mu\left(\Theta_i\right), \quad Y_i = (1, \ldots, 1)'.$$

Example 8.1 (simple linear regression)

$$E\left[X_{ij} \mid \Theta_i\right] = \beta_0\left(\Theta_i\right) + j\beta_1\left(\Theta_i\right)$$

or in matrix notation

$$E\left[\mathbf{X}_i|\, \Theta_i\right] = \underbrace{\begin{pmatrix} 1 & 1 \\ 1 & 2 \\ \vdots & 3 \\ \vdots & \vdots \\ 1 & n \end{pmatrix}}_{= Y_i} \underbrace{\begin{pmatrix} \beta_0\left(\Theta_i\right) \\ \beta_1\left(\Theta_i\right) \end{pmatrix}}_{= \boldsymbol{\beta}(\Theta_i)}.$$

Next we derive the formula of the credibility estimator by use of the results from multidimensional credibility.

Theorem 8.3 (data compression). *Under Model Assumptions 8.2 (standard case) we have:*
i) The best linear and individually unbiased estimator of $\boldsymbol{\beta}(\Theta_i)$ *is*

$$\mathbf{B}_i = \left(Y_i' W_i Y_i\right)^{-1} Y_i' W_i \mathbf{X}_i. \tag{8.12}$$

ii) The quadratic loss matrix of $\mathbf{B}_i$ *is*

$$E\left[\left(\mathbf{B}_i - \boldsymbol{\beta}(\Theta_i)\right) \cdot \left(\mathbf{B}_i - \boldsymbol{\beta}(\Theta_i)\right)'\right] = \sigma^2 \cdot \left(Y_i' W_i Y_i\right)^{-1}. \tag{8.13}$$

Proof: Given Θ_i, we may first ask what is the best unbiased linear estimator from classical regression analysis. Theorem 8.1, resp. formula (8.5) tells us that (8.12) is already the best unbiased linear estimator of $\boldsymbol{\beta}(\Theta_i)$. Since the matrices of (8.12) do not depend on Θ_i, it must be the optimal data compression. From (8.6), we also get that

$$\operatorname{Cov}\left(\mathbf{B}_i, \mathbf{B}_i' \middle| \Theta_i\right) = \sigma^2(\Theta_i) \cdot \left(Y_i' W_i Y_i\right)^{-1}. \tag{8.14}$$

Since

$$\operatorname{Cov}\left(\mathbf{B}_i, \mathbf{B}_i' \middle| \Theta_i\right) = E\left[\left(\boldsymbol{\beta}_i - \boldsymbol{\beta}(\Theta_i)\right)\left(\boldsymbol{\beta}_i - \boldsymbol{\beta}(\Theta_i)\right)' \middle| \Theta_i\right] \tag{8.15}$$

and by taking the expected value of the right-hand side of (8.14), we obtain (8.13). This ends the proof of Theorem 8.3. □

Theorem 8.4 (credibility formula standard case). *Under Model Assumptions 8.2 we get that the credibility estimator for* $\boldsymbol{\beta}(\Theta_i)$ *satisfies*

$$\widehat{\widehat{\boldsymbol{\beta}(\Theta_i)}} = A_i \mathbf{B}_i + (I - A_i)\boldsymbol{\beta},$$

where

$$A_i = T\left(T + \sigma^2 \left(Y_i' W_i Y_i\right)^{-1}\right)^{-1}.$$

The quadratic loss matrix is given by

$$E\left[\left(\widehat{\widehat{\boldsymbol{\beta}(\Theta_i)}} - \boldsymbol{\beta}(\Theta_i)\right) \cdot \left(\widehat{\widehat{\boldsymbol{\beta}(\Theta_i)}} - \boldsymbol{\beta}(\Theta_i)\right)'\right] = (I - A_i)T.$$

Proof: Theorem 8.4 follows directly from Theorem 7.13 and Theorem 7.5 applied to the compressed data given by Theorem 8.3. □

8.3.2 The General Regression Case (Hachemeister)

The only difference from the standard regression case (Model Assumptions 8.2) is that the conditional independence assumption between the components of $\mathbf{X}_i$ is dropped and that we allow any conditional covariance structure within $\mathbf{X}_i$.

Model Assumptions 8.5 (Hachemeister model)
The risk i is characterized by an individual risk profile ϑ_i, which is itself the realization of a random variable Θ_i. We make the following assumptions:

H1: Conditionally, given Θ_i, the entries X_{ij}, $j = 1,2,\ldots,n$, are independent and we have

$$\underbrace{E\left[\mathbf{X}_i \mid \Theta_i\right]}_{n \times 1} = \underbrace{Y_i}_{n \times p} \cdot \underbrace{\boldsymbol{\beta}\left(\Theta_i\right)}_{p \times 1}, \tag{8.16}$$

$$\begin{aligned} \textit{where } \boldsymbol{\beta}\left(\Theta_i\right) &= \textit{regression vector of length } p \leq n, \\ &\quad \textit{with linearly independent components,} \\ Y_i &= \textit{known “design” matrix of rank } p, \end{aligned}$$

$$\operatorname{Cov}\left(\mathbf{X}_i, \mathbf{X}_i' \mid \Theta_i\right) = \Sigma_i\left(\Theta_i\right). \tag{8.17}$$

H2: The pairs $(\Theta_1,\mathbf{X}_1)$, $(\Theta_2,\mathbf{X}_2)$,... are independent, and Θ_1, Θ_2,... are independent and identically distributed.

Remarks:

- Of course, the Hachemeister model includes the standard regression as a special case.
- (8.17) implies that we have as structural parameters the matrices

$$S_i = E\left[\Sigma_i\left(\Theta_i\right)\right] \quad (i = 1, 2, \ldots, I)$$

  instead of $\sigma^2 W_i^{-1}$ with only one parameter σ^2 in the standard regression case.
- Usually one assumes

$$\Sigma_i\left(\Theta_i\right) = W_i^{-1/2} \Sigma\left(\Theta_i\right) W_i^{-1/2}$$

  and hence

$$S_i = W_i^{-1/2} S \, W_i^{-1/2},$$

  where

$$W_i^{-1/2} = \begin{pmatrix} w_{i1}^{-1/2} & & 0 \\ & \ddots & \\ 0 & & w_{in}^{-1/2} \end{pmatrix}.$$

Even under this assumption, we still have many more structural parameters than in the standard regression case.
Indeed the matrix

$$S = E\left[\Sigma\left(\Theta_i\right)\right]$$

contains $n\left(n+1\right)/2$ structural parameters instead of only one parameter σ^2 in the standard regression model.

Again we want to find the credibility estimator and again we first focus on the data compression.

Conditionally, given Θ_i, the optimal linear unbiased estimator resulting from classical statistics (cf. Theorem 8.1) is

$$\widehat{\boldsymbol{\beta}\left(\Theta_i\right)} = \left(Y_i' \Sigma_i \left(\Theta_i\right)^{-1} Y_i\right)^{-1} Y_i' \Sigma_i \left(\Theta_i\right)^{-1} \mathbf{X}_i \tag{8.18}$$

and the covariance matrix is

$$\Sigma_{\widehat{\beta\left(\Theta_i\right)}} = \left(Y_i' \Sigma_i \left(\Theta_i\right)^{-1} Y_i\right)^{-1}. \tag{8.19}$$

Unfortunately, the covariance matrix $\Sigma_i\left(\Theta_i\right)$ appearing on the right-hand side of (8.18) depends on the unknown Θ_i and is therefore itself unknown. Remembering that the structural parameter entering in the credibility weight of the Bühlmann–Straub model is $\sigma^2 = E\left[\sigma^2\left(\Theta_i\right)\right]$, it suggests that the unknown covariance matrix (8.18) should be replaced by the structural parameter matrix $S_i = E\left[\Sigma_i\left(\Theta_i\right)\right]$. The next theorem shows that indeed we then obtain the optimal data compression.

Theorem 8.6 (data compression). *Under Model Assumptions 8.5 (Hachemeister) we have:*
i) The best linear and individually unbiased estimator of $\boldsymbol{\beta}\left(\Theta_i\right)$ *based on* $\mathbf{X}_i$ *is*

$$\mathbf{B}_i = \left(Y_i' S_i^{-1} Y_i\right)^{-1} Y_i' S_i^{-1} \mathbf{X}_i. \tag{8.20}$$

ii) The quadratic loss matrix of $\mathbf{B}_i$ *is*

$$E\left[\left(\mathbf{B}_i - \boldsymbol{\beta}\left(\Theta_i\right)\right) \cdot \left(\mathbf{B}_i - \boldsymbol{\beta}\left(\Theta_i\right)\right)'\right] = \left(Y_i' S_i^{-1} Y_i\right)^{-1}. \tag{8.21}$$

Remarks:

- $\mathbf{B}_i$ is individually unbiased, since

$$\begin{aligned} E\left[\mathbf{B}_i \middle| \Theta_i\right] &= \left(Y_i' S_i^{-1} Y_i\right)^{-1} Y_i' S_i^{-1} E\left[\mathbf{X}_i \middle| \Theta_i\right] \\ &= \left(Y_i' S_i^{-1} Y_i\right)^{-1} Y_i' S_i^{-1} Y_i \boldsymbol{\beta}\left(\Theta_i\right) \\ &= \boldsymbol{\beta}\left(\Theta_i\right). \end{aligned}$$

Proof of Theorem 8.6: We must show that the orthogonality condition

$$\boldsymbol{\beta}(\Theta_i) - \mathbf{B}_i \perp A\mathbf{X}_i - \mathbf{B}_i \tag{8.22}$$

is satisfied for all matrices A with

$$E\left[A\mathbf{X}_i|\,\Theta_i\right] = \boldsymbol{\beta}(\Theta_i). \tag{8.23}$$

The orthogonality condition (8.22) can be written as

$$E\left[\mathrm{Cov}\left(\mathbf{B}_i, (A\mathbf{X}_i)'|\,\Theta_i\right)\right] = E\left[\mathrm{Cov}\left(\mathbf{B}_i, \mathbf{B}_i'|\,\Theta_i\right)\right]. \tag{8.24}$$

From (8.23) and Model Assumptions 8.5, formula (8.16), we have

$$AY_i = I,$$

since $E\left[A\mathbf{X}_i|\,\Theta_i\right] = AY_i\boldsymbol{\beta}(\Theta_i)$ for all values of $\boldsymbol{\beta}(\Theta_i)$. Using (8.20) we find that the left-hand side of (8.24) is equal to

$$\left(Y_i'S_i^{-1}Y_i\right)^{-1} Y_i'S_i^{-1}S_iA' = \left(Y_i'S_i^{-1}Y_i\right)^{-1} Y_i'A' = \left(Y_i'S_i^{-1}Y_i\right)^{-1}.$$

Since the left-hand side of (8.24) is independent of A, it must also hold that

$$E\left[\mathrm{Cov}\left(\mathbf{B}_i, \mathbf{B}_i'|\,\Theta_i\right)\right] = \left(Y_i'S_i^{-1}Y_i\right)^{-1}, \tag{8.25}$$

which proves Theorem 8.6. □

The following theorem follows directly from Theorem 7.13 and Theorem 7.5 from Chapter 7 (Multidimensional Credibility) applied to the compression $\mathbf{B}_i$.

Theorem 8.7 (Hachemeister formula). *Under Model Assumptions 8.5 we get that the credibility estimator for $\boldsymbol{\beta}(\Theta_i)$ satisfies*

$$\begin{aligned} \widehat{\widehat{\boldsymbol{\beta}(\Theta_i)}} &= A_i\mathbf{B}_i + (I - A_i)\,\boldsymbol{\beta}, \qquad (8.26)\\ \text{where } A_i &= T\left(T + \left(Y_i'S_i^{-1}Y_i\right)^{-1}\right)^{-1},\\ \mathbf{B}_i &= \left(Y_i'S_i^{-1}Y_i\right)^{-1} Y_i'S_i^{-1}\mathbf{X}_i,\\ S_i &= E\left[\Sigma_i\left(\Theta_i\right)\right] = E\left[\mathrm{Cov}\left(\mathbf{X}_i, \mathbf{X}_i'|\,\Theta_i\right)\right],\\ T &= \mathrm{Cov}\left(\boldsymbol{\beta}\left(\Theta_i\right), \boldsymbol{\beta}\left(\Theta_i\right)'\right),\\ \boldsymbol{\beta} &= E\left[\boldsymbol{\beta}\left(\Theta_i\right)\right]. \end{aligned}$$

The quadratic loss matrix is given by

$$E\left[\left(\widehat{\widehat{\boldsymbol{\beta}(\Theta_i)}} - \boldsymbol{\beta}(\Theta_i)\right)\cdot\left(\widehat{\widehat{\boldsymbol{\beta}(\Theta_i)}} - \boldsymbol{\beta}(\Theta_i)\right)'\right] = (I - A_i)\,T.$$

8.3.3 Homogeneous Credibility Estimator and Quadratic Loss

The homogeneous credibility estimator as well as the quadratic loss of the inhomogeneous and the homogeneous credibility estimator are given by the results presented in Section 7.1 (Theorem 7.3, Theorem 7.5, Theorem 7.6) and by inserting there the specific credibility matrix and the structural parameter found in the standard regression case (Subsection 8.3.1) or in the general regression case (Subsection 8.3.2).

8.4 The Simple Linear Regression Case (Linear Trend Model)

The simple linear regression model with time as covariable (linear trend) was the basis for Hachemeister's original problem. In practice, this problem is still the most important application of the Hachemeister regression model. For this reason, we will examine this case more closely here.

In the simple linear regression model the regression equation is

$$E\left[X_{ij}\middle|\,\Theta_i\right] = \beta_0\left(\Theta_i\right) + j \cdot \beta_1\left(\Theta_i\right),$$

or in matrix notation

$$E\left[\mathbf{X}_i\middle|\,\Theta_i\right] = \underbrace{\begin{pmatrix} 1 & 1 \\ 1 & 2 \\ \vdots & 3 \\ \vdots & \vdots \\ 1 & n \end{pmatrix}}_{=\mathbf{Y}_i} \underbrace{\begin{pmatrix} \beta_0\left(\Theta_i\right) \\ \\ \beta_1\left(\Theta_i\right) \end{pmatrix}}_{=\boldsymbol{\beta}(\Theta_i)},$$

where $\beta_0\left(\Theta_i\right)$ is the intercept and $\beta_1\left(\Theta_i\right)$ is the slope.

Further we assume that we are in the standard regression case, resp. that

$$S_i = E\left[\mathrm{Cov}(\mathbf{X}_i, \mathbf{X}_i \mid \Theta_i)\right] = \sigma^2 \cdot W_i^{-1},$$

$$\text{where } W_i = \begin{pmatrix} w_{i1} & 0 & \ldots & 0 \\ 0 & w_{i2} & \ldots & 0 \\ \vdots & \vdots & \ddots & \vdots \\ 0 & 0 & \ldots & w_{in} \end{pmatrix}.$$

Theoretically everything seems to be in good order. The model fits exactly into the framework of the credibility regression model. However, when Hachemeister applied this model to his bodily injury data, he had an unwelcome surprise. In some states, the obtained results were rather strange and contrary to commonsense judgement. For example in state number 4 (see Figure 8.1) the slope of the credibility adjusted regression line was smaller than that in the collective as well as that for the individual risk and not somewhere between the two, as one would expect.

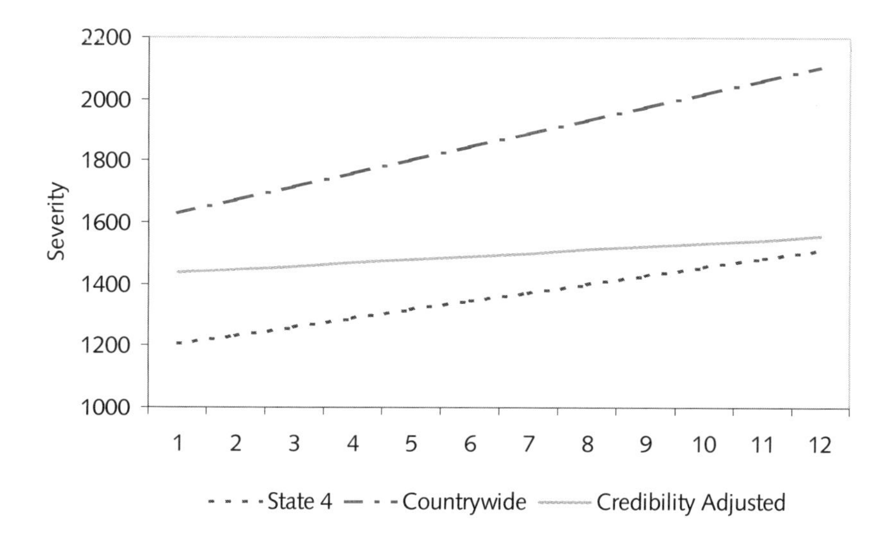

Fig. 8.1. State 4

For this reason the predictions based on the credibility regression line, say for the next four periods (the Hachemeister data were quarterly data) were very strange. They were less than what one would predict using either the individual regression line for the state under consideration as well as being less than the predictions one would make using all the states together. No practising actuary would trust such a prediction. Hachemeister had found a nice theory and an appealing theoretical result, but there still remained a question mark: the method didn't work well with the practical problem he was faced with.

It is amazing that a satisfactory solution to this problem was found only 20 years later in [BG97]. As we will see, the "trick" consists of modelling the intercept not in the time origin but rather in the centre of the time range.

First, however, we want to work with the original model with the intercept at the time origin in order to understand where the difficulties arise.

Take the intuitively reasonable case, where the intercept and the slope are independent of each other. This implies that the covariance matrix T has the following form:

$$T = \mathrm{Cov}(\boldsymbol{\beta}(\Theta_i), \boldsymbol{\beta}(\Theta_i)') = \begin{pmatrix} \tau_0^2 & 0 \\ 0 & \tau_1^2 \end{pmatrix}. \tag{8.27}$$

To simplify the notation, in the following we will omit the subscript i. Further we introduce the quantity

$$V := Y'WY.$$

Further, also to simplify the notation we define the following quantities. We regard the relative weights $w_j/w_\bullet$ as "sample weights" and denote with $E^{(s)}$ or $\mathrm{Var}^{(s)}$ the moments with respect to this distribution. We will also apply these "operators" to quantities which are not random variables. Examples of this notation are

$$E^{(s)}[j] = \sum_{j=1}^{n} j \frac{w_j}{w_\bullet},$$

$$E^{(s)}[X_j] = \sum \frac{w_j}{w_\bullet} X_j,$$

$$\mathrm{Var}^{(s)}[j] = \sum_{j=1}^{n} j^2 \frac{w_j}{w_\bullet} - \left(\sum_{j=1}^{n} j \frac{w_j}{w_\bullet} \right)^2 = E^{(s)}[j^2] - \left(E^{(s)}[j] \right)^2.$$

In the simple linear regression case we get

$$V = \begin{pmatrix} 1 & 1 & \dots & 1 \\ 1 & 2 & \dots & n \end{pmatrix} \begin{pmatrix} w_1 & 0 & \dots & 0 \\ 0 & w_2 & \dots & 0 \\ \vdots & \vdots & \ddots & \vdots \\ 0 & 0 & \dots & w_n \end{pmatrix} \begin{pmatrix} 1 & 1 \\ 1 & 2 \\ \vdots & \vdots \\ 1 & n \end{pmatrix}$$

$$= w_\bullet \begin{pmatrix} \sum_{j=1}^{n} \frac{w_j}{w_\bullet} & \sum_{j=1}^{n} j \frac{w_j}{w_\bullet} \\ \sum_{j=1}^{n} j \frac{w_j}{w_\bullet} & \sum_{j=1}^{n} j^2 \frac{w_j}{w_\bullet} \end{pmatrix} \tag{8.28}$$

$$= w_\bullet \begin{pmatrix} 1 & E^{(s)}[j] \\ E^{(s)}[j] & E^{(s)}[j^2] \end{pmatrix}.$$

For the best linear and individually unbiased estimator we get

$$\mathbf{B} = V^{-1}Y'W\mathbf{X}$$

$$= \begin{pmatrix} 1 & E^{(s)}[j] \\ E^{(s)}[j] & E^{(s)}[j^2] \end{pmatrix}^{-1} \begin{pmatrix} \sum_j \frac{w_j}{w_\bullet} X_j \\ \sum_j \frac{w_j}{w_\bullet} jX_j \end{pmatrix}$$

$$= \frac{1}{\mathrm{Var}^{(s)}[j]} \cdot \begin{pmatrix} E^{(s)}[j^2] & -E^{(s)}[j] \\ -E^{(s)}[j] & 1 \end{pmatrix} \begin{pmatrix} E^{(s)}[X_j] \\ E^{(s)}[jX_j] \end{pmatrix}$$

$$= \frac{1}{\mathrm{Var}^{(s)}[j]} \cdot \begin{pmatrix} E^{(s)}[j^2] \cdot E^{(s)}[X_j] - E^{(s)}[j] \cdot E^{(s)}[jX_j] \\ E^{(s)}[jX_j] - E^{(s)}[j] \cdot E^{(s)}[X_j] \end{pmatrix}. \tag{8.29}$$

Next we want to calculate the credibility matrix

$$A = T\left(T + \sigma^2 V^{-1}\right)^{-1}.$$

In order to reduce our calculations, we first rewrite the matrix A:

$$\begin{aligned} A &= T\left(T + \sigma^2 V^{-1}\right)^{-1} \\ &= \left(I + \sigma^2 V^{-1} T^{-1}\right)^{-1} \\ &= \left(V + \sigma^2 T^{-1}\right)^{-1} V. \end{aligned} \tag{8.30}$$

For the inverse of T we get

$$\begin{aligned} T^{-1} &= \frac{1}{\tau_0^2 \tau_1^2} \begin{pmatrix} \tau_1^2 & 0 \\ 0 & \tau_0^2 \end{pmatrix} \\ &= \frac{1}{\sigma^2} \begin{pmatrix} \kappa_0 & 0 \\ 0 & \kappa_1 \end{pmatrix}, \\ &\text{where } \kappa_0 = \frac{\sigma^2}{\tau_0^2}, \quad \kappa_1 = \frac{\sigma^2}{\tau_1^2}, \end{aligned}$$

and thus

$$V + \sigma^2 T^{-1} = \begin{pmatrix} w_\bullet + \kappa_0 & w_\bullet E^{(s)}[j] \\ w_\bullet E^{(s)}[j] & w_\bullet E^{(s)}[j^2] + \kappa_1 \end{pmatrix}.$$

From this we get

$$(V+\sigma^2 T^{-1})^{-1} = \frac{1}{N} \cdot \begin{pmatrix} w_\bullet E^{(s)}[j^2] + \kappa_1 & -(w_\bullet E^{(s)}[j]) \\ -(w_\bullet E^{(s)}[j]) & w_\bullet + \kappa_0 \end{pmatrix},$$

$$\text{where } N = (w_\bullet + \kappa_0)\left(w_\bullet E^{(s)}[j^2] + \kappa_1\right) - (w_\bullet E^{(s)}[j])^2$$

$$= w_\bullet^2 \mathrm{Var}^{(s)}[j] + \kappa_0 w_\bullet E^{(s)}[j^2] + w_\bullet \kappa_1 + \kappa_0 \kappa_1 .$$

Substituting these in (8.30) gives

$$A = \frac{w_\bullet}{N} \cdot \begin{pmatrix} w_\bullet \mathrm{Var}^{(s)}[j] + \kappa_1 & \kappa_1 E^{(s)}[j] \\ \kappa_0 E^{(s)}[j] & w_\bullet \mathrm{Var}^{(s)}[j] + \kappa_0 E^{(s)}[j^2] \end{pmatrix}. \tag{8.31}$$

The credibility matrix is a very complicated expression and it is difficult to interpret it. We see that the credibility matrix is not diagonal, even though S and T are diagonal.

The fact that the credibility matrix is non-diagonal is the source of all the unpleasantness and can lead to implausible results, as Hachemeister experienced in his application of this method. Moreover, the parameters are difficult to interpret and their effect is hard to predict, which is a major disadvantage in practice.

However, there is an astoundingly simple way of removing this difficulty: one should take the intercept not at the origin of the time axis but rather at the "centre of gravity" of the observed time range, which is defined by

$$j_0 = E^{(s)}[j] = \sum_{j=1}^{n} j \, \frac{w_j}{w_\bullet}.$$

The regression equation is then

$$\mu_j(\Theta) = \beta_0(\Theta) + (j - E^{(s)}[j]) \cdot \beta_1(\Theta),$$

and the design matrix becomes

$$Y = \begin{pmatrix} 1 & 1 - E^{(s)}[j] \\ 1 & 2 - E^{(s)}[j] \\ \vdots & \vdots \\ 1 & n - E^{(s)}[j] \end{pmatrix}.$$

Remark:

- It is well known that a linear transformation of the time axis (or more generally of any covariable) has no effect on the credibility estimator (this follows from the linearity property of the credibility estimator). But what we have done here is more than just a transformation of the time axis. We have changed the original model in that $\beta_0(\Theta_i)$ now has a slightly different meaning, namely the intercept at the "centre of gravity" of the time variable (rather than at the zero of the time variable). We model therefore a "new" vector and we assume that the covariance matrix T is diagonal with respect to this "new" vector.

Repeating the calculations in (8.28) and (8.29) with the "new time" $k - j - E^{(s)}[j]$ and using $E^{(s)}[k] = 0$ plus the fact that all variances and covariances involving j are unaffected by the time change, expression (8.29) – expressed in the time variable j – becomes

$$\mathbf{B} = \begin{pmatrix} E^{(s)}[X_j] \\ \\ \dfrac{\mathrm{Cov}^{(s)}(j, X_j)}{\mathrm{Var}^{(s)}(j)} \end{pmatrix} \tag{8.32}$$

and

$$A = \begin{pmatrix} a_{11} & 0 \\ \\ 0 & a_{22} \end{pmatrix}, \tag{8.33}$$

where
$$a_{11} = \frac{w_\bullet}{w_\bullet + \kappa_0} = \frac{w_\bullet}{w_\bullet + \frac{\sigma^2}{\tau_0^2}}, \tag{8.34}$$

$$a_{22} = \frac{w_\bullet \mathrm{Var}^{(s)}(j)}{w_\bullet \mathrm{Var}^{(s)}(j) + \kappa_1} = \frac{w_\bullet \cdot \mathrm{Var}^{(s)}(j)}{w_\bullet \cdot \mathrm{Var}^{(s)}(j) + \frac{\sigma^2}{\tau_1^2}}. \tag{8.35}$$

Notice that the credibility estimators for the components of $\boldsymbol{\beta}(\Theta)$ are given by

$$\widehat{\widehat{\beta_0(\Theta)}} = a_{11} B_0 + (1 - a_{11})\beta_0, \tag{8.36}$$

$$\widehat{\widehat{\beta_1(\Theta)}} = a_{22} B_1 + (1 - a_{22})\beta_1, \tag{8.37}$$

where B_0 and B_1 were defined in (8.32).

Under these assumptions, the credibility formula in the simple linear regression model can then be split into two one-dimensional credibility formulae. These have the simple form familiar from the Bühlmann–Straub model. In particular, notice that the credibility weights a_{11} and a_{22} have the same structure as the credibility weights in the Bühlmann–Straub model. The volumes are $w_\bullet$ for a_{11} and $w_\bullet \mathrm{Var}^{(s)}[j]$ for a_{22}, and the credibility coefficients are σ^2/τ_0^2 for a_{11} and σ^2/τ_1^2 for a_{22}.

These are formulae that one can use directly in practice. The results that we get with them make sense and reflect our intuitive expectations. The intercept and slope of the credibility regression line lie between the corresponding values for the individual and the collective regression models. Moreover, the intercept and the slope can be separately estimated using already known techniques from the Bühlmann–Straub model. This applies also to the estimator of the structural parameters, which we will describe below. The structural parameters have a direct meaning and we can see the influence that changes in these will have on the result. If, for example, τ_1^2 is small compared to $\sigma^2 / \left(w_\bullet \mathrm{Var}^{(s)}(j) \right)$, then the slope of the credibility line will lie close to the slope of the collective line.

Mathematically, what we have done here, is to make the columns of the design matrix Y mutually orthogonal. That is, the inner product of the column of 1's and the centred column $(1 - E^{(s)}[j], \ldots, n - E^{(s)}[j])'$, defined by the $E^{(s)}$ operator, is zero. As a consequence, the credibility matrix A is diagonal, and the credibility adjustment to β_0 (respectively β_1), has no effect on the credibility adjustment to β_1 (respectively β_0). Moreover, with this model specification, we have a more coherent view of our a priori beliefs: Classical statistics tells us, when the columns of Y are mutually orthogonal, then the estimators for the fixed parameter β_0 and β_1 are stochastically uncorrelated (conditional on Θ, in this case). Assuming that the stochastic parameters β_0 and β_1 are mutually independent, therefore seems more reasonable when the columns of Y are mutually orthogonal, than when they are not.

One point has been "brushed over" until now: the centre of gravity of the time variable may be different from risk to risk. In practice, one very often sees that the underlying volumes of the different risks don't vary too much over time and that they show a similar development pattern. In other words, the centres of gravity lie in most cases close together. For example, in the Hachemeister data, the observation period was 12 quarters and the centres of gravity varied in a range between 6.41 and 6.70. The question remains, however, as to how we should define the centre of gravity, and the natural answer is that we take the centre of gravity of the collective, i.e.

$$j_0 = \sum_{j=1}^{n} \frac{w_{\bullet j}}{w_{\bullet\bullet}}\, j.$$

In all situations, where the individual centres of gravity lie close together, we recommend taking the collective centre of gravity and applying the credibility formulae (8.36) and (8.37) developed above. The difference between this result and the exact credibility result will be negligible in most cases.

Figure 8.2 shows the result for state number 4, when the technique described above is applied to the Hachemeister data.

The credibility lines now look reasonable. The predictions make sense. The method delivers results which are useful. It seems to us that the Hachemeister problem is solved and the method is developed sufficiently to be used in practice for the estimation of trends by means of credibility.

For the sake of completeness, we give the exact expression for the credibility matrix when the centre of gravity of the time variable varies from one risk to

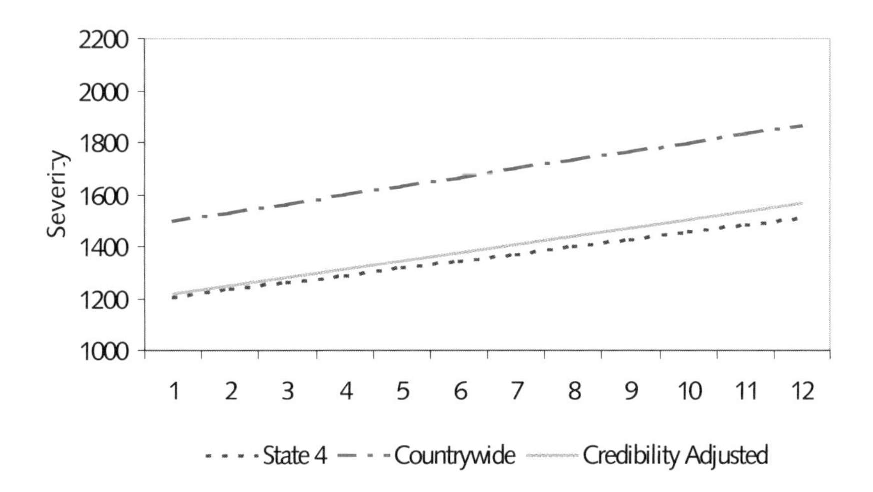

Fig. 8.2. State number 4 with intercept of centre of gravity

another. For the derivation, see the original article [BG97].

$$A_i = \frac{w_{i\bullet}}{N_i} \cdot \begin{pmatrix} w_{i\bullet} \cdot \mathrm{Var}^{(s_i)}[j] + \frac{\sigma^2}{\tau_1^2} + \Delta_i^2 \frac{\sigma^2}{\tau_0^2} & \Delta_i \mathrm{Var}^{(s_i)}[j] \frac{\sigma^2}{\tau_0^2} \\ \Delta_i \frac{\sigma^2}{\tau_0^2} & (w_{i\bullet} + \frac{\sigma^2}{\tau_0^2}) \mathrm{Var}^{(s_i)}[j] \end{pmatrix},$$

where $N_i = (w_{i\bullet} + \dfrac{\sigma^2}{\tau_0^2})(w_{i\bullet} \cdot \mathrm{Var}^{(s_i)}[j] + \dfrac{\sigma^2}{\tau_1^2}) + w_{i\bullet} \Delta_i^2 \dfrac{\sigma^2}{\tau_0^2},$

$$\Delta_i = E^{(s_i)}[j] - j_0$$

(deviation of the individual from the collective centre of gravity).

The credibility matrix is again quite complicated and is not diagonal. As we have already said, in most cases in practice, the deviations Δ_i are small. In this case the influence of the non-diagonal elements is negligible. Notice also, that for $\Delta_i = 0$ the above credibility matrix is identical with (8.33). For small Δ_i the results that one gets with the above exact credibility formula are very close to those that one gets with the far simpler one-dimensional credibility formulae (8.36) and (8.37). In practice then, one can work with the simple one-dimensional model and get results which are exact enough for practical purposes.

In order to apply the methodology, we still have to estimate the structural parameters. In the following we give estimators for the structural parameters which are analogous to those in the Bühlmann–Straub model.

Consider

$$\widehat{\sigma}_i^2 = \frac{1}{n-2} \sum_{j=1}^{n} w_{ij} \left(X_{ij} - \widehat{\mu}_{ij}\right)^2,$$

where $\widehat{\mu}_{ij}$ = fitted values from the individual regression lines.

It is well known from classical statistics that the $\widehat{\sigma}_i^2$'s are conditionally unbiased estimators for the $\sigma^2(\Theta_i)$'s.

A natural estimator for σ^2 is therefore

$$\widehat{\sigma}^2 = \frac{1}{I} \sum_{i=1}^{I} \widehat{\sigma}_i^2.$$

From (8.6) and (8.28) we get that with time origin at the individual centre of gravity, the conditional covariance matrix of the individual regression parameters is diagonal and given by

$$\operatorname{Cov}\left(\mathbf{B}_i, \mathbf{B}_i' \middle| \Theta_i\right) = \sigma^2(\Theta_i) \begin{pmatrix} w_{i\bullet} & 0 \\ 0 & w_{i\bullet} d_i \end{pmatrix}^{-1}, \tag{8.38}$$

$$\text{where } d_i = \operatorname{Var}^{(s_i)}(j) = \sum_{j=1}^{n} \frac{w_{ij}}{w_{i\bullet}} \left(j - j_0^{(i)}\right)^2, \tag{8.39}$$

$$j_0^{(i)} = \sum\nolimits_{j=1}^{n} \tfrac{w_{ij}}{w_{i\bullet}}\, j.$$

In most cases in practice, the individual centres of gravity $j_0^{(i)}$ vary little from one risk to another. Then (8.38) and (8.39) are still approximately fulfilled, if the individual centres of gravity $j_0^{(i)}$ are replaced by the global centre of gravity

$$j_0 = \sum_{j=1}^{n} \frac{w_{ij}}{w_{\bullet\bullet}}\, j.$$

The conditional variance of the elements of $\mathbf{B}_i$ in (8.38) have the same structure as the conditional variance of X_i in the Bühlmann–Straub model. If we take the estimator which is analogous to the estimator in the Bühlmann–Straub model, we get the following unbiased estimators:

$$\widehat{\tau}_0^2 = c_0 \cdot \left\{ \frac{1}{I-1} \sum_{i=1}^{I} \frac{w_{i\bullet}}{w_{\bullet\bullet}} \left(B_{0i} - \overline{B_0}\right)^2 - \frac{I\,\widehat{\sigma^2}}{w_{\bullet\bullet}} \right\},$$

where $c_0 = \frac{I-1}{I} \left\{ \sum_{i=1}^{I} \frac{w_{i\bullet}}{w_{\bullet\bullet}} \left(1 - \frac{w_{i\bullet}}{w_{\bullet\bullet}}\right) \right\}^{-1}$,

$$\overline{B_0} = \sum_{i=1}^{I} \frac{w_{i\bullet}}{w_{\bullet\bullet}} B_{0i}.$$

$$\widehat{\tau}_1^2 = c_1 \left\{ \frac{1}{I-1} \sum_{i=1}^{I} \frac{w_{i\bullet}^*}{w_{\bullet\bullet}^*} \left(B_{1i} - \overline{B_1}\right)^2 - \frac{I\,\widehat{\sigma^2}}{w_{\bullet\bullet}^*} \right\},$$

where $w_{i\bullet}^* = d_i \cdot w_{i\bullet}$,

$$c_1 = \frac{I-1}{I} \left\{ \sum_{i=1}^{I} \frac{w_{i\bullet}^*}{w_{\bullet\bullet}^*} \left(1 - \frac{w_{i\bullet}^*}{w_{\bullet\bullet}^*}\right) \right\}^{-1},$$

$$\overline{B_1} = \sum_{i=1}^{I} \frac{w_{i\bullet}^*}{w_{\bullet\bullet}^*} B_{1i}.$$

8.5 Exercises

Exercise 8.1

To get a deeper understanding of the simple regression credibility model (see Section 8.4), consider the following fictitious situation: the number of observation years n is equal to 5 and the weights w_{ij} are all equal to 1. For the collective regression line we obtained $\widehat{\beta}_0 = 130$ for the intercept (at the centre of gravity, which is 3 in this case) and $\widehat{\beta}_1 = 10$ for the slope. For the individual regression line we got $B_0 = 90$ for the intercept and $B_1 = 0$ for the slope.

We assume that (8.27) is fulfilled, i.e. $T = \mathrm{Cov}\left(\boldsymbol{\beta}(\Theta_i), \boldsymbol{\beta}(\Theta_i)'\right) = \begin{pmatrix} \tau_0^2 & 0 \\ 0 & \tau_1^2 \end{pmatrix}$.

a) Calculate the credibility regression line for the following structural parameters:

i)	$\sigma = 20$	$\tau_0 = 5$	$\tau_1 = 3$
ii)	$\sigma = 20$	$\tau_0 = 5$	$\tau_1 = 0.00001$
ii)	$\sigma = 20$	$\tau_0 = 0.00001$	$\tau_1 = 3$

b) Display the different regression lines (individual , collective and credibility) in a graph and interpret the results.

Exercise 8.2

The following table shows the claim number and the average claim amounts of four risk groups of a motor liability portfolio.

	risk group				risk group			
	1	2	3	4	1	2	3	4
year	claim number				average claim amount			
1	6 244	4 067	2 455	951	6 182	5 462	6 233	3 214
2	6 583	4 166	2 250	1 086	6 104	5 668	6 997	4 297
3	6 156	3 526	2 068	883	6 574	6 143	6 680	4 112
4	6 297	3 432	1 974	786	6 561	5 942	8 088	3 987
5	6 217	3 213	1 720	744	6 479	7 112	6 232	7 132
6	6 363	3 151	1 633	760	6 821	6 547	7 562	5 490

Table: *average claim amounts of motor liability portfolio*

a) Calculate the centre of gravity of time for each of the four risk groups and check that they are close to each other. In the following neglect the differences in the centre of gravity.
b) Calculate the (empirical) credibility lines for the average claim amounts for each risk group by applying formulae (8.36) and (8.37) and display them together with the individual and the collective regression lines.
c) Calculate for each risk group the values of the (empirical) estimator of the average claim amount for year 8.
d) As in c), but now under the a priori assumption that the slope is the same for all risk groups.

Exercise 8.3

Write the Bühlmann–Straub model as a credibility regression model and verify that Theorem 8.4 yields the credibility formula of the Bühlmann–Straub model.

9

Evolutionary Credibility Models and Recursive Calculation

9.1 Motivation

In most of the credibility models which we have dealt with until now, e.g. the simple Bühlmann model (Subsection 3.1.4) or the Bühlmann–Straub model (Chapter 4), the best linear estimator for the individual premium $\mu(\Theta)$ was of the form

$$P^{\text{cred}} = \widehat{\widehat{\mu(\Theta)}} = \alpha \overline{X} + (1-\alpha) P^{\text{coll}},$$

a weighted average of the collective premium $P^{\text{coll}} = \mu := E[\mu(\Theta)]$ and the individual, observed average $\overline{X}$.

In the multidimensional case (Chapter 7) we also found that

$$\widehat{\widehat{\boldsymbol{\mu}(\Theta)}} = A\mathbf{B} + (I - A)\boldsymbol{\mu},$$

where $\mathbf{B}$ can be interpreted as the observed average (best linear individually unbiased estimator).

In many applications, the components of the observation vector $\mathbf{X}' = (X_1, X_2, \ldots, X_n)$ represent the observed aggregate claim amounts over the past n observation years. Formulae of the above type impose that in order to calculate the credibility premium, one needs to have information about the complete individual claims history of the last n years.

On the other hand, one of the first and most well-known applications of credibility in insurance practice was the development of Bonus–Malus systems for motor insurance (see Section 2.4.1). Most Bonus–Malus systems used in practice have the property that the premium in period $n+1$ only depends on the premium in period n and the newly observed claim experience (mostly the number of claims) in year n. Such a system has the big advantage that the "claim history" over many years is not required. This makes the system for setting premiums easier both for the customer to understand and also for the insurers to administer (data storage, IT systems, etc.).

Also, the methodology used earlier by the rating bureaus of the United States (e.g. by the National Bureau of Casualty Underwriters or by the ISO) calculated the new premiums based on old premiums, but adjusted for the new claim experience, i.e.

$$\begin{aligned} P_{n+1} &= P_n + \alpha_n \left(X_n - P_n\right), \qquad (9.1)\\ \text{where } P_n &= \text{pure risk premium in period } n,\\ X_n &= \text{claim experience in period } n,\\ \alpha_n &= \text{weight or ``credibility measure''}\\ &\quad\ \text{for the most recent observation.} \end{aligned}$$

Formulae of the kind in (9.1) are called *recursive premium formulae.*

9.2 Recursive Credibility

In the following we consider a particular risk over time and we let:
P_n^{cred} be the credibility premium for the nth period based on the observations from the years $1, 2, \ldots, n-1$,
$X_1, X_2, \ldots, X_n, X_{n+1}, \ldots$ be the observed aggregate claim amounts in the years $j = 1, 2, \ldots, n, n+1, \ldots$ and
$P_1^{\text{cred}}, P_2^{\text{cred}}, \ldots, P_n^{\text{cred}}, P_{n+1}^{\text{cred}}, \ldots$ be the associated premiums for the corresponding years.

Definition 9.1. *The credibility premium is recursive, if*

$$\begin{aligned} P_{n+1}^{\text{cred}} &= \alpha_n X_n + (1-\alpha_n)\, P_n^{\text{cred}} \qquad (9.2)\\ &= P_n^{\text{cred}} + \alpha_n \left(X_n - P_n^{\text{cred}}\right), \end{aligned}$$

where P_n^{cred} denotes the credibility premium for year n based on the observations $X_1, X_2, \ldots, X_{n-1}$.

In the following we discuss in which cases the credibility premiums are recursive, i.e. in which cases the credibility premium can be calculated recursively over time by averaging the "old" credibility premium and the latest observation.

From a practical point of view the problem is that of how to calculate, over time, each year's premium (e.g. for fire insurance or motor insurance) for an existing policy. If the credibility premium is recursive, then according to (9.2) we have the very useful interpretation:

$$\begin{aligned} \text{new premium} = \text{old premium } &+ \alpha_n \cdot \text{ (deviation of the observed aggregate}\\ &\text{claim amount } X_n \text{ from the old premium } P_n). \end{aligned}$$

Example 9.1 (simple Bühlmann model)

For the credibility premium in year $n+1$we found that (Subsection 3.1.1, Theorem 3.2)

$$P_{n+1}^{\text{cred}} = \beta_n X^{(n)} + (1-\beta_n)\,\mu, \tag{9.3}$$

$$\text{where} \quad \beta_n = \frac{n}{n+\sigma^2/\tau^2},$$

$$X^{(n)} = \frac{1}{n}\sum_{j=1}^{n} X_j.$$

One can easily verify that (9.3) can also be written recursively,

$$P_{n+1}^{\text{cred}} = \alpha_n X_n + (1-\alpha_n)\,P_n^{\text{cred}},$$

$$\text{where} \quad \alpha_n = \frac{1}{n+\frac{\sigma^2}{\tau^2}}.$$

Example 9.2 (Bühlmann–Straub model)

For the credibility premium in year $n+1$ we found (Section 4.3, Theorem 4.2)

$$P_{n+1}^{\text{cred}} = \beta_n X^{(n)} + (1-\beta_n)\,\mu, \tag{9.4}$$

$$\text{where} \quad \beta_n = \frac{w_\bullet^{(n)}}{w_\bullet^{(n)} + \frac{\sigma^2}{\tau^2}},$$

$$w_\bullet^{(n)} = \sum_{j=1}^{n} w_j,$$

$$X^{(n)} = \sum_{j=1}^{n} \frac{w_j}{w_\bullet^{(n)}} X_j.$$

Formula (9.4) can also be written recursively as follows:

$$P_{n+1}^{\text{cred}} = \alpha_n X_n + (1-\alpha_n)\,P_n^{\text{cred}},$$

$$\text{where } \alpha_n = \frac{w_n}{w_\bullet^{(n)} + \frac{\sigma^2}{\tau^2}}.$$

In both Example 9.1 and Example 9.2, the weights α_n tend to zero with increasing observation period n and with bounded volume w_n. Concluding that premium adjustments for large n don't achieve anything, however, is to throw the baby out with the bathwater! The fact that in Examples 9.1 and 9.2 the weights α_n tend to zero with a growing observation period is a direct consequence of the underlying model assumptions, where it is assumed that the correct individual premium remains constant over time, i.e. that

$$P^{\text{ind}} = \mu(\Theta) = E[X_j|\Theta] \quad \text{for all } j.$$

In order to study situations in which the α_n do not tend to zero with increasing n,we must therefore study models in which the individual premium $E[X_j|\Theta]$is not the same for every j, but rather changes with the passage of time. Indeed the assumption of unchanging individual premiums over the years in the simple Bühlmann model and in the Bühlmann–Straub model is a simplified approach to reality (see also the remarks on Assumption BS1 on page 80).

Before turning to the consideration of how we should model situations where the correct individual premium changes over time, we take a look at the case where the recursive premium formula has constant weights $\alpha_n = \alpha$.

Example 9.3 ("geometric"premium formula)

We consider a recursive premium formula with constant weights $\alpha_n = \alpha$,with $0 < \alpha < 1$,i.e.

$$\begin{aligned} P_1 &= \mu_0, \\ P_{n+1} &= \alpha X_n + (1-\alpha) P_n \quad \text{for } n \geq 1. \end{aligned} \tag{9.5}$$

It is illustrative to consider the development of the premium P_n over time as a function of the observations X_j. We get

$$\begin{aligned} P_1 &= \mu_0, \\ P_2 &= \alpha X_1 + (1-\alpha)\mu_0, \\ P_3 &= \alpha X_2 + (1-\alpha) P_2 = \alpha X_2 + \alpha(1-\alpha) X_1 + (1-\alpha)^2 \mu_0, \\ &\vdots \\ P_{n+1} &= \sum_{k=0}^{n-1} \alpha(1-\alpha)^k X_{n-k} + (1-\alpha)^n \mu_0. \end{aligned} \tag{9.6}$$

Comparing this premium formula with the credibility premium in the simple Bühlmann model is very enlightening. With the above recursive premium formula, the a priori value μ_0, and "older" claim experience is weighted less and less as time goes on. It is intuitively obvious that the claim experience from more distant years should be weighted less than more recent claim experience in calculating this year's premium. In contrast, the credibility formula (9.3) gives the same weight to each of the observations. In practice, we accept this because mostly (9.3) is used only over a very limited observation period, e.g. over the last five years. The choice of the length of this observation period is somewhat arbitrary and has the unpleasant consequence that e.g. a five-year-old observation has the same weight as the newest observation, while the observation from six years back is totally ignored. In the geometric premium formula, older observations get successively less weight, removing the need to impose an artificial limit on the observation period. In particular, we note that the starting value $P_1 = \mu_0$ gets the smallest weight. For n sufficiently large, i.e. for a process that runs for a very long time, this starting value has a practically negligible influence on P_{n+1}^{cred}.

Let us now address the modelling aspect: In the simple Bühlmann model, the reason that all observations have the same weight is due to the assumption that the correct individual premium does not change over time. In contrast, lower weights for "older" observations as in the premium formula (9.6) can only be theoretically justified when the risk changes with the passage of time and older observations are less relevant than newer ones in the rating of the current risk. In the next section we will propose a model for the situation where the correct individual premium changes with time. It should be noted that in the example with the geometric premium formula we just have written down a recursive premium formula and that as yet, we have not demonstrated any connection between this and the credibility premium as defined in Chapter 3 (formula (3.17)). The question is, whether there exists a credibility model with a resulting recursive credibility formula as in Example 9.3 and what this model looks like. Amongst others, this question will be answered in Section 9.5 (Example 9.4).

9.3 Evolutionary Credibility Models

We will now describe situations in which the correct individual premium changes with the passage of time.

We have already seen one such situation in the regression case considered in Chapter 8. In the regression model, the individual premium in year j is determined by the formula

$$E\left[X_j \middle| \Theta\right] = \mathbf{y}_j' \,\boldsymbol{\beta}\left(\Theta\right).$$

It depends on the design vector $\mathbf{y}_j$, which represents the values of the covariates in year j. In this way we can model dependencies on known covariates such as time, inflation, age, sex, etc., in the credibility framework. In the regression approach the individual risk is described by the parameter vector $\boldsymbol{\beta}(\Theta)$, which stays constant with the passage of time. This approach has been treated in Chapter 8 and is not the subject of this or the following section. We will, however, come back to it in Section 9.6.

In many situations in insurance practice, the mechanics of the changes in the correct individual premiums are unknown and cannot be described by means of covariates in a regression model, be it because the information needed to do this has not been collected by the insurer or it is not used (see the comments under the thesis on page 2), or because the changes are a result of random events which cannot be quantified a priori. Indeed, most real risks are continually changing. For example, changes in living conditions like starting a family, the birth of a child, divorce, change of profession, job loss, can have an impact on the risk behaviour of the individual driver. For a variety of reasons, such events are not recorded by the insurer and are not directly used in the calculation of the premium and are therefore, from the point of view of the insurer, of an unknown nature. In collective health and accident insurance, the risk profile of the business insured changes with fluctuations in personnel and the different risk behaviour and recreational activities of the individual employees. In addition, we have to consider changes in the business organization and in management, which may have an effect on things like the work atmosphere, controls, etc., and therefore are not without influence on the insured risk. It is obvious that all of these potential explanatory factors cannot be observed by the insurer, let alone be quantified.

Such changes in the risk premiums, which cannot be explained by covariates which have been collected by the insurer and used in the rating of risks, are most easily understood as *stochastic* in nature. This approach leads us to *evolutionary* models.

In the following we consider a particular individual risk. Let $\mathbf{X}' = (X_1, \dots)$ be the observed aggregate claim amounts of this risk in years $j = 1, 2, \dots$. This risk is characterized by its individual risk profile, which is not directly observed by the insurer but which, in this approach, is allowed to *change stochastically with the passage of time.* This situation is described mathematically by letting the risk profile itself be a stochastic process, i.e. a sequence of random variables

$$\boldsymbol{\Theta} = \{\Theta_1, \Theta_2, \dots, \Theta_j, \dots\}. \tag{9.7}$$

The components Θ_j are to be understood as the risk profile in year j. We assume that

$$E[X_j | \boldsymbol{\Theta}] = E[X_j | \Theta_j] = \mu(\Theta_j), \tag{9.8}$$

$$\operatorname{Var}[X_j | \boldsymbol{\Theta}] = \operatorname{Var}[X_j | \Theta_j] = \frac{\sigma^2(\Theta_j)}{w_j}, \tag{9.9}$$

where the w_j are appropriate weights.

Notice that $E[X_j | \boldsymbol{\Theta}]$ (resp. $\operatorname{Var}[X_j | \boldsymbol{\Theta}]$) is the conditional expectation (resp. the conditional variance) given the sequence $\boldsymbol{\Theta}$, while $E[X_j | \Theta_j]$ (resp. $\operatorname{Var}[X_j | \Theta_j]$) is the conditional expected value (resp. the conditional variance) given the component Θ_j. Equations (9.8) and (9.9) express the fact that the first two moments of X_j depend only on the risk profile Θ_j in the year j and not on the other components $\{\Theta_k : k \neq j\}$.

In this way, the correct individual premium becomes a real-valued stochastic process

$$\boldsymbol{\mu}(\boldsymbol{\Theta}) = \{\mu(\Theta_1), \mu(\Theta_2), \dots, \mu(\Theta_j) \dots\}. \tag{9.10}$$

Further, we assume as always that the first two moments exist and are finite.

Definition 9.2. *By evolutionary credibility models we mean models with the structure given by (9.7)–(9.10).*

Remarks on the evolutionary credibility models:

- The sequence (9.10) stands for the sequence of correct individual premiums in time order. In contrast to regression, where one can imagine that the regression vector $\boldsymbol{\beta}(\Theta)$ is chosen once and for all, and that the changes over time for the correct individual premium are determined only by changes in the covariables, the elements of the sequence (9.10) change *stochastically.*
- Hachemeister's credibility regression approach can be combined with the ideas of evolutionary credibility in that we consider the components of the regression vector $\boldsymbol{\beta}(\Theta)$ to be non-constant over time, but rather to form a stochastic process. This is an extension of the Hachemeister regression model and will be treated later.
- Purely formally, the simple Bühlmann model and the Bühlmann–Straub model can be considered in the framework defined here for the evolutionary model. In this case, the stochastic process (9.10) has a "degenerate" form, i.e. the sequence $\boldsymbol{\Theta}$ is a repetition of the same random variable.
- We have limited ourselves here to the description of a model for an individual risk. In order to embed this in a collective of independent risks $i = 1, 2, \ldots, I$ we can proceed as we did in the previous chapters for the "classical" models:
  Each risk $i\,(i = 1, 2, \ldots, I)$ is characterized by its own risk profile
  $$\boldsymbol{\Theta}_i = \{\Theta_{i1}, \Theta_{i2}, \ldots, \Theta_{ij}, \ldots\}$$
  and the process of observations
  $$\mathbf{X}_i = \{X_{i1}, X_{i2}, \ldots, X_{ij}, \ldots\}.$$
  We assume that the pairs $\{(\mathbf{X}_1, \boldsymbol{\Theta}_1), (\mathbf{X}_2, \boldsymbol{\Theta}_2), \ldots, (\mathbf{X}_I, \boldsymbol{\Theta}_I)\}$ are independent of each other and that $\{\boldsymbol{\Theta}_1, \boldsymbol{\Theta}_2, \ldots, \boldsymbol{\Theta}_I\}$ are independent and identically distributed.
  With this embedding, nothing changes in our customary interpretation of the collective from the non-evolutionary situation. As in the previous chapters, it holds that in this case the credibility estimator for the ith contract depends only on the observations from this contract and not on those from the other contracts. The data from the collective are merely used to estimate the structural parameters in the credibility estimator. The credibility estimators that we derive in the following sections are therefore, as earlier, the credibility formulae for the individual risks in the case of a collective of independent risks. In certain applications, this interpretation may be useful.
- On the other hand, it must also be mentioned that in evolutionary models, the assumption that the risk profiles of the different risks are independent of each other is very incisive. In reality, it is easy to see situations where

this independence is not satisfied. This is especially the case when certain changes (changes in the law, economic circumstances, weather, etc.) have an effect on all risks in the risk collective at the same time. Modelling such situations is possible, but for this we need the multidimensional evolutionary model which we will describe in Chapter 10.

As an alternative one might, however, try to first filter out the "overall premium development" (which affects all the risks together over time) and then to apply the evolutionary model to the resulting "as if" data: Having filtered out the overall development, the assumption of independent, a priori equal risks is again realistic.

9.4 Evolutionary Models and Recursive Credibility

In this section we want to determine for which evolutionary models the credibility premiums are recursive (cf. Definition 9.1).

It is very useful, conceptually, to split up the path from P_n^{cred} to P_{n+1}^{cred} into two steps, namely:

a) Updating
Improve the estimation of $\mu(\Theta_n)$ on the basis of the newest information X_n,
b) Parameter movement
Change the estimation due to the switch of the parameter from $\mu(\Theta_n)$ to $\mu(\Theta_{n+1})$.

In this connection it is useful to introduce the usual terminology and notation from state space models (see e.g. Abraham and Ledolter [AL83]):

$$\begin{aligned}
\mu_{n|n-1} &:= \operatorname{Pro}(\mu(\Theta_n)|\, L(1, X_1, \dots, X_{n-1})),\\
\mu_{n|n} &:= \operatorname{Pro}(\mu(\Theta_n)|\, L(1, X_1, \dots, X_n)).
\end{aligned}$$

Notice that in this new notation

$$P_n^{\text{cred}} = \widehat{\widehat{\mu(\Theta_n)}} = \mu_{n|n-1},$$

and that the steps described above entail:

a) Updating: Step from $P_n^{\text{cred}} = \mu_{n|n-1}$ to $\mu_{n|n}$;
b) Parameter movement: Step from $\mu_{n|n}$ to $\mu_{n+1|n} = P_{n+1}^{\text{cred}}$.

In consideration of the recursiveness, we ask the following questions:

a) Under which conditions is the step from $\mu_{n|n-1}$ to $\mu_{n|n}$ recursive?
b) Under which conditions do we have recursiveness in the step from $\mu_{n|n}$ to $\mu_{n+1|n}$?

The following result gives the answer to question a).

Theorem 9.3. *The updating of* $\mu_{n|n-1}$ *to* $\mu_{n|n}$ *is recursive, i.e.*

$$\mu_{n|n} = \alpha_n X_n + (1-\alpha_n)\,\mu_{n|n-1}, \tag{9.11}$$

if and only if

$$E\left[\operatorname{Cov}\left(X_k, X_j \middle| \boldsymbol{\Theta}\right)\right] = 0 \quad \textit{for all } k \neq j. \tag{9.12}$$

Remarks:

- Formula (9.12) means that X_k and X_j *are on the average conditionally uncorrelated.* In particular, (9.12) is fulfilled, if the components of the observation vector are conditionally independent (standard assumption).

Proof of Theorem 9.3:

i) "⇒": From (9.11) it follows that (9.12) holds.
Let

$$\mu_{n|n} = a_{n0} + \sum_{j=1}^{n} a_{nj} X_j,$$

$$\mu_{n|n-1} = b_{n0} + \sum_{j=1}^{n-1} b_{nj} X_j.$$

From (9.11) we get

$$a_{nj} = (1-\alpha_n)\,b_{nj} \qquad j = 0,1,\dots,n-1, \tag{9.13}$$

$$a_{nn} = \alpha_n. \tag{9.14}$$

The normal equations (Corollary 3.17) for $\mu_{n|n}$ and $\mu_{n|n-1}$ are

$$\operatorname{Cov}\left(\mu\left(\Theta_n\right), X_k\right) = \sum_{j=1}^{n} a_{nj} \operatorname{Cov}\left(X_j, X_k\right) \quad \text{for } k \leq n, \tag{9.15}$$

$$\operatorname{Cov}\left(\mu\left(\Theta_n\right), X_k\right) = \sum_{j=1}^{n-1} b_{nj} \operatorname{Cov}\left(X_j, X_k\right) \quad \text{for } k \leq n-1. \tag{9.16}$$

Substituting the right-hand sides of (9.13) and (9.14) in (9.15) gives

$$\operatorname{Cov}\left(\mu\left(\Theta_n\right), X_k\right) = (1-\alpha_n) \sum_{j=1}^{n-1} b_{nj} \operatorname{Cov}\left(X_j, X_k\right) + \alpha_n \operatorname{Cov}\left(X_n, X_k\right).$$

For $k \leq n-1$ and applying (9.16) we get

$$\operatorname{Cov}\left(\mu\left(\Theta_n\right), X_k\right) = (1-\alpha_n) \operatorname{Cov}\left(\mu\left(\Theta_n\right), X_k\right) + \alpha_n \operatorname{Cov}\left(X_n, X_k\right),$$

and with that

$$\operatorname{Cov}\left(\mu\left(\Theta_n\right), X_k\right)=\operatorname{Cov}\left(X_n, X_k\right) \quad \text{for } k \leq n-1.$$

Using the fact that $E\left[X_k \mid \boldsymbol{\Theta}\right]=\mu\left(\Theta_k\right)$ and the well-known identity

$$\operatorname{Cov}\left(X_n, X_k\right)=\operatorname{Cov}\left(\mu\left(\Theta_n\right), \mu\left(\Theta_k\right)\right)+E\left[\operatorname{Cov}\left(X_n, X_k \mid \boldsymbol{\Theta}\right)\right],$$

we conclude that

$$E\left[\operatorname{Cov}\left(X_n, X_k \mid \boldsymbol{\Theta}\right)\right]=0. \tag{9.17}$$

Formula (9.17) must be true for any n and $k \leq n-1$, which is equivalent to (9.12), which is what we had to show.

ii) "$\Leftarrow$": From (9.12) it follows that (9.11) holds.
From (9.15) we get for $k \leq n$

$$\operatorname{Cov}\left(\mu\left(\Theta_n\right), X_k\right)=\sum_{j=1}^{n-1} a_{nj} \operatorname{Cov}\left(X_j, X_k\right)+a_{nn} \operatorname{Cov}(X_n, X_k),$$

and thus from (9.12) for $k \leq n-1$

$$\left(1-a_{nn}\right) \operatorname{Cov}\left(\mu\left(\Theta_n\right), X_k\right)=\sum_{j=1}^{n-1} a_{nj} \operatorname{Cov}\left(X_j, X_k\right).$$

From the uniqueness of the solution for the system of equations (9.16) we then have

$$a_{nj}=\left(1-a_{nn}\right) b_{nj} \quad \text{for } j=1,2, \ldots, n-1. \tag{9.18}$$

It remains to show that (9.18) is also true for $j=0$.

Because of the collective unbiasedness of the estimators $\mu_{n|n}$ and $\mu_{n|n-1}$ we have

$$E\left[a_{n0}+\sum_{j=1}^{n} a_{nj} X_j\right]=E\left[\mu\left(\Theta_n\right)\right],$$

$$E\left[a_{n0}+\sum_{j=1}^{n-1} a_{nj} X_j\right]=\left(1-a_{nn}\right) E\left[\mu\left(\Theta_n\right)\right],$$

$$E\left[b_{n0}+\sum_{j=1}^{n-1} b_{nj} X_j\right]=E\left[\mu\left(\Theta_n\right)\right].$$

From (9.18) it follows from the above system of equations that $a_{n0}=\left(1-a_{nn}\right) b_{n0}$.

This completes the proof of Theorem 9.3. □

We turn now to the parameter movement of $\mu(\Theta_n)$ to $\mu(\Theta_{n+1})$, or equivalently the step b) of $\mu_{n|n}$ to $\mu_{n+1|n}$. In order to get recursiveness also for the parameter movement, the following assumption is very natural:

$$\operatorname{Pro}(\mu(\Theta_{n+1})|\, L(1, \mu(\Theta_1), \ldots, \mu(\Theta_n))) = \mu(\Theta_n). \tag{9.19}$$

This means that the best linear estimator of $\mu(\Theta_{n+1})$, given all earlier "states" $\mu(\Theta_1), \ldots, \mu(\Theta_n)$, is equal to the last known state $\mu(\Theta_n)$. Equation (9.19) is equivalent to saying that the process $\{\mu(\Theta_1), \mu(\Theta_2), \ldots\}$ has orthogonal increments, i.e. that

$$\operatorname{Pro}(\mu(\Theta_{n+1}) - \mu(\Theta_n)\, | L(1, \mu(\Theta_1), \ldots, \mu(\Theta_n))) = 0 \quad \text{for } n \geq 1. \tag{9.20}$$

This implies that the increments are uncorrelated and have expectation zero, so that

$$\operatorname{Cov}(\mu(\Theta_{j+1}) - \mu(\Theta_j), \mu(\Theta_{k+1}) - \mu(\Theta_k)) = 0 \quad \text{for all } j \neq k, \tag{9.21}$$

$$E[\mu(\Theta_{j+1}) - \mu(\Theta_j)] = 0 \quad \text{for all } j. \tag{9.22}$$

Note that (9.20) holds if $\{\mu(\Theta_1), \mu(\Theta_2), \ldots\}$ is a *martingale* sequence.

Theorem 9.4. *If the X_j are, on average, conditionally uncorrelated and if the process $\{\mu(\Theta_1), \mu(\Theta_2), \ldots\}$ has orthogonal increments, then we have that*

$$\mu_{n+1|n} = \mu_{n|n}.$$

Proof: Because of the linearity property of projections we have that

$$\begin{aligned}\mu_{n+1|n} &= \operatorname{Pro}(\mu(\Theta_n)|\, L(1, X_1, \ldots, X_n)) \\ &\quad + \operatorname{Pro}(\mu(\Theta_{n+1}) - \mu(\Theta_n)|\, L(1, X_1, \ldots, X_n)) \\ &= \mu_{n|n} + \operatorname{Pro}(\mu(\Theta_{n+1}) - \mu(\Theta_n)|\, L(1, X_1, \ldots, X_n)).\end{aligned}$$

From (9.20) we have that

$$E[(\mu(\Theta_{n+1}) - \mu(\Theta_n)) \cdot 1] = 0.$$

It thus remains to show that

$$\mu(\Theta_{n+1}) - \mu(\Theta_n) \perp L(X_1, \ldots, X_n). \tag{9.23}$$

If we condition on $\boldsymbol{\Theta}$, we get

$$\begin{aligned}E[(\mu(\Theta_{n+1}) - \mu(\Theta_n))\, X_k] &= E[(\mu(\Theta_{n+1}) - \mu(\Theta_n))\, \mu(\Theta_k)] \\ &= 0 \quad \text{for } k = 1, 2, \ldots, n,\end{aligned}$$

where the second equality follows from (9.20). With this (9.23) is shown, which completes the proof of Theorem 9.4. □

We get an important extension to our collection of models when we allow linear transformations on the right-hand side of (9.19).

Corollary 9.5. *If instead of (9.19) the condition*

$$\operatorname{Pro}\left(\mu\left(\Theta_{n+1}\right) \mid L\left(1, \mu\left(\Theta_1\right) \ldots, \mu\left(\Theta_n\right)\right)\right)=a_n \mu\left(\Theta_n\right)+b_n \tag{9.24}$$

is satisfied, we then have

$$\mu_{n+1 \mid n}=a_n \mu_{n \mid n}+b_n . \tag{9.25}$$

Remarks:

- (9.24) means that the process $\{\mu(\Theta_j) : j=1,2,\ldots\}$ satisfies the linear Markov property.
- (9.24) is the most general assumption, under which we can hope to have recursiveness for updating.
- The condition (9.24) implies that the elements of the sequence
$$\left\{\mu\left(\Theta_{n+1}\right)-a_n \mu\left(\Theta_n\right)-b_n: n=1,2, \ldots\right\}$$
are uncorrelated and have expected value zero. Things like inflation measured according to some given index can therefore be taken into account by means of a linear transformation (change of scale and shift of origin).
- In particular, (9.24) is satisfied when the process $\{\mu(\Theta_j) : j=1,2,\ldots\}$ is an autoregressive process of order 1 with centered i.i.d. innovations $\Delta_2, \Delta_3, \ldots$. Then we have
$$\mu\left(\Theta_{n+1}\right)=\rho \mu\left(\Theta_n\right)+(1-\rho) \mu+\Delta_{n+1} \quad \text{with } |\rho|<1,$$
and thus $a_n=\rho$ and $b_n=(1-\rho)\mu$.

Proof: The corollary follows directly from Theorem 9.4 and the linearity property of projections. □

9.5 Recursive Calculation Method (Kalman Filter)

In accordance with the general Kalman Filter method, we name the recursive calculation algorithm described in Section 9.4, which is split into the two steps "updating" and "parameter movement", the *Kalman Filter procedure.* The original application of the Kalman Filter can be found in the engineering literature. The usual formulation leads to a more general formula (see Chapter 10) than the one found in this section. But the essential component of this general form is also the splitting of the calculation into the two steps of updating and parameter movement.

Analogously to the notation in Section 9.4, we introduce the following quantities for the expected quadratic loss of the corresponding estimators:

$$\begin{aligned}
q_{n \mid n-1} & :=E\left[\left(\mu_{n \mid n-1}-\mu\left(\Theta_n\right)\right)^2\right], \\
q_{n \mid n} & :=E\left[\left(\mu_{n \mid n}-\mu\left(\Theta_n\right)\right)^2\right].
\end{aligned}$$

It is also useful at this point to introduce the following notation for the structural variance paramcters, namely

$$\begin{aligned}\sigma_j^2 &:= E\left[\sigma^2(\Theta_j)\right], \\ \tau_j^2 &:= \operatorname{Var}\left[\mu(\Theta_j)\right].\end{aligned}$$

It is convenient, to set without loss of generality

$$\sigma_j^2 = \frac{\sigma^2}{w_j},$$

where w_j is an appropriate weight.

Theorem 9.6 (recursion formula in Kalman Filter notation). *Under the assumption that the X_j are, on the average, conditionally uncorrelated, the following recursion formulae hold:*

i) Anchoring (Period 1)

$$\mu_{1|0} = E\left[\mu(\Theta_1)\right] \quad \textit{and} \quad q_{1|0} = \tau_1^2. \tag{9.27}$$

ii) Recursion ($n \geq 1$)

a) Updating

$$\mu_{n|n} = \alpha_n X_n + (1-\alpha_n)\,\mu_{n|n-1}, \tag{9.28}$$

$$\textit{where} \quad \alpha_n = \frac{q_{n|n-1}}{q_{n|n-1} + \frac{\sigma^2}{w_n}} = \frac{w_n}{w_n + \frac{\sigma^2}{q_{n|n-1}}}, \tag{9.29}$$

$$q_{n|n} = (1-\alpha_n)\, q_{n|n-1}. \tag{9.30}$$

b) Change from $\mu(\Theta_n)$ to $\mu(\Theta_{n+1})$ (parameter movement)

Case I: The process $\{\mu(\Theta_j) : j = 1, 2, \ldots\}$ has orthogonal increments, i.e. $\operatorname{Pro}\left(\mu(\Theta_{n+1}) \middle| L(1, \mu(\Theta_1) \ldots, \mu(\Theta_n)\right) = \mu(\Theta_n)$.
Then it holds that

$$\mu_{n+1|n} = \mu_{n|n}, \tag{9.31}$$

$$q_{n+1|n} = q_{n|n} + \delta_{n+1}^2, \tag{9.32}$$

$$\textit{where } \delta_{n+1}^2 = \operatorname{Var}[\mu(\Theta_{n+1}) - \mu(\Theta_n)].$$

Case II: Linear transformations are also allowed, i.e. $\operatorname{Pro}\left(\mu(\Theta_{n+1}) \middle| L(1, \mu(\Theta_1) \ldots, \mu(\Theta_n)\right) = a_n \mu(\Theta_n) + b_n$.
Then it holds that

$$\mu_{n+1|n} = a_n \mu_{n|n} + b_n \quad \textit{for } n = 1, 2, \ldots, \tag{9.33}$$

$$q_{n+1|n} = a_n^2 \cdot q_{n|n} + \delta_{n+1}^2, \tag{9.34}$$

$$\textit{where} \quad \delta_{n+1}^2 = \operatorname{Var}\left[\mu(\Theta_{n+1}) - a_n \mu(\Theta_n)\right].$$

Remarks:

- Notice that, in the updating step, the quadratic loss always gets smaller. Notice also, that in Case I, in the second step, the quadratic loss is increased because of the change in the risk profile, but that the credibility estimator stays the same. This is a natural consequence of the assumption that the $\mu(\Theta_n)$ have orthogonal increments.
- Notice that, in the updating step, the estimator (9.28) and its associated quadratic loss (9.30) have the same form as for the credibility estimator, respectively, the quadratic loss in the Bühlmann–Straub model. The term $\mu_{n|n-1} = P_n^{\text{cred}}$ plays the role of the collective average and $q_{n|n-1}$ plays the role of the mean squared error of the individual premium to be estimated. In a fully parameterized model, in which $\mu_{n|n-1} = P_n^{\text{cred}}$ is equal to the conditional expectation $E[\mu(\Theta_n)|X_1, \ldots, X_{n-1}]$ (*exact credibility* case), the quantity $q_{n|n-1}$ corresponds to the conditional variance of $\mu(\Theta_n)$, given the observations $X_1, \ldots, X_{n-1}$.
- (9.28) is a weighted average of $\mu_{n|n-1}$ and X_n, where the weights are again proportional to the precisions (= reciprocal values of the mean squared error) of the single components. The "general intuitive principle" (Principle 3.3) also holds here.

Proof of Theorem 9.6: According to Theorem 9.3, $\mu_{n|n}$ is of the form (9.28). We have yet to determine the value of α_n.

For the quadratic loss of $\mu_{n|n}$ we get

$$\begin{aligned} q_{n|n} &= E\left[\left(\mu(\Theta_n) - \alpha_n X_n - (1-\alpha_n)\,\mu_{n|n-1}\right)^2\right] \\ &= \alpha_n^2 E\left[\left(\mu(\Theta_n) - X_n\right)^2\right] + (1-\alpha_n)^2 E\left[\left(\mu(\Theta_n) - \mu_{n|n-1}\right)^2\right] \\ &= \alpha_n^2 \frac{\sigma^2}{w_n} + (1-\alpha_n)^2 q_{n|n-1}. \end{aligned} \tag{9.35}$$

To get the second equality, we use the fact that the X_j's are on average conditionally uncorrelated, which leads to the disappearance of the cross-term.

Setting the derivative of (9.35) with respect to α_n to zero (minimizing the quadratic loss) we get immediately that the optimal choice of α_n satisfies equation (9.29). Substituting α_n in (9.35), we get directly that

$$q_{n|n} = (1-\alpha_n)\, q_{n|n-1}.$$

Equation (9.31) follows directly from Theorem 9.4.

For the quadratic loss of $\mu_{n+1|n}$ we get in Case I

$$
\begin{aligned}
q_{n+1|n} &= E\left[\left(\mu_{n|n} - \mu(\Theta_n) + \mu(\Theta_n) - \mu(\Theta_{n+1})\right)^2\right] \\
&= E\left[\left(\mu_{n|n} - \mu(\Theta_n)\right)^2\right] + E\left[\left(\mu(\Theta_n) - \mu(\Theta_{n+1})\right)^2\right] \\
&= (1-\alpha_n)\, q_{n|n-1} + \delta_{n+1}^2. \qquad (9.36)
\end{aligned}
$$

This is exactly equation (9.32).

In Case II the calculation is analogous and will therefore not be explained in more detail here. The proof of Theorem 9.6 is thus complete. □

In Theorem 9.6 we consciously kept the two steps "updating" and "movement in parameter space" separate from one another. This is useful, firstly, for the purpose of interpretation and secondly, because in this way we can understand the extension in ii b) of Case I to Case II. As we have already stated, we call this two-step procedure *Kalman Filter.* The recursion formula can however be written more compactly by combining the two steps. We name this combined procedure the *recursive calculation* of the credibility premium. This is the subject of the following corollary that follows directly from Theorem 9.6 using the notation

$$
\begin{aligned}
P_n^{\text{cred}} &:= \mu_{n|n-1}, \\
q_n &:= q_{n|n-1}.
\end{aligned}
$$

Corollary 9.7 (recursive calculation of the credibility premium). *Under the same assumptions as in Theorem 9.6 we have:*

i) Anchoring (Period 1)

$$
P_1^{\text{cred}} = E\left[\mu(\Theta_1)\right] \quad \text{and} \quad q_1 = \tau_1^2. \qquad (9.37)
$$

ii) Recursion ($n \geq 1$)

Case I: $\{\mu(\Theta_j) : j = 1, 2, \ldots\}$ *has orthogonal increments. Then it holds that*

$$
P_{n+1}^{\text{cred}} = \alpha_n X_n + (1-\alpha_n) P_n^{\text{cred}}, \qquad (9.38)
$$

$$
\text{where } \alpha_n = \frac{q_n}{q_n + \frac{\sigma^2}{w_n}} = \frac{w_n}{w_n + \frac{\sigma^2}{q_n}};
$$

$$
q_{n+1} = (1-\alpha_n)\, q_n + \delta_{n+1}^2, \qquad (9.39)
$$

$$
\text{where } \delta_{n+1}^2 = \operatorname{Var}\left[\mu(\Theta_{n+1}) - \mu(\Theta_n)\right].
$$

Case II: $\operatorname{Pro}\left(\mu(\Theta_{n+1}) \middle| L(1, \mu(\Theta_1) \ldots, \mu(\Theta_n))\right) = a_n \mu(\Theta_n) + b_n$. *Then it holds that*

$$P_{n+1}^{\text{cred}} = a_n \left(\alpha_n X_n + (1-\alpha_n) P_n^{\text{cred}}\right) + b_n, \tag{9.40}$$
$$\textit{where } \alpha_n = \frac{q_n}{q_n + \frac{\sigma^2}{w_n}} = \frac{w_n}{w_n + \frac{\sigma^2}{q_n}};$$

$$q_{n+1} = a_n^2 \left(1-\alpha_n\right) q_n + \delta_{n+1}^2, \tag{9.41}$$
$$\textit{where } \delta_{n+1}^2 = \operatorname{Var}\left[\mu\left(\Theta_{n+1}\right) - a_n \mu\left(\Theta_n\right)\right].$$

Remarks:

- In Case I the credibility formula (9.38) is recursive (see Definition 9.1).
- When, additionally, linear transformations are allowed (Case II), then according to (9.40), the new credibility premium (in period $n+1$) continues to depend only on the newest observation X_n and the current credibility premium (in period n). However, it is no longer of the strict recursive form given in Definition 9.1.

Example 9.4 (Gerber and Jones model)

In this example treated in [GJ75b], it is assumed that $\mu(\Theta_n) = \mu(\Theta_1) + \Delta_2 + \ldots + \Delta_n$ for $n \geq 2$. Moreover, $\Delta_2, \Delta_3, \ldots$ are independent with $E[\Delta_j] = 0$ and $\operatorname{Var}[\Delta_j] = \delta^2$. Further, the random variables X_j are conditionally independent, given $\boldsymbol{\Theta}$, with $\operatorname{Var}[X_j|\boldsymbol{\Theta}] = \sigma^2$.

From Corollary 9.7 we get

$$P_1^{\text{cred}} = \mu_0,$$
$$q_1 = \tau_1^2,$$

$$P_{n+1}^{\text{cred}} = \alpha_n X_n + (1-\alpha_n) P_n^{\text{cred}},$$
$$\text{where } \alpha_n = \frac{1}{1 + \frac{\sigma^2}{q_n}},$$
$$q_{n+1} = (1-\alpha_n) q_n + \delta^2 = \frac{\sigma^2 q_n}{\sigma^2 + q_n} + \delta^2.$$

One can show that the last equation defines a contraction with limit

$$q_n \overset{n\to\infty}{\longrightarrow} q = \frac{\delta^2 \left(1 + \sqrt{1 + 4\frac{\sigma^2}{\delta^2}}\right)}{2}. \tag{9.42}$$

If $\tau_1^2 = \operatorname{Var}[\mu(\Theta_1)]$ is equal to the right-hand side of (9.42), then this is true, not only asymptotically, but for every n. Notice that in this case all the weights α_n are the same, i.e.

$$\alpha_n = \alpha = \frac{1}{1 + \frac{\sigma^2}{q}}.$$

Thus we have constructed a model which gives a geometric premium formula as postulated in Example 9.3. The assumption that

$$\tau_1^2 = \frac{\delta^2 \left(1 + \sqrt{1 + 4\frac{\sigma^2}{\delta^2}}\right)}{2},$$

looks very artificial at first. However, when we conceive that the underlying process has been running for a long time before the time-point zero, it makes sense to take this value from the stationary distribution.

General remarks on Example 9.4:

- The model of Gerber and Jones may be appropriate when our interest is in a single risk, the premium of which we need to calculate.
- However, if we want to apply this model to a collective of independent and "similar" risks (see also the two last points in the remarks on the evolutionary credibility model on page 225), this model may be inappropriate for many cases in practice. The reason for this is that, in the above model

$$\mathrm{Var}\left[\mu\left(\Theta_n\right)\right] = \tau_1^2 + (n-1)\,\delta^2. \tag{9.43}$$

  The variance of $\mu(\Theta_n)$ therefore grows linearly with time. If we consider a whole portfolio of contracts, independent of each other, which can be described by this model, then (9.43) means that we are working with a portfolio, whose heterogeneity increases constantly with time, that is to say, a portfolio that somehow is spreading out. However, this feature does not correspond to what is usually observed in practice. This critique applies not only to the Gerber and Jones model, but to all models which satisfy the requirements of Theorem 9.6 for Case I. This is because for all such models, the orthogonal increments lead to the quantity $\mathrm{Var}\left[\mu\left(\Theta_j\right)\right]$ growing with increasing j.
- As the next example shows, we can avoid this growing heterogeneity by allowing for linear transformations, which leads to models corresponding to Case II of Theorem 9.6. In Section 10.4 we shall discuss a further approach to dealing with this problem, namely that of multidimensional evolutionary credibility (see also the discussion in the last point of the remarks on page 225).

Example 9.5 (autoregressive process of order 1)

Let the process $\{\mu\left(\Theta_j\right) : j = 1, 2, \ldots\}$ be an autoregressive process of order 1, i.e.

$$\mu\left(\Theta_{n+1}\right) = \rho\mu\left(\Theta_n\right) + (1-\rho)\,\mu + \Delta_{n+1}, \tag{9.44}$$

where $E\left[\mu(\Theta_1)\right] = \mu$ and the Δ_j are again uncorrelated with $E\left[\Delta_j\right] = 0$ and $\mathrm{Var}\left[\Delta_j\right] = \delta^2$. We also assume that $|\rho| < 1$, so that $\mathrm{Var}\left[\mu(\Theta_j)\right]$ is bounded.

Further, we make the usual assumption that the X_j, conditionally, given Θ, are uncorrelated with $\operatorname{Var}[X_j|\Theta] = \sigma^2$.

As a consequence of Corollary 9.7 (Case II) we get the following recursion formula for the credibility estimator and its quadratic loss:

$$\begin{aligned} P_1^{\text{cred}} &= \mu, \\ q_1 &= \tau_1^2, \end{aligned}$$

$$\begin{aligned} P_{n+1}^{\text{cred}} &= \rho\left(\beta_n X_n + (1-\beta_n)\, P_n^{\text{cred}}\right) + (1-\rho)\,\mu, \\ &\qquad \text{where } \beta_n = \frac{1}{1+\frac{\sigma^2}{q_n}}, \\ q_{n+1} &= \rho^2\,(1-\beta_n)\, q_n + \delta^2. \end{aligned}$$

Remarks:

- For the process $\{\mu(\Theta_j) : j = 1, 2, \ldots\}$ we have

$$\begin{aligned} E\left[\mu(\Theta_j)\right] &= \mu \quad \text{for } j \geq 1, \\ \tau_n^2 = \operatorname{Var}\left[\mu(\Theta_n)\right] &\overset{n\to\infty}{\longrightarrow} \tau^2 := \frac{\delta^2}{1-\rho^2}. \end{aligned} \tag{9.45}$$

  If we further assume that

$$\operatorname{Var}\left[\mu(\Theta_1)\right] = \tau^2 = \frac{\delta^2}{1-\rho^2}, \tag{9.46}$$

  then the process $\{\mu(\Theta_j) : j = 1, 2, \ldots\}$ is stationary with

$$\begin{aligned} \operatorname{Var}\left[\mu(\Theta_n)\right] &= \tau^2, \\ \operatorname{Cov}\left[\mu(\Theta_{n+k}), \mu(\Theta_n)\right] &= \rho^k\, \tau^2. \end{aligned}$$

- One can show that – irrespective of the behaviour of τ_n^2 – asymptotically

$$q_n \overset{n\to\infty}{\longrightarrow} q = \frac{\delta^2 - (1-\rho^2)\,\sigma^2 + \sqrt{\left(\delta^2 - (1-\rho^2)\,\sigma^2\right)^2 + 4\sigma^2\delta^2}}{2}. \tag{9.47}$$

  In the case where $E\left[\left(P_1^{\text{cred}} - \mu(\Theta_1)\right)^2\right]$ is already the same as the right-hand side of (9.47), this is true, not only asymptotically, but for all n, and for the weights α_n, it then holds that

$$\alpha_n = \alpha = \frac{q}{q+\sigma^2}.$$

- The conditions which lead to constant weights $\alpha_n = \alpha$, and to a constant quadratic loss $q_n = q$ for the credibility estimator, look very artificial at first glance. However, with the notion of the credibility estimator as a yearly "rating revision" and a conceptually very long lasting process, the recursion formula

$$P_{n+1}^{\text{cred}} = \rho\left\{\alpha X_n + (1-\alpha)\, P_n^{\text{cred}}\right\} + (1-\rho)\mu \quad for\ n = 1, 2, \ldots \tag{9.48}$$

makes sense. We assume that our window of observation covers a time period in which we are close enough to the asymptotic state In other words the time at which our observation period begins is not the same as that at which the AR(1)-process (9.44) begins, and the process of the credibility estimator P_n^{cred} as well as that of the $\mu(\Theta_n)$ have already stabilized, so that the "starting conditions" of the original processes no longer play a role. The recursion (9.48) starts then with

$$P_1^{\text{cred}} = \text{"premium in period 1"},$$

for which we have

$$E\left[\left(P_1^{\text{cred}} - \mu(\Theta_1)\right)^2\right] = q. \tag{9.49}$$

For the anchoring we must be careful not to use $P_1^{\text{cred}} = \mu = E[\mu(\Theta_1)]$ (collective mean), but rather the "premium in period 1" (which is assumed to reflect the claims history before period one and to satisfy equation (9.49)). The recursion formula (9.48) is then given by

$$P_{n+1}^{\text{cred}} = \rho\left\{\frac{q}{\sigma^2+q} X_n + \frac{\sigma^2}{\sigma^2+q} P_n^{\text{cred}}\right\} + (1-\rho)\mu, \tag{9.50}$$

$$\text{where} \quad q = E\left[\left(P_n^{\text{Cred}} - \mu(\Theta_n)\right)^2\right] \quad \text{for } n \geq 1,$$

$$\sigma^2 = E\left[\left(X_n - \mu(\Theta_n)\right)^2\right].$$

- For a portfolio of independent contracts obeying this model, $\text{Var}[\mu(\Theta_j)]$ is a measure of the heterogeneity of the portfolio at time j. In contrast to Example 9.4, the portfolio does not spread out over time. Instead, in this example we are modelling a portfolio with constant heterogeneity, a property which seems reasonable to assume for many cases in practice.
- If this model is to be used for a portfolio of independent, but similar, contracts, one should note that the model describes a situation where, over the passage of time, the correct individual premium of each contract randomly varies around the *same* expected value. A more realistic model would be one where the long-term average varied from one contract to another and is drawn at random for each contract i. We will describe such a model in the next section.

9.6 The Evolutionary Credibility Regression Model

As in the previous sections, we consider a particular risk over time, and we let $X_1, X_2, \ldots, X_n, \ldots$ be the observed claim data (e.g. aggregate claim amount, claim ratio, claim frequency, etc.) of this risk in the years $j = 1, 2, \ldots, n, \ldots$. Again, the goal is to calculate the pure risk premium for a future year.

It is often useful to relate the pure risk premium to a number of explanatory parameters. Indeed, in most branches the premium depends on observed rating criteria (e.g. in motor insurance these might be age, sex, region, kilometres driven, engine size).

In the credibility regression model (Chapter 8) we saw how we could use a credibility regression model to combine a priori criteria and credibility. However, in the regression model, the regression parameter vector $\boldsymbol{\beta}(\Theta)$ remains constant over time. We now extend the ideas of the evolutionary credibility model from Sections 9.4 and 9.5 to the credibility regression case.

In the credibility regression model in Chapter 8, we have for the risk premium in year j the regression equation

$$E[X_j|\Theta] = \mathbf{y}_j' \boldsymbol{\beta}(\Theta), \tag{9.51}$$

where $\mathbf{y}_j'$ contains the values of the covariables in year j and $\boldsymbol{\beta}(\Theta)$ is the regression parameter vector, the value of which depends on the risk profile Θ. In the notation of Chapter 8 (see formula (8.7)), $\mathbf{y}_j'$ corresponds to the jth row vector of the design matrix Y. The risk profile Θ and thus also $\boldsymbol{\beta}(\Theta)$ stay unchanged over time.

In contrast, in the evolutionary case, we assume that the risk profile can change over time. The notation and the assumptions (9.7)–(9.9) can be taken directly from Section 9.3, i.e.

$$\begin{aligned}
\boldsymbol{\Theta} &= \{\Theta_1, \Theta_2, \ldots\}, \\
E[X_j|\boldsymbol{\Theta}] = E[X_j|\Theta_j] &= \mu(\Theta_j), \\
\operatorname{Var}[X_j|\boldsymbol{\Theta}] = \operatorname{Var}[X_j|\Theta_j] &= \frac{\sigma^2(\Theta_j)}{w_j},
\end{aligned}$$

where the w_j are appropriate weights.

In addition, the regression equation

$$E[X_j|\boldsymbol{\Theta}] = \mu(\Theta_j) = \mathbf{y}_j'\boldsymbol{\beta}(\Theta_j) \tag{9.52}$$

is also assumed to hold. Note the difference from the regression model of Chapter 8 (difference between (9.51) and (9.52)): instead of a parameter vector $\boldsymbol{\beta}(\Theta)$ which does not change with time, we have the sequence

$$\boldsymbol{\beta}(\boldsymbol{\Theta}) = \{\boldsymbol{\beta}(\Theta_1), \boldsymbol{\beta}(\Theta_2), \ldots\}.$$

Using the terminology of state-space models, we call $\boldsymbol{\beta}(\Theta_j)$ the state vector in the period j.

Analogously to Section 9.4 we introduce the following notation:

$$\begin{aligned}
\boldsymbol{\beta}_{n|n-1} &:= \operatorname{Pro}\left(\boldsymbol{\beta}\left(\Theta_n\right)|L\left(1, X_1, \ldots, X_{n-1}\right)\right), \\
\boldsymbol{\beta}_{n|n} &:= \operatorname{Pro}\left(\boldsymbol{\beta}\left(\Theta_n\right)|L\left(1, X_1, \ldots, X_n\right)\right), \\
Q_{n|n-1} &:= E\left[\left(\boldsymbol{\beta}\left(\Theta_n\right)-\boldsymbol{\beta}_{n|n-1}\right)\left(\boldsymbol{\beta}\left(\Theta_n\right)-\boldsymbol{\beta}_{n|n-1}\right)^{\prime}\right], \\
Q_{n|n} &:= E\left[\left(\boldsymbol{\beta}\left(\Theta_n\right)-\boldsymbol{\beta}_{n|n}\right)\left(\boldsymbol{\beta}\left(\Theta_n\right)-\boldsymbol{\beta}_{n|n}\right)^{\prime}\right].
\end{aligned}$$

Notice that

$$\boldsymbol{\beta}_{n|n-1}=\widehat{\widehat{\boldsymbol{\beta}\left(\Theta_n\right)}}=\text{credibility estimator of } \boldsymbol{\beta}\left(\Theta_n\right),$$

and that $Q_{n|n-1}$, respectively $Q_{n|n}$ are the quadratic loss matrices of the vector $\boldsymbol{\beta}_{n|n-1}$, respectively of the vector $\boldsymbol{\beta}_{n|n}$.

We will now show that the results from Sections 9.4 and 9.5 can be carried over to the regression case.

9.7 Recursive Calculation in the Evolutionary Regression Model

As in Theorem 9.6, we assume that the X_j's are, on average, uncorrelated, i.e.

$$E\left[\operatorname{Cov}\left(X_j, X_k|\boldsymbol{\Theta}\right)\right]=0 \quad \text{for } j \neq k. \tag{9.53}$$

In most cases in practice, the condition (9.53) is a reasonable assumption, because in most cases one can assume that, conditional on the risk profile, the observations in the different years are independent, i.e. that $\operatorname{Cov}\left(X_j, X_k|\boldsymbol{\Theta}\right)=0$ for $j \neq k$.

In Section 9.4 we then imposed conditions on the process $\{\mu\left(\Theta_j\right): j=1,2,\ldots\}$. We saw that the credibility premium was recursive, as defined in Definition 9.1, if this process had orthogonal increments, which is equivalent to saying that $\operatorname{Pro}\left(\mu\left(\Theta_{n+1}\right)|\, L\left(1, \mu\left(\Theta_1\right), \ldots, \mu\left(\Theta_n\right)\right)\right)=\mu\left(\Theta_n\right)$. Then we weakened this condition somewhat, in that we allowed, in addition, for linear transformations, i.e. we assumed that $\operatorname{Pro}\left(\mu\left(\Theta_{n+1}\right)|\, L\left(1, \mu\left(\Theta_1\right), \ldots, \mu\left(\Theta_n\right)\right)\right) = a_n \mu\left(\Theta_n\right)+b_n$ for known constants a_n and b_n. Here, we extend these conditions to the sequence $\{\boldsymbol{\beta}\left(\Theta_j\right): j=1,2,\ldots\}$ in an appropriate manner. The corresponding assumptions are

$$\operatorname{Pro}\left(\boldsymbol{\beta}\left(\Theta_{n+1}\right)|L(1, \boldsymbol{\beta}\left(\Theta_1\right), \boldsymbol{\beta}\left(\Theta_2\right), \ldots, \boldsymbol{\beta}\left(\Theta_n\right))\right)=\boldsymbol{\beta}\left(\Theta_n\right), \tag{9.54}$$

for Case I, respectively

$$\operatorname{Pro}\left(\boldsymbol{\beta}\left(\Theta_{n+1}\right)|L(1, \boldsymbol{\beta}\left(\Theta_1\right), \boldsymbol{\beta}\left(\Theta_2\right), \ldots, \boldsymbol{\beta}\left(\Theta_n\right))\right)=A_n \boldsymbol{\beta}\left(\Theta_n\right)+\mathbf{b}_n, \tag{9.55}$$

for Case II.

Note that (9.54) is equivalent to saying that the process $\{\boldsymbol{\beta}(\Theta_j) : j = 1, 2, \ldots\}$ has orthogonal increments, i.e. that

$$\boldsymbol{\beta}(\Theta_{n+1}) - \boldsymbol{\beta}(\Theta_n) \perp L(1, \boldsymbol{\beta}(\Theta_1), \ldots, \boldsymbol{\beta}(\Theta_n)), \tag{9.56}$$

or that

$$\begin{aligned} E[\boldsymbol{\beta}(\Theta_{n+1}) - \boldsymbol{\beta}(\Theta_n)] &= \mathbf{0} \quad \text{and} \\ E\left[(\boldsymbol{\beta}(\Theta_{n+1}) - \boldsymbol{\beta}(\Theta_n)) \cdot \boldsymbol{\beta}(\Theta_k)'\right] &= \mathbf{0} \quad \text{for } k = 1, 2, \ldots, n, \end{aligned}$$

where on the right-hand side, by $\mathbf{0}$ we mean a vector or matrix full of zeroes.

As defined in Section 9.5 we write without loss of generality

$$\sigma_j^2 := E[\mathrm{Var}[X_j | \Theta_j]] = \frac{\sigma^2}{w_j},$$

where the w_j's are appropriate weights.

Theorem 9.8 *Under the assumptions (9.53)–(9.55) we have the following recursion formulae (Kalman Filter form):*

i) Anchoring (period 1)

$$\begin{aligned} \boldsymbol{\beta}_{1|0} &= E[\boldsymbol{\beta}(\Theta_1)], \\ Q_{1|0} &= E\left[\left(\boldsymbol{\beta}(\Theta_1) - \boldsymbol{\beta}_{1|0}\right) \cdot \left(\boldsymbol{\beta}(\Theta_1) - \boldsymbol{\beta}_{1|0}\right)'\right]. \end{aligned}$$

ii) Recursion ($n \geq 1$)

a) Updating

$$\boldsymbol{\beta}_{n|n} = \boldsymbol{\beta}_{n|n-1} + \boldsymbol{\gamma}_n \left(X_n - \mathbf{y}_n' \boldsymbol{\beta}_{n|n-1}\right), \tag{9.57}$$

$$\text{where } \boldsymbol{\gamma}_n = \frac{Q_{n|n-1} \mathbf{y}_n}{\mathbf{y}_n' Q_{n|n-1} \mathbf{y}_n + \frac{\sigma^2}{w_n}}, \tag{9.58}$$

$$Q_{n|n} = (I - \boldsymbol{\gamma}_n \mathbf{y}_n') Q_{n|n-1}. \tag{9.59}$$

b) Change from Θ_n to Θ_{n+1} (movement in parameter space)
Case I:

$$\boldsymbol{\beta}_{n+1|n} = \boldsymbol{\beta}_{n|n}, \tag{9.60}$$

$$Q_{n+1|n} = Q_{n|n} + D_{n+1}, \tag{9.61}$$

$$\text{where } D_{n+1} = \mathrm{E}\big[(\boldsymbol{\beta}(\Theta_{n+1}) - \boldsymbol{\beta}(\Theta_n)) \cdot (\boldsymbol{\beta}(\Theta_{n+1}) - \boldsymbol{\beta}(\Theta_n))' \big].$$

Case II:

$$\boldsymbol{\beta}_{n+1|n} = A_n \boldsymbol{\beta}_{n|n} + \mathbf{b}_n, \tag{9.62}$$

$$Q_{n+1|n} = A_n Q_{n|n} A_n' + D_{n+1}, \tag{9.63}$$

$$\text{where } D_{n+1} = \mathrm{E}\big[\left(\boldsymbol{\beta}(\Theta_{n+1}) - A_n \boldsymbol{\beta}(\Theta_n) - \mathbf{b}_n\right) \cdot \left(\boldsymbol{\beta}(\Theta_{n+1}) - A_n \boldsymbol{\beta}(\Theta_n) - \mathbf{b}_n\right)' \big].$$

Analogously to Section 9.5, the recursive calculation of the credibility premium and its quadratic loss can be written more compactly by combining the two steps "updating" and "movement in parameter space", which have been deliberately separated in the statement of the Theorem 9.8. This leads to the following corollary which is an immediate consequence of Theorem 9.8 and where we use the identities

$$\widehat{\widehat{\boldsymbol{\beta}(\Theta_{n+1})}} := \boldsymbol{\beta}_{n+1|n},$$
$$Q_n := Q_{n|n-1}.$$

Corollary 9.9 (recursive calculation of the credibility estimator of the regression parameter). *Under the same conditions as in Theorem 9.8, we can calculate the credibility estimator $\widehat{\widehat{\boldsymbol{\beta}(\Theta_n)}}$ and its quadratic loss recursively as follows:*

i) Anchoring (period 1)

$$\widehat{\widehat{\boldsymbol{\beta}(\Theta_1)}} = E\left[\boldsymbol{\beta}(\Theta_1)\right],$$

$$Q_1 = E\left[\left(\boldsymbol{\beta}(\Theta_1) - \widehat{\widehat{\boldsymbol{\beta}(\Theta_1)}}\right)\left(\boldsymbol{\beta}(\Theta_1) - \widehat{\widehat{\boldsymbol{\beta}(\Theta_1)}}\right)'\right].$$

ii) Recursion ($n \geq 1$)
Case I:

$$\widehat{\widehat{\boldsymbol{\beta}(\Theta_{n+1})}} = \left\{\widehat{\widehat{\boldsymbol{\beta}(\Theta_n)}} + \boldsymbol{\gamma}_n \left(X_n - \mathbf{y}_n' \widehat{\widehat{\boldsymbol{\beta}(\Theta_n)}}\right)\right\}, \tag{9.64}$$

$$\text{where } \boldsymbol{\gamma}_n = \frac{Q_n \mathbf{y}_n}{\mathbf{y}_n' Q_n \mathbf{y}_n + \frac{\sigma^2}{w_n}}, \tag{9.65}$$

$$\begin{aligned} Q_{n+1} &= \left(I - \boldsymbol{\gamma}_n \mathbf{y}_n'\right) Q_n + D_{n+1} \\ &= Q_n \left(I - \mathbf{y}_n \boldsymbol{\gamma}_n'\right) + D_{n+1}, \end{aligned} \tag{9.66}$$

$$\text{where } D_{n+1} = E\big[\left(\boldsymbol{\beta}(\Theta_{n+1}) - \boldsymbol{\beta}(\Theta_n)\right) \cdot \left(\boldsymbol{\beta}(\Theta_{n+1}) - \boldsymbol{\beta}(\Theta_n)\right)' \big].$$

Case II:

$$\widehat{\widehat{\boldsymbol{\beta}(\Theta_{n+1})}} = A_n \left\{ \widehat{\widehat{\boldsymbol{\beta}(\Theta_n)}} + \boldsymbol{\gamma}_n \cdot \left(X_n - \mathbf{y}_n' \widehat{\widehat{\boldsymbol{\beta}(\Theta_n)}} \right) \right\} + \mathbf{b}_n, \tag{9.67}$$

$$\text{where } \boldsymbol{\gamma}_n = \frac{Q_n \mathbf{y}_n}{\mathbf{y}_n' Q_n \mathbf{y}_n + \frac{\sigma^2}{w_n}}, \tag{9.68}$$

$$\begin{aligned} Q_{n+1} &= A_n \left(I - \boldsymbol{\gamma}_n \mathbf{y}_n'\right) Q_n A_n' + D_{n+1} \\ &= A_n Q_n \left(I - \mathbf{y}_n \boldsymbol{\gamma}_n'\right) A_n' + D_{n+1}, \\ &\quad \text{where } D_{n+1} = E\big[\left(\boldsymbol{\beta}(\Theta_{n+1}) - A_n \boldsymbol{\beta}(\Theta_n) - \mathbf{b}_n\right) \\ &\qquad\qquad \cdot \left(\boldsymbol{\beta}(\Theta_{n+1}) - A_n \boldsymbol{\beta}(\Theta_n) - \mathbf{b}_n\right)' \big]. \end{aligned} \tag{9.69}$$

Remarks:

- Notice the analogy to Theorem 9.6 and Corollary 9.7.
- One can easily show that the credibility estimator $\widehat{\widehat{\mu(\Theta_n)}} := \mathbf{y}_n' \widehat{\widehat{\boldsymbol{\beta}(\Theta_n)}}$ is also the credibility estimator (predictor) $\widehat{\widehat{X_n}}$ of X_n. Notice that the second summand in (9.57), which represents the "correction" or adjustment of $\widehat{\widehat{\boldsymbol{\beta}(\Theta_n)}}$ resulting from the newly arrived observation X_n, is of the form

$$\boldsymbol{\gamma}_n \left(X_n - \widehat{\widehat{X_n}} \right). \tag{9.70}$$

  The "direction vector" $\boldsymbol{\gamma}_n$ gives the (shrinking) direction of the correction, and the deviation of X_n from the predicted value $\widehat{\widehat{X_n}}$ gives the size of the correction.
- The "non-evolutionary" credibility regression model (see Chapter 8, Model Assumptions 8.2) is included in Case I). Because in this case $\boldsymbol{\beta}(\Theta_n) = \boldsymbol{\beta}(\Theta)$ for all n (the risk profile doesn't change), the matrix D_{n+1} falls out of (9.66). As in the Bühlmann–Straub model, the credibility estimator of the standard regression model (cf. Model Assumptions 8.2) can also be written recursively. In contrast to the Bühlmann—Straub model, the transition from the non-recursive to the recursive formula is, however, not so immediate. It is important to realize that the recursive formula is more general, in that it allows one to also find the credibility estimator in cases where there is only one or very few observations and $(Y'Y)$ is not of full rank, where Y is the design matrix as defined in (8.7).
- By an appropriate choice of the components of $\boldsymbol{\beta}(\Theta_n)$ (state space), as well as application of linear transformations (movement equations), the evolutionary regression model allows the modelling of a large number of evolutionary phenomena. For example, in this way we can allow the components of $\boldsymbol{\beta}(\Theta_n)$ to be the correct individual premiums in a limited number of earlier periods, so that

$$\boldsymbol{\beta}(\Theta_n)' = \left(\mu(\Theta_n), \mu(\Theta_{n-1}), \ldots, \mu(\Theta_{n-p+1})\right). \tag{9.71}$$

Formally, the argument of $\boldsymbol{\beta}$ in (9.71) depends on $(\Theta_n, \Theta_{n-1}, \ldots, \Theta_{n-p+1})$. Nevertheless we have written $\boldsymbol{\beta}(\Theta_n)$ as before.
In particular, we can model in this way all situations where the process $\{\mu(\Theta_j) : j = 1, 2, \ldots\}$ is an ARMA process (autoregressive moving average). We can also model the situation described in Chapter 9 of a stationary AR(1) process with random mean $\mu(\Theta)$. After the proof of Theorem 9.8 we will give a number of examples from this rich family of time series models.

Proof of Theorem 9.8: It is enough to prove the theorem for Case I. Case II then follows directly from Case I and the linearity property of projections.

Let $\mu_{n|n-1}$, respectively $\mu_{n|n}$, be estimators of $\mu(\Theta_n)$ based on the first $n-1$, respectively the first n observations and $q_{n|n-1}$, respectively $q_{n|n}$, be the corresponding quadratic loss. From Theorem 9.3 and Theorem 9.6 we have

$$\mu_{n|n} = \alpha_n X_n + (1 - \alpha_n)\,\mu_{n|n-1}, \tag{9.72}$$

$$\text{where } \alpha_n = \frac{q_{n|n-1}}{q_{n|n-1} + \frac{\sigma^2}{w_n}},$$

or written somewhat differently

$$\mu_{n|n} = \mu_{n|n-1} + \frac{q_{n|n-1}}{q_{n|n-1} + \frac{\sigma^2}{w_n}} \left(X_n - \mu_{n|n-1} \right). \tag{9.73}$$

From the regression assumption (9.52) and the linearity property of projections we have also that

$$\mu_{n|n} = \mathbf{y}_n' \boldsymbol{\beta}_{n|n}, \tag{9.74}$$

$$\mu_{n|n-1} = \mathbf{y}_n' \boldsymbol{\beta}_{n|n-1}, \tag{9.75}$$

$$q_{n|n-1} = \mathbf{y}_n' Q_{n|n-1} \mathbf{y}_n. \tag{9.76}$$

Substituting (9.74)–(9.76) into (9.73) we get

$$\mathbf{y}_n' \boldsymbol{\beta}_{n|n} = \mathbf{y}_n' \boldsymbol{\beta}_{n|n-1} + \frac{\mathbf{y}_n' Q_{n|n-1} \mathbf{y}_n}{\mathbf{y}_n' Q_{n|n-1} \mathbf{y}_n + \frac{\sigma^2}{w_n}} \left(X_n - \mathbf{y}_n' \boldsymbol{\beta}_{n|n-1} \right),$$

or

$$\mathbf{y}_n' \boldsymbol{\beta}_{n|n} = \mathbf{y}_n' \left(\boldsymbol{\beta}_{n|n-1} + \boldsymbol{\gamma}_n \left(X_n - \mathbf{y}_n' \boldsymbol{\beta}_{n|n-1} \right) \right). \tag{9.77}$$

where $\boldsymbol{\gamma}_n$ was defined in (9.58). Because (9.77) holds for every $\mathbf{y}_n$, we conjecture that

$$\boldsymbol{\beta}_{n|n} = \boldsymbol{\beta}_{n|n-1} + \boldsymbol{\gamma}_n \left(X_n - \mathbf{y}_n' \boldsymbol{\beta}_{n|n-1} \right). \tag{9.78}$$

Note that (9.77) does not imply the conjecture, since $\mathbf{y}_n$ appears also inside the parentheses.

We prove the validity of (9.78) by checking that the right-hand side of (9.78), which we denote by $\boldsymbol{\beta}^*_{n|n}$, satisfies the normal equations (Corollary 3.17). We have then to show that

$$E\left[\boldsymbol{\beta}(\Theta_n) - \boldsymbol{\beta}^*_{n|n}\right] = \mathbf{0}, \tag{9.79}$$

$$E\left[\left(\boldsymbol{\beta}(\Theta_n) - \boldsymbol{\beta}^*_{n|n}\right) X_j\right] = \mathbf{0} \quad \text{for } j = 1, 2, \dots, n. \tag{9.80}$$

i) From

$$E\left[\boldsymbol{\beta}(\Theta_n) - \boldsymbol{\beta}_{n|n-1}\right] = \mathbf{0}$$

and

$$\begin{aligned} E\left[X_n - \mathbf{y}_n' \boldsymbol{\beta}_{n|n-1}\right] &= E\left[E\left[X_n | \Theta_n\right] - \mathbf{y}_n' \boldsymbol{\beta}_{n|n-1}\right] \\ &= \mathbf{y}_n' E\left[\boldsymbol{\beta}(\Theta_n) - \boldsymbol{\beta}_{n|n-1}\right] = \mathbf{0}, \end{aligned}$$

(9.79) follows directly.

ii) Because $\boldsymbol{\beta}_{n|n-1}$ is the credibility estimator of $\boldsymbol{\beta}(\Theta_n)$ based on the observations until period $n-1$, it holds for $j = 1, 2, \dots, n-1$ that

$$\begin{aligned} E\left[\left(\boldsymbol{\beta}(\Theta_n) - \boldsymbol{\beta}_{n|n-1}\right) X_j\right] &= \mathbf{0}, \\ E\left[\left(X_n - \mathbf{y}_n' \boldsymbol{\beta}_{n|n-1}\right) X_j\right] &= E\left[\mathbf{y}_n' \left(\boldsymbol{\beta}(\Theta_n) - \boldsymbol{\beta}_{n|n-1}\right) X_j\right] \\ &= \mathbf{y}_n' E\left[\left(\boldsymbol{\beta}(\Theta_n) - \boldsymbol{\beta}_{n|n-1}\right) X_j\right] \\ &= \mathbf{0}. \end{aligned}$$

Thus (9.80) is also satisfied for $j = 1, 2, \dots, n-1$.

iii) It remains to show that

$$\begin{aligned} &E\left[\left(\boldsymbol{\beta}(\Theta_n) - \boldsymbol{\beta}^*_{n|n}\right) X_n\right] \\ &\quad = E\left[\left(\boldsymbol{\beta}(\Theta_n) - \boldsymbol{\beta}^*_{n|n}\right) \left(X_n - \mu_{n|n-1}\right)\right] = \mathbf{0}, \end{aligned} \tag{9.81}$$

where the insertion of $\mu_{n|n-1}$ is allowable because of i) and ii). This is the same as saying that

$$E\left[\left(\boldsymbol{\beta}(\Theta_n) - \boldsymbol{\beta}_{n|n-1}\right) \left(X_n - \mu_{n|n-1}\right)\right] = \gamma_n E\left[\left(X_n - \mu_{n|n-1}\right)^2\right].$$

It holds that

$$E\left[\left(\boldsymbol{\beta}(\Theta_n) - \boldsymbol{\beta}_{n|n-1}\right)\left(X_n - \mu_{n|n-1}\right)\right]$$
$$= E\left[\left(\boldsymbol{\beta}(\Theta_n) - \boldsymbol{\beta}_{n|n-1}\right)\left(\boldsymbol{\beta}(\Theta_n) - \boldsymbol{\beta}_{n|n-1}\right)'\mathbf{y}_n\right]$$
$$= Q_{n|n-1}\,\mathbf{y}_n.$$

$$\gamma_n E\left[\left(X_n - \mu(\Theta_n) + \mu(\Theta_n) - \mu_{n|n-1}\right)^2\right]$$
$$= \gamma_n\left(\frac{\sigma^2}{w_n} + \mathbf{y}_n' Q_{n|n-1}\mathbf{y}\right) = Q_{n|n-1}\,\mathbf{y}_n.$$

and thus (9.81) is shown.

This ends the proof for the updating formula (9.57) of Theorem 9.8. □

For the quadratic loss matrix, by explicit calculation we get

$$\begin{aligned}
Q_{n|n} = E\Big[&\left(\boldsymbol{\beta}(\Theta_n) - \boldsymbol{\beta}_{n|n-1} - \gamma_n\left(X_n - \mu_{n|n-1}\right)\right) \\
&\cdot\left(\boldsymbol{\beta}(\Theta_n) - \boldsymbol{\beta}_{n|n-1} - \gamma_n\left(X_n - \mu_{n|n-1}\right)\right)'\Big] \\
= {}& E\left[\left(\boldsymbol{\beta}(\Theta_n) - \boldsymbol{\beta}_{n|n-1}\right)\cdot\left(\boldsymbol{\beta}(\Theta_n) - \boldsymbol{\beta}_{n|n-1}\right)'\right] \\
&- E\left[\left(\boldsymbol{\beta}(\Theta_n) - \boldsymbol{\beta}_{n|n-1}\right)\cdot\left(\boldsymbol{\beta}(\Theta_n) - \boldsymbol{\beta}_{n|n-1}\right)'\right]\mathbf{y}_n\gamma_n' \\
&- \gamma_n\mathbf{y}_n' E\left[\left(\boldsymbol{\beta}(\Theta_n) - \boldsymbol{\beta}_{n|n-1}\right)\cdot\left(\boldsymbol{\beta}(\Theta_n) - \boldsymbol{\beta}_{n|n-1}\right)'\right] \\
&+ \gamma_n\mathbf{y}_n' E\left[\left(\boldsymbol{\beta}(\Theta_n) - \boldsymbol{\beta}_{n|n-1}\right)\cdot\left(\boldsymbol{\beta}(\Theta_n) - \boldsymbol{\beta}_{n|n-1}\right)'\right]\mathbf{y}_n\gamma_n' \\
&+ \gamma_n E\left[\left(X_n - \mu(\Theta_n)\right)^2\right]\gamma_n' \\
= {}& Q_{n|n-1} - Q_{n|n-1}\mathbf{y}_n\gamma_n' - \gamma_n\mathbf{y}_n' Q_{n|n-1} \\
&+ \gamma_n\left(\mathbf{y}_n' Q_{n|n-1}\mathbf{y}_n + \frac{\sigma^2}{w_n}\right)\gamma_n'.
\end{aligned} \tag{9.82}$$

Note that, for the term

$$\mathbf{y}_n' Q_{n|n-1}\mathbf{y}_n + \frac{\sigma^2}{w_n},$$

which occurs in the numerator of γ_n (see (9.58)), it holds that

$$\gamma_n\left(\mathbf{y}_n' Q_{n|n-1}\mathbf{y}_n + \frac{\sigma^2}{w_n}\right)\gamma_n' = Q_{n|n-1}\mathbf{y}_n\gamma_n'$$

and

$$\gamma_n \left(\mathbf{y}_n' Q_{n|n-1} \mathbf{y}_n + \frac{\sigma^2}{w_n} \right) \gamma_n' = \gamma_n \mathbf{y}_n' Q_{n|n-1}.$$

Substituting into (9.82) we get

$$Q_{n|n} = \left(I - \gamma_n \mathbf{y}_n' \right) Q_{n|n-1} = Q_{n|n-1} \left(I - \mathbf{y}_n \gamma_n' \right),$$

whereby the part "updating" of Theorem 9.8 is proved.

The validity of (9.60) and (9.61) can be proved exactly as in Theorem 9.4 and is a consequence of the requirement that the sequence $\{\boldsymbol{\beta}(\Theta_j) : j = 1, \dots\}$ has orthogonal increments. We first show that $\boldsymbol{\beta}_{n+1|n} = \boldsymbol{\beta}_{n|n}$. From the linearity property of projections we have that

$$\boldsymbol{\beta}_{n+1|n} = \boldsymbol{\beta}_{n|n} + \text{Pro}\left(\boldsymbol{\beta}(\Theta_{n+1}) - \boldsymbol{\beta}(\Theta_n) \,|\, L(1, X_1, \dots, X_n) \right).$$

Thus, we have to show that

$$\boldsymbol{\beta}(\Theta_{n+1}) - \boldsymbol{\beta}(\Theta_n) \perp L(1, X_1, \dots, X_n). \tag{9.83}$$

From (9.56) it follows directly that

$$E\left[\left(\boldsymbol{\beta}(\Theta_{n+1}) - \boldsymbol{\beta}(\Theta_n) \right) \cdot 1 \right] = \mathbf{0}.$$

Further, we have that

$$\begin{aligned} E\left[\left(\boldsymbol{\beta}(\Theta_{n+1}) - \boldsymbol{\beta}(\Theta_n) \right) X_j \right] &= E\left[\left(\boldsymbol{\beta}(\Theta_{n+1}) - \boldsymbol{\beta}(\Theta_n) \right) \boldsymbol{\beta}'(\Theta_j) \right] \mathbf{y}_j \\ &= \mathbf{0} \quad \text{for } j = 1, 2, \dots, n. \end{aligned}$$

And so (9.83) and (9.60) are shown.

For the quadratic loss of $\boldsymbol{\beta}_{n+1|n}$ we get

$$\begin{aligned} Q_{n+1|n} &= E\left[\left(\boldsymbol{\beta}_{n+1|n} - \boldsymbol{\beta}(\Theta_{n+1}) \right)^2 \right] \\ &= E\left[\left(\boldsymbol{\beta}_{n|n} - \boldsymbol{\beta}(\Theta_n) + \boldsymbol{\beta}(\Theta_n) - \boldsymbol{\beta}(\Theta_{n+1}) \right)^2 \right] \\ &= \left[E\left(\boldsymbol{\beta}_{n|n} - \boldsymbol{\beta}(\Theta_n) \right)^2 \right] + E\left[\left(\boldsymbol{\beta}(\Theta_{n+1}) - \boldsymbol{\beta}(\Theta_n) \right)^2 \right] \\ &= Q_{n|n} + D_{n+1}. \end{aligned}$$

In the second equation we used the already proved fact, that $\boldsymbol{\beta}_{n+1|n} = \boldsymbol{\beta}_{n|n}$. By conditioning on $\boldsymbol{\Theta}$ and keeping in mind that $E\left[\boldsymbol{\beta}_{n|n} \middle| \boldsymbol{\Theta} \right] = \boldsymbol{\beta}(\Theta_n)$, it becomes clear why the cross-term disappears in the third equation. With this (9.61) is shown. This completes the proof of Theorem 9.8. □

In the following, for illustrative purposes, we present a few examples from the family of ARMA processes. In each of these examples, we assume that

$$E\left[\text{Cov}\left(X_k, X_j \middle| \boldsymbol{\Theta} \right) \right] = \sigma^2 \delta_{kj}$$

holds.

Example 9.6 (autoregressive process of order 1 with random mean)

We refer to Example 9.5. However, now the mean of the autoregressive process is assumed not to be a constant μ, but a random variable, the value of which depends on the individual risk profile. In this way a collective of independent and similar risks can be described, in which the individual, correct premium of each contract varies randomly about a mean, which depends on the individual risk profile.

The model assumptions for the pure risk premiums of a particular risk are given by

$$\mu(\Theta_{n+1}) = \rho\mu(\Theta_n) + (1-\rho)\,\mu(\Theta_0) + \Delta_{n+1} \quad \text{for } n = 1,2,\ldots, \tag{9.84}$$

where the Δ_j are uncorrelated with $E[\Delta_j] = 0$ and $\mathrm{Var}[\Delta_j] = \delta^2$. We further assume that, conditional on $\mu(\Theta_0)$, the process $\{\mu(\Theta_1), \mu(\Theta_2), \ldots\}$ is a stationary process with

$$\begin{aligned} E[\mu(\Theta_j)|\mu(\Theta_0)] &= \mu(\Theta_0), \\ \mathrm{Var}[\mu(\Theta_j)|\mu(\Theta_0)] &= \frac{\delta^2}{1-\rho^2} \qquad \text{(cf. (9.46))}. \end{aligned}$$

$\mu(\Theta_0)$ is therefore the stochastically drawn "mean", around which the AR(1) process of the correct individual premiums varies. For the unconditioned process we have then that

$$\begin{aligned} E[\mu(\Theta_j)] &= E[\mu(\Theta_0)] = \mu \quad \text{for } j = 1,2,\ldots, \\ \mathrm{Var}[\mu(\Theta_j)] &= \frac{\delta^2}{1-\rho^2} + \tau^2, \\ \tau^2 &= \mathrm{Var}[\mu(\Theta_0)]. \end{aligned} \tag{9.85}$$

Note: If we use this model to model a portfolio of independent risks, then (9.85) means that the heterogeneity of the portfolio stays constant over the passage of time – an assumption which is reasonable.

In order to find the credibility estimator for $\mu(\Theta_j)$, we want to use the Kalman Filter technique (Theorem 9.8). To do this, we must first define the state or regression vector $\boldsymbol{\beta}(\Theta_n)$. In our case it is given by

$$\boldsymbol{\beta}(\Theta_n) = \begin{pmatrix} \mu(\Theta_n) \\ \mu(\Theta_0) \end{pmatrix}.$$

On the basis of (9.84) we get

$$\boldsymbol{\beta}(\Theta_{n+1}) = \begin{pmatrix} \rho & 1-\rho \\ 0 & 1 \end{pmatrix} \boldsymbol{\beta}(\Theta_n) + \begin{pmatrix} \Delta_{n+1} \\ 0 \end{pmatrix},$$

and the regression equation is

$$\mu\left(\Theta_{n+1}\right)=(1,0)\,\boldsymbol{\beta}\left(\Theta_{n+1}\right).$$

We can now apply Theorem 9.8 and get:

i) Anchoring

$$\boldsymbol{\beta}_{1|0}=\boldsymbol{\beta}_0=\begin{pmatrix}\mu\\ \mu\end{pmatrix} \quad\text{and}\quad Q_{1|0}=\begin{pmatrix}\tau^2+\frac{\delta^2}{1-\rho^2} & \tau^2\\ \tau^2 & \tau^2\end{pmatrix}.$$

ii) Recursion

a) Updating

$$\begin{aligned}
\mathbf{y}'_n &= (1,0) \text{ for all } n\ ,\\
\boldsymbol{\beta}_{n|n} &= \boldsymbol{\beta}_{n|n-1}+\gamma_n\left(X_n-(1,0)\ \boldsymbol{\beta}_{n|n-1}\right),\\
\text{where } \gamma_n &= \frac{Q_{n|n-1}\ (1,0)'}{(1,0)\ Q_{n|n-1}\ (1,0)'+\sigma^2},\\
&= \frac{\text{first column of } Q_{n|n-1}}{q^{(11)}_{n|n-1}+\sigma^2},\\
&\quad\ \text{where } q^{(11)}_{n|n-1} = \text{upper left corner of } Q_{n|n-1},\\
&\quad\ \sigma^2 = E\left[\mathrm{Var}\left[X_j|\boldsymbol{\Theta}\right]\right].
\end{aligned}$$

$$Q_{n/n} = \left(I-\frac{Q_{n|n-1}\begin{pmatrix}1&0\\0&0\end{pmatrix}}{q^{(11)}_{n|n-1}+\sigma^2}\right)Q_{n|n-1}.$$

After some calculations one also obtains that

$$Q_{n/n}=\frac{\sigma^2}{q^{(11)}_{n|n-1}+\sigma^2}\left(Q_{n|n-1}+\begin{pmatrix}0&0\\0&\frac{\Delta}{\sigma^2}\end{pmatrix}\right),$$
$$\text{where } \Delta=\det\left(Q_{n|n-1}\right).$$

b) Movement in parameter space

$$\begin{aligned}
\boldsymbol{\beta}_{n+1|n} &= \begin{pmatrix}\rho & 1-\rho\\ 0 & 1\end{pmatrix}\boldsymbol{\beta}_{n|n},\\
Q_{n+1|n} &= \begin{pmatrix}\rho & 1-\rho\\ 0 & 1\end{pmatrix}Q_{n|n}\begin{pmatrix}\rho & 0\\ 1-\rho & 1\end{pmatrix}+\begin{pmatrix}\delta^2 & 0\\ 0 & 0\end{pmatrix}.
\end{aligned}$$

Example 9.7 (autoregressive process of order p with random mean)

All assumptions are analogous to Example 9.6, only instead of (9.84) we have

$$\mu(\Theta_{n+1}) = \rho_1 (\mu(\Theta_n) - \mu(\Theta_0)) + \cdots \rho_p (\mu(\Theta_{n-p+1}) - \mu(\Theta_0)) + \mu(\Theta_0) + \Delta_{n+1}.$$

We choose

$$\boldsymbol{\beta}(\Theta_n) = \begin{pmatrix} \mu(\Theta_n) \\ \vdots \\ \mu(\Theta_{n-p+1}) \\ \mu(\Theta_0) \end{pmatrix}$$

and get as movement equations

$$\boldsymbol{\beta}(\Theta_{n+1}) = \begin{pmatrix} \rho_1 & \rho_2 & \cdots & \cdots & \rho_p & 1 - \sum_{k=1}^{p} \rho_k \\ 1 & 0 & \cdots & \cdots & 0 & 0 \\ 0 & 1 & 0 & & 0 & 0 \\ \vdots & \ddots & \ddots & \ddots & \vdots & 0 \\ 0 & & 0 & 1 & 0 & 0 \\ 0 & \cdots & \cdots & 0 & 0 & 1 \end{pmatrix} \boldsymbol{\beta}(\Theta_n) + \begin{pmatrix} \Delta_{n+1} \\ 0 \\ \vdots \\ \vdots \\ \vdots \\ \vdots \\ 0 \end{pmatrix}.$$

Thus all of the conditions for Theorem 9.8 are satisfied and we can again use the Kalman Filter to calculate the credibility estimator.

Remark on Examples 9.6 and 9.7:

- These are very natural examples to illustrate that, for the evolutionary credibility regression model, we may have a situation where the dimension of the regression parameter is higher than the number of observations (see fourth bullet of the remarks beginning on page 242).

Example 9.8 (Moving average process of order q)

Here also the assumptions are analogous to the previous examples, but instead of (9.84) we have

$$\mu(\Theta_{n+1}) = \mu(\Theta_n) + \Delta_{n+1} + \rho_1 \Delta_n + \cdots + \rho_q \Delta_{n-q}.$$

We define

$$\boldsymbol{\beta}(\Theta_n) = \begin{pmatrix} \mu(\Theta_n) \\ \Delta_n \\ \vdots \\ \Delta_{n-q} \end{pmatrix}$$

and get the movement equations

$$\boldsymbol{\beta}\left(\Theta_{n+1}\right)=\begin{pmatrix} 1 & \rho_1 & \cdots & \cdots & \cdots & \rho_q \\ 0 & 0 & \cdots & \cdots & \cdots & 0 \\ 0 & 1 & 0 & & & \vdots \\ \vdots & 0 & 1 & \ddots & & \vdots \\ \vdots & & \ddots & \ddots & \ddots & \vdots \\ 0 & \cdots & \cdots & 0 & 1 & 0 \end{pmatrix}\boldsymbol{\beta}\left(\Theta_n\right)+\begin{pmatrix} 0 \\ \Delta_{n+1} \\ 0 \\ \vdots \\ \vdots \\ 0 \end{pmatrix}.$$

Thus we can also apply the Kalman Filter algorithm here.

10

Multidimensional Evolutionary Models and Recursive Calculation

10.1 Introduction

We have already introduced multidimensional credibility in Chapter 7. It is characterized by the fact that the quantity $\boldsymbol{\mu}(\Theta)$ to be estimated is multidimensional. The observations or a suitable compression of the observations are then described by vectors $\mathbf{X}$, having the same dimension as $\boldsymbol{\mu}(\Theta)$ (see abstract multidimensional credibility model, Section 7.2).

In Chapter 7 we discussed, as a typical example, the case where for each risk i, the components of the vector $\mathbf{X}_i$ represent the claim frequencies of normal and large claims. Other examples mentioned there were claim frequency as one component and average claim amount as the other component or claims observations of different layers or different claim types. In Chapter 9, in the last point of the remarks beginning on page 225, we indicated that we can also interpret the abstract vector $\mathbf{X}$ in another way, namely for each time point j as vector $\mathbf{X}_j$ summarizing the individual observations of a whole collective of risks at that time. As mentioned there, this last possibility has a particular relevance in the evolutionary case and lends the evolutionary multidimensional model, so to speak, an extra dimension, in that it allows the modelling of the development over time of a collective in its entirety. We will deal with this point more closely in the next section.

Let us now consider the vectors $\mathbf{X}_j$. In the non-evolutionary case, discussed in Chapter 7, we assumed that the risk profile Θ remained constant over time, i.e. that

$$\boldsymbol{\mu}(\Theta) := E[\mathbf{X}_j|\Theta] \quad \text{for all } j.$$

In contrast, in the evolutionary model, we assume analogously to chapter 9 that the risks change over time and that the risk profile Θ is a stochastic process

$$\boldsymbol{\Theta} = \{\Theta_1, \Theta_2, \ldots\}, \tag{10.1}$$

where the components Θ_j represent the risk profiles in the corresponding years $j = 1, 2, \ldots$. The conditional expectations also change with time, i.e.

$$\boldsymbol{\mu}(\boldsymbol{\Theta}) = \{\boldsymbol{\mu}(\Theta_1), \boldsymbol{\mu}(\Theta_2), \ldots\}.$$

Remarks:

- As discussed above, depending on the application, the $\mathbf{X}_j$ can represent the multidimensional observation of a single risk or the entirety of the single observations from a whole collective in year j. In the first case, we have for each risk i, a multidimensional observation $\mathbf{X}_{ij}$. Its embedding in a collective of similar risks can be modelled as it was done in Section 9.3. The credibility estimator then depends only on the observation vectors of the risk i under consideration, and not on the observations from the other risks. The data from the collective are, however, used to determine the structural parameter. In the second case, for each year only one observation vector $\mathbf{X}_j$ is available. The credibility estimator for the ith component, i.e. for the ith risk, then depends also on the other data from the collective. On the other hand, because the whole collective is described by one single vector, we must find other ways to estimate the structural parameter.
- Because in the following the letter I is reserved for the identity matrix, throughout this chapter we denote the total number of risks by N, i.e. the index i takes the values $i = 1, 2, \ldots, N$.

10.2 On the Embedding of the Individual Risk in a Collective

In Chapter 9, Section 9.3, (see the second last point in the remarks on the evolutionary credibility model, page 226) we have characterized each risk i $(i = 1, 2, \ldots, N)$ by its risk profile

$$\boldsymbol{\Theta}_i = \{\Theta_{i1}, \Theta_{i2}, \ldots, \Theta_{ij}, \ldots\}$$

and the process of observations

$$\mathbf{X}_i = \{X_{i1}, X_{i2}, \ldots, X_{ij}, \ldots\}.$$

The embedding in the collective, which we have carried out as in the non-evolutionary case, has then led to the assumptions that the pairs $\{(\mathbf{X}_1, \boldsymbol{\Theta}_1), (\mathbf{X}_2, \boldsymbol{\Theta}_2), \ldots (\mathbf{X}_N, \boldsymbol{\Theta}_N)\}$ are independent and that $\{\boldsymbol{\Theta}_1, \boldsymbol{\Theta}_2, \ldots, \boldsymbol{\Theta}_N\}$ are independent and identically distributed.

The problems associated with these assumptions have already been indicated (page 225ff). In many cases, it may be true that the risk profiles of the different individual risks are effected by *common* (and *not independent*) changes. We also noted in Chapter 9 that some of the models handled there (e.g. Gerber–Jones) led to collectives which drifted apart, i.e. whose heterogeneity could become arbitrarily large. We rescued ourselves from this difficulty by the introduction of appropriate linear transformations (or more generally by the elimination of common trends).

In this section, we choose a more radical procedure: we remove the above assumptions. Instead, we describe and analyse all the risks $i = 1, 2, \ldots, N$ of the collective *simultaneously*, i.e. we consider the *multidimensional* sequences

$$\begin{aligned} \boldsymbol{\Theta} &= \{\boldsymbol{\Theta}_1, \boldsymbol{\Theta}_2, \ldots\}, \text{ where } \boldsymbol{\Theta}_j = (\Theta_{ij}), \quad i = 1, 2, \ldots, N, \\ \mathbf{X} &= \{\mathbf{X}_1, \mathbf{X}_2, \ldots\}, \text{ where } \mathbf{X}_j = (X_{ij}), \quad i = 1, 2, \ldots, N. \end{aligned} \tag{10.2}$$

In this way, we allow arbitrary dependencies *within* (i.e. between the components of) the multidimensional quantities $\boldsymbol{\Theta}_j, \mathbf{X}_j$.

Notice that (10.2) fits well with the general description (10.1), because the Θ_j in (10.1) can be arbitrary random variables. In the case which we discuss here, it is sensible to consider Θ_j as a vector with the components Θ_{ij} $(i = 1, 2, \ldots, N)$, where Θ_{ij} is the risk profile of the ith risk in year j. Therefore, in view of the general meaning of the parameter Θ_i, we may use the notation Θ_j instead of $\boldsymbol{\Theta}_j$ also in cases where Θ_j represents a vector, which we will do from now on.

Analogously to Section 9.3 it should also be true in the multidimensional case that

$$\begin{aligned} E\left[\mathbf{X}_j | \boldsymbol{\Theta}\right] &= E\left[\mathbf{X}_j | \Theta_j\right] =: \boldsymbol{\mu}\left(\Theta_j\right), \\ E\left[\operatorname{Cov}\left(\mathbf{X}_j, \mathbf{X}_j' | \boldsymbol{\Theta}\right)\right] &= E\left[\operatorname{Cov}\left(\mathbf{X}_j, \mathbf{X}_j' | \Theta_j\right)\right] =: E_j, \end{aligned}$$

i.e. the first two moments of the multidimensional quantity $\mathbf{X}_j$ depend only on the current value of the risk profile Θ_j. We also call Θ_j (non-observable) *state variables at time j*.

Analogously to (9.12) we require that

$$E\left[\operatorname{Cov}\left(\mathbf{X}_j, \mathbf{X}_k' | \boldsymbol{\Theta}\right)\right] = 0 \quad \text{for } j \neq k,$$

i.e. the observation vectors at different time points should, on the average, be uncorrelated.

10.3 Multidimensional Evolutionary Models

Analogous to the differentiation between Cases I and II in Theorem 9.6, respectively between the assumptions "orthogonal increments" (formula 9.19) and "linear Markov property" (formula 9.23), we want to differentiate between these two cases here.

- **Case I:** The process $\{\boldsymbol{\mu}(\Theta_j) : j = 1, 2, \ldots\}$ has orthogonal increments. This means that

$$\operatorname{Pro}\left[\boldsymbol{\mu}\left(\Theta_{j+1}\right) | L\left(\mathbf{1}, \boldsymbol{\mu}\left(\Theta_1\right), \ldots, \boldsymbol{\mu}\left(\Theta_j\right)\right)\right] = \boldsymbol{\mu}\left(\Theta_j\right), \tag{10.3}$$

which implies that the increments have expected value zero and that the covariance of increments at different times is zero, i.e.

$$\mathrm{Cov}\left(\boldsymbol{\mu}(\Theta_{j+1})-\boldsymbol{\mu}(\Theta_j),(\boldsymbol{\mu}(\Theta_{k+1})-\boldsymbol{\mu}(\Theta_k))'\right)=\mathbf{0} \quad \text{for } j\neq k,$$
$$E\left[\boldsymbol{\mu}(\Theta_{j+1})-\boldsymbol{\mu}(\Theta_j)\right]=\mathbf{0} \quad \text{for all } j.$$

- **Case II:** The process $\{\boldsymbol{\mu}(\Theta_j): j=1,2,\dots\}$ satisfies the linear Markov property.
  This means that

$$\mathrm{Pro}\left[\boldsymbol{\mu}(\Theta_{j+1})\,|L(\mathbf{1},\boldsymbol{\mu}(\Theta_1),\dots,\boldsymbol{\mu}(\Theta_j))\right]=A_{j+1}\,\boldsymbol{\mu}(\Theta_j)+\mathbf{b}_{j+1}. \tag{10.4}$$

  Note that Case II is a generalization of Case I, in that additionally linear transformations are allowed (transformation matrix A_{j+1}, shift vector $\mathbf{b}_{j+1}$).

In the following we use the notation introduced in Chapter 9.

$$\begin{aligned}\boldsymbol{\mu}_{n|k}:=&\ \mathrm{Pro}\left[\boldsymbol{\mu}(\Theta_n)\,|L(1,\mathbf{X}_1,\dots,\mathbf{X}_k)\right]\\ =&\ \text{projection on data until time } k\ (k=n \text{ or } k=\ n-1).\\ Q_{n|k}:=&\ E\left[\left(\boldsymbol{\mu}(\Theta_n)-\boldsymbol{\mu}_{n|k}\right)\cdot\left(\boldsymbol{\mu}(\Theta_n)-\boldsymbol{\mu}_{n|k}\right)'\right]\\ =&\ \text{loss matrix using data until time } k.\end{aligned}$$

Analogous to Theorem 9.6 the following recursive formulae hold.

Theorem 10.1. *(multidimensional Kalman-Filter algorithm)*

i) Anchoring (period 1)

$$\begin{aligned}\boldsymbol{\mu}_{1|0}&=E\left[\boldsymbol{\mu}(\Theta_1)\right],\\ Q_{1|0}&=\mathrm{Cov}\left(\boldsymbol{\mu}(\Theta_1),\boldsymbol{\mu}'(\Theta_1)\right).\end{aligned}$$

ii) Recursion (periods $n\geq 1$)

a) Updating

$$\boldsymbol{\mu}_{n|n}=\boldsymbol{\mu}_{n|n-1}+Z_n\left(\mathbf{X}_n-\boldsymbol{\mu}_{n|n-1}\right), \tag{10.5}$$
$$\text{where } Z_n=Q_{n|n-1}\left(Q_{n|n-1}+S_n\right)^{-1},$$
$$S_n=E\left[\mathrm{Cov}\left(\mathbf{X}_n,\mathbf{X}_n'|\Theta_n\right)\right],$$
$$Q_{n|n}=(I-Z_n)\,Q_{n|n-1}. \tag{10.6}$$

b) Change from Θ_n to Θ_{n+1} (movement in parameter space)
Case I: $\{\boldsymbol{\mu}(\Theta_j): j=1,2,\dots\}$ has orthogonal increments (see (10.3)). It then holds that

$$\begin{aligned}\boldsymbol{\mu}_{n+1|n}&=\boldsymbol{\mu}_{n|n}, \qquad (10.7)\\ Q_{n+1|n}&=Q_{n|n}+D_{n+1},\end{aligned}$$

where

$$D_{n+1} = E\left[\boldsymbol{\mu}\left(\Theta_{n+1}\right) - \boldsymbol{\mu}\left(\Theta_n\right)\left(\boldsymbol{\mu}\left(\Theta_{n+1}\right) - \boldsymbol{\mu}\left(\Theta_n\right)\right)'\right].$$

Case II: The process $\{\boldsymbol{\mu}(\Theta_j) : j = 1, 2, \ldots\}$ satisfies the linear Markov property (see (10.4)).
It then holds that

$$\boldsymbol{\mu}_{n+1|n} = A_{n+1}\boldsymbol{\mu}_{n|n} + \mathbf{b}_{n+1}, \tag{10.9}$$

$$Q_{n+1|n} = A_{n+1}Q_{n|n}A'_{n+1} + D_{n+1}, \tag{10.10}$$

where

$$D_{n+1} = E\left[\left(\boldsymbol{\mu}\left(\Theta_{n+1}\right) - A_{n+1}\boldsymbol{\mu}\left(\Theta_n\right) - \mathbf{b}_{n+1}\right) \cdot \left(\boldsymbol{\mu}\left(\Theta_{n+1}\right) - A_{n+1}\boldsymbol{\mu}\left(\Theta_n\right) - \mathbf{b}_{n+1}\right)'\right]. \tag{10.11}$$

Remark:

- The matrices of the observation errors, i.e. the matrices E_j $(j = 1, 2, \ldots)$ will have a diagonal form in many applications. The matrices of the innovation errors of the risk profile process, i.e. the D_j $(j = 1, 2, \ldots)$ are typically *not* diagonal. If this were the case (and if in addition $Q_{1|0}$ is diagonal), then the multidimensional algorithm just reduces to the individual algorithm applied to each individual risk as presented in Chapter 9.

Proof of Theorem 10.1:
The proof follows the argument presented in Chapter 7. In the estimator $\boldsymbol{\mu}_{n|n}$ (formula (10.5)) $\boldsymbol{\mu}_{n|n-1}$ assumes the role of the "collective average" and $Q_{n|n-1}$ assumes the role of the loss matrix (deviations from the collective average). □

10.4 Modelling a Collective with Both Joint and Individual Movements

We saw in Sections 10.1 and 10.2 that the multidimensional model allows us to model the development over time of the entire collective, and also to model changes which affect all risks in the collective in a similar manner. In the following, we present two examples in which we first show how such a model might look, and then how, using multidimensional techniques, we can find the credibility estimator.

As we already mentioned, in real contexts, it is frequently the case that the risk profile of an individual depends both on individual specific factors and on global factors which affect all risks in the collective. For example, consider the share prices of a firm: they depend on both global business conditions and market trends, as well as developments which are specific to the firm itself. In the insurance business, global factors could include things like changes in the law, prevailing weather conditions, cyclical effects, all of which will affect all risks in the collective at the same time.

In order to model such situations, it is convenient to divide the innovations (changes to the correct individual premiums) into a *collective* and an *individual* part. This is what we will do in the following two examples. These are the counterparts to Examples 9.4 (Gerber and Jones model) and 9.5 (autoregressive process of order 1) with the difference that now we also have movements which affect all risks in the collective at the same time.

We write

$$\mathbf{X}_j = \boldsymbol{\mu}(\Theta_j) + \boldsymbol{\varepsilon}_j, \tag{10.12}$$

i.e. $\boldsymbol{\varepsilon}_j$ is the vector of deviations of the observation vector $\mathbf{X}_j$ from the vector $\boldsymbol{\mu}(\Theta_j)$ of the associated individual premiums.

Example 10.1 (Counterpart to Example 9.4)

Assumptions:

- Movement

$$\begin{aligned}\boldsymbol{\mu}(\Theta_{n+1}) &= \boldsymbol{\mu}(\Theta_n) + \boldsymbol{\delta}_{n+1} \\ &= \boldsymbol{\mu}(\Theta_n) + \xi_{n+1}\mathbf{1} + \boldsymbol{\eta}_{n+1}.\end{aligned} \tag{10.13}$$

  The innovation of the ith risk (ith component of the vector $\boldsymbol{\delta}_{n+1}$) is composed of a component ξ_{n+1} common to all risks in the collective and a component $\eta_{i,n+1}$ which is individual specific.
  Remark on the notation: in Chapter 9 we used Δ_{n+1} for the innovations.
- Independence between the components

  $\{\xi_j, \eta_{ij}, \varepsilon_{ij} : j = 1, 2, \ldots;\ i = 1, 2, \ldots, N\}$ are independent random variables and are also independent of the components of $\boldsymbol{\mu}(\Theta_{j-1})$.

- Moments

$$\begin{aligned}E[\xi_j] &= E[\eta_{ij}] = E[\varepsilon_{ij}] = 0 \quad \text{for all } i \text{ and } j, \\ \operatorname{Var}[\xi_j] &= \omega^2 \quad \text{for } j = 1, 2, \ldots, \\ \operatorname{Var}[\varepsilon_{ij}] &= \sigma^2 \quad \text{for all } i \text{ and } j, \\ \operatorname{Var}[\eta_{ij}] &= \tau^2 \quad \text{for all } i \text{ and } j,\end{aligned}$$

For the matrices S_n and D_{n+1}, which appeared in Theorem 10.1, we get from the above assumptions

$$\begin{aligned}S_n &= \operatorname{Var}[\boldsymbol{\varepsilon}_n] = \sigma^2 I, \\ D_{n+1} &= \operatorname{Cov}(\boldsymbol{\delta}_{n+1}, \boldsymbol{\delta}'_{n+1}) = D \text{ (independent of } n),\end{aligned}$$

where

$$D = \begin{pmatrix} \tau^2+\omega^2 & \omega^2 & \dots & \omega^2 \\ \omega^2 & \tau^2+\omega^2 & \dots & \vdots \\ \vdots & & \ddots & \vdots \\ \omega^2 & \omega^2 & \dots & \tau^2+\omega^2 \end{pmatrix}. \tag{10.14}$$

It is useful to write the matrix D as follows:

$$D = \tau^2 I + \omega^2 E,$$

where I is the identity matrix and E is a matrix of 1's.

We will see that in Example 10.1 (and also in Example 10.2), we can in general restrict ourselves to matrices of the form

$$D_{p,q} = qI + pE, \quad q, p \in \mathbb{R}.$$

We will refer to a matrix of this form as being of the *$D_{p,q}$-form*. It is obvious that we should also use a matrix of the $D_{p,q}$-form for the initial matrix $Q_{1|0}$ in Theorem 10.1.

From Theorem 10.1 we get the following recursion formulae:

i) Anchoring (period 1)

$$\boldsymbol{\mu}_{1|0} = E\left[\boldsymbol{\mu}\left(\Theta_1\right)\right] = \boldsymbol{\mu}_0, \tag{10.15}$$

$$Q_{1|0} = q_{1|0} I + p_{1|0} E. \tag{10.16}$$

ii) Recursion

a) Updating

$$\boldsymbol{\mu}_{n|n} = \boldsymbol{\mu}_{n|n-1} + Z_n \left(\mathbf{X}_n - \boldsymbol{\mu}_{n|n-1}\right), \tag{10.17}$$

$$\text{where } Z_n = Q_{n|n-1}\left(Q_{n|n-1} + \sigma^2 I\right)^{-1}, \tag{10.18}$$

$$Q_{n|n} = \left(I - Z_n\right) Q_{n|n-1} \,. \tag{10.19}$$

b) Movement

$$\boldsymbol{\mu}_{n+1|n} = \boldsymbol{\mu}_{n|n}, \tag{10.20}$$

$$Q_{n+1|n} = Q_{n|n} + D. \tag{10.21}$$

The recursion formulae can be written out explicitly using the fact that all of the matrices therein have the $D_{p,q}$-form, e.g. $Q_{n|n-1} = q_{n|n-1} I + p_{n|n-1} E$. We write this explicit representation componentwise, and we write $\mu_{n|n}^{(i)}$ for the ith component of $\boldsymbol{\mu}_{n|n}$. We will show that the following holds:

a) Updating

$$\mu_{n|n}^{(i)} = \mu_{n|n-1}^{(i)} + \beta_n \left(X_{in} - \mu_{n|n-1}^{(i)} \right) + (1 - \beta_n)\, \alpha_n \left(\overline{X}_n - \overline{\mu}_{n|n-1} \right), \tag{10.22}$$

$$\text{where} \quad \overline{X}_n = \frac{1}{N} \sum_{i=1}^{N} X_{in},$$

$$\overline{\mu}_{n|n-1} = \frac{1}{N} \sum_{i=1}^{N} \mu_{n|n-1}^{(i)}.$$

$$\beta_n = \frac{q_{n|n-1}}{q_{n|n-1} + \sigma^2}, \tag{10.23}$$

$$\alpha_n = \frac{N p_{n|n-1}}{N p_{n|n-1} + q_{n|n-1} + \sigma^2}. \tag{10.24}$$

$$Q_{n|n} = q_{n|n} I + p_{n|n} E, \tag{10.25}$$

$$\text{where } q_{n|n} = \frac{\sigma^2}{q_{n|n-1} + \sigma^2}\, q_{n|n-1} = (1 - \beta_n)\, q_{n|n-1}, \tag{10.26}$$

$$p_{n|n} = \frac{\sigma^2}{q_{n|n-1} + \sigma^2} \cdot \frac{\sigma^2}{q_{n|n-1} + \sigma^2 + N p_{n|n-1}} \cdot p_{n|n-1} = (1 - \beta_n)^2 (1 - \alpha_n)\, p_{n|n-1}. \tag{10.27}$$

b) Movement in parameter space

$$\mu_{n+1|n}^{(i)} = \mu_{n|n}^{(i)}, \tag{10.28}$$

$$Q_{n+1|n} = q_{n+1|n} I + p_{n|n+1} E, \tag{10.29}$$

where

$$q_{n|n+1} = q_{n|n} + \tau^2, \tag{10.30}$$

$$p_{n+1|n} = p_{n|n} + \omega^2. \tag{10.31}$$

Remarks:

- (10.22) is a very intuitive result. The first correction term gives the correction resulting from the latest individual claim experience. Included in that is the estimation for the correction to the collective mean as a result of that individual observation. The second term is the correction term based on the collective experience and reflects the estimation of the collective innovation based on all the data from the collective. The factor $(1-\beta_n)$ takes account of the fact that this collective innovation is already included in the first correction term with the credibility weight β_n.
- Both in the updating step and in the step "movement in parameter space", the corresponding loss matrices $Q_{n|n}$, respectively $Q_{n|n-1}$, can be calculated by updating the scalar quantities q and p using the formulae given in (10.26) and (10.27), respectively (10.30) and (10.31).

In order to show the validity of (10.22)–(10.31), we need the following lemma.

Lemma 10.2.

$$\begin{aligned}(qI+pE)^{-1} &= \frac{1}{q}\left(I-kE\right),\\ \textit{where}\ \ k &= \frac{p}{(q+pN)},\\ N &= \textit{dimension of the matrix.}\end{aligned}$$

Proof:
The result is very easy to verify by multiplying out and using the identity $E^2 = N \cdot E$. □

We carry out the calculations for the recursion formulae (10.17)–(10.19) under the assumption that $Q_{n|n-1}$ has the $D_{p,q}$-form, i.e. that

$$Q_{n|n-1} = q_{n|n-1}I + p_{n|n-1}E. \tag{10.32}$$

We will show at the end that $Q_{n|n-1}$ does indeed have this form.

From (10.32) and Lemma 10.2, we get

$$\begin{aligned}\left(Q_{n|n-1}+\sigma^2 I\right)^{-1} &= \frac{1}{q_{n|n-1}+\sigma^2}\,I\\ &\quad - \frac{p_{n|n-1}}{\left(q_{n|n-1}+\sigma^2\right)\left(q_{n|n-1}+\sigma^2+Np_{n|n-1}\right)}\,E,\end{aligned} \tag{10.33}$$

$$
\begin{aligned}
Z_n &= Q_{n|n-1}\left(Q_{n|n-1}+\sigma^2 I\right)^{-1} \\
&= \frac{q_{n|n-1}}{q_{n|n-1}+\sigma^2} I \\
&\quad + \left(\frac{\sigma^2}{q_{n|n-1}+\sigma^2}\right)\left(\frac{p_{n|n-1}}{q_{n|n-1}+\sigma^2+Np_{n|n-1}}\right) E.
\end{aligned}
$$

This can also be written as

$$
Z_n = \beta_n I + \frac{1}{N}\left(1-\beta_n\right)\alpha_n E,
$$

$$
\text{where} \quad \beta_n = \frac{q_{n|n-1}}{q_{n|n-1}+\sigma^2}, \tag{10.34}
$$

$$
\alpha_n = \frac{Np_{n|n-1}}{Np_{n|n-1}+q_{n|n-1}+\sigma^2}. \tag{10.35}
$$

Plugging into (10.17) we thus get the update formula (10.22), which we have written componentwise.

The next step in the recursion formula is the updating of the loss matrix using (10.19). We get

$$
(I-Z_n) = \left(1-\beta_n\right)\left(I-\frac{\alpha_n}{N}E\right),
$$

and after some calculation

$$
\begin{aligned}
(I-Z_n)\,Q_{n/n-1} &= \left(1-\beta_n\right)\left(q_{n|n-1}I + \left(\left(1-\alpha_n\right)p_{n|n-1} - \frac{\alpha_n}{N}q_{n|n-1}\right)E\right) \\
&= \frac{\sigma^2 q_{n|n-1}}{q_{n|n-1}+\sigma^2} I \\
&\quad + \frac{\sigma^2}{q_{n|n-1}+\sigma^2} \cdot \frac{\sigma^2 p_{n|n-1}}{q_{n|n-1}+\sigma^2+Np_{n|n-1}} E.
\end{aligned}
$$

From this we get directly (10.25)–(10.27).

Finally we get formulae (10.28)–(10.31) for the movement in parameter space from (10.14),(10.20), (10.21) as well as (10.25)–(10.27).

The starting point for the above derivation was that the matrix $Q_{n|n-1}$ has the $D_{p,q}$ model form. We have shown that if this is the case then $Q_{n+1|n}$ also has the $D_{p,q}$ model form. We have yet to show that it is indeed the case that $Q_{n|n-1}$ has the $D_{p,q}$ model form. But this follows by induction from the assumption that $Q_{1|0}$ has the $D_{p,q}$ model form. □

Remarks on Example 10.1:

- If the individual components of the innovations are non-degenerate, i.e. if $\tau^2 > 0$, this example also describes a situation where, analogous to the Gerber and Jones model, the heterogeneity of the portfolio grows without limit, which does not correspond to anything one usually sees in practice. Equally unrealistic is the situation where $\tau^2 = 0$, which would imply that the size of the changes to the correct individual premiums is the same for all the risks.
- In order to avoid the drifting-apart phenomenon and at the same time to allow for individual changes, we may consider an autoregressive model instead of a model with orthogonal increments. This leads to the next example.

Example 10.2 (Counterpart to Example 9.5)

The difference between this and Example 10.1 is that instead of the movement equation (10.13) we have an autoregressive process of order 1, i.e. we have that

$$\boldsymbol{\mu}(\Theta_{n+1}) = \boldsymbol{\mu}_0 + \rho\left(\boldsymbol{\mu}(\Theta_n) - \boldsymbol{\mu}_0\right) + \boldsymbol{\delta}_{n+1}, \tag{10.36}$$

$$\text{where } |\rho| < 1, \tag{10.37}$$

$$\boldsymbol{\mu} = E\left[\boldsymbol{\mu}(\Theta_1)\right]$$

The *innovations* $\boldsymbol{\delta}_j$ and the deviations $\boldsymbol{\varepsilon}_j$ are defined exactly as in Example 10.1 and satisfy the same conditions as there. As in Example 10.1, we further assume that $\text{Cov}\left(\boldsymbol{\mu}(\Theta_1), \boldsymbol{\mu}'(\Theta_1)\right)$ has the $D_{p,q}$ model form.

Remarks:

- Conceptually, one can also think of the individual premium as being divided into an overall "collective component" and an "individual component". Assumption (10.36) is then equivalent to saying that the collective, as well as the individual components are autoregressive with the same contraction parameter ρ. We will consider the case where the contraction parameters for the individual and collective process are not the same, in Section 10.6. The extension of Example 10.2 to this case will follow in Example 10.4.
- The conditions (10.36) and (10.37) together with the conditions about the innovations $\boldsymbol{\delta}_n$ imply that (10.36), under "appropriate" assumptions for $\text{Var}\left(\boldsymbol{\mu}(\Theta_1)\right)$, is a stationary process (cf. (9.46) in Example 9.5). Note also, that $E\left[\boldsymbol{\mu}(\Theta_j)\right] = \boldsymbol{\mu}$ for all j.
- In contrast to Example 10.1, in this example the portfolio does not drift apart.

The recursion formulae are derived using exactly the same considerations as in Example 10.1. From the anchoring to the updating, nothing changes.

The corresponding formulae (10.15)–(10.19), resp. (10.22)–(10.27) derived in Example 10.1 can be used without change.

Only the movement equations look different. Instead of (10.28) - (10.31), we have

$$\mu_{n+1|n}^{(i)} = \mu_i + \rho\left(\mu_{n|n}^{(i)} - \mu_i\right), \tag{10.38}$$

$$Q_{n+1|n} = q_{n+1|n} I + p_{n+1|n} E, \tag{10.39}$$

$$\text{where} \quad q_{n+1|n} = \rho^2\, q_{n|n} + \tau^2, \tag{10.40}$$

$$p_{n+1|n} = \rho^2\, p_{n|n} + \omega^2. \tag{10.41}$$

These formulae follow from (10.36) and equations (10.9)–(10.10) in Theorem 10.1.

10.5 Multidimensional Evolutionary Regression Models

In contrast to the set-up in Section 10.3, where we directly modelled the multidimensional process $\boldsymbol{\mu}(\Theta_1), \boldsymbol{\mu}(\Theta_2), \ldots, \boldsymbol{\mu}(\Theta_n), \ldots$, we take the regression approach (multidimensional version of the procedure in Theorem 9.6 and Corollary 9.7). We assume that

$$\boldsymbol{\mu}(\Theta_k) = H_k\, \boldsymbol{\beta}(\Theta_k), \tag{10.42}$$

$$\begin{aligned} \text{where } \boldsymbol{\mu}(\Theta_k) &= N \times 1 \text{ vector}, \\ \boldsymbol{\beta}(\Theta_k) &= p \times 1 \text{ vector}, \\ H_k &= N \times p \text{ matrix}. \end{aligned}$$

We model the process of the regressors

$$\{\boldsymbol{\beta}(\Theta_1), \boldsymbol{\beta}(\Theta_2), \ldots, \boldsymbol{\beta}(\Theta_n), \ldots\}.$$

This generalization allows a greater flexibility and variety of models. This is because we have so much freedom in the choice of the regressors.

From the point of view of the statistician, since these regressors have to be estimated from the data, we need to be careful in our choice of regressors. While a very highly parameterized model may lead to a good fit, it may have the undesirable feature of having a large estimation error and therefore be a poor predictor.

For the process of regressors, we make the following assumptions, where as before (cf. Section 10.3), we consider two cases:

- **Case I:** The process $\{\boldsymbol{\beta}(\Theta_j) : j = 1, 2, \ldots\}$ is a multidimensional process with orthogonal increments, i.e.

$$\operatorname{Pro}\left[\boldsymbol{\beta}(\Theta_{n+1}) | L(1, \boldsymbol{\beta}(\Theta_1), \ldots, \boldsymbol{\beta}(\Theta_n))\right] = \boldsymbol{\beta}(\Theta_n).$$

- **Case II:** The process $\{\boldsymbol{\beta}(\Theta_j) : j = 1, 2, \ldots\}$ is a multidimensional process having the linear Markov property, i.e.

$$\text{Pro}\left[\boldsymbol{\beta}(\Theta_{n+1}) \,|L(\mathbf{1}, \boldsymbol{\beta}(\Theta_1), \ldots, \boldsymbol{\beta}(\Theta_n))\right] = A_{n+1}\,\boldsymbol{\beta}(\Theta_n) + \mathbf{b}_{n+1}.$$

Theorem 10.3 is the multidimensional version of the Theorem 10.1. This "general form of the Kalman Filter" corresponds to the form which is usually referred to as the Kalman Filter in the literature.

Theorem 10.3 (General form of the Kalman Filter).

i) Anchoring (period 1)

$$\boldsymbol{\beta}_{1|0} = E\left[\boldsymbol{\beta}(\Theta_1)\right], \tag{10.43}$$

$$Q_{1|0} = \text{Cov}\left(\boldsymbol{\beta}(\Theta_1), \boldsymbol{\beta}'(\Theta_1)\right). \tag{10.44}$$

ii) Recursion (periods $n \geq 1$)

a) Updating

$$\boldsymbol{\beta}_{n|n} = \boldsymbol{\beta}_{n|n-1} + G_n\left(\mathbf{X}_n - H_n\boldsymbol{\beta}_{n|n-1}\right), \tag{10.45}$$

$$\text{where } G_n = Q_{n|n-1}H_n'\left(H_nQ_{n|n-1}H_n' + S_n\right)^{-1}, \tag{10.46}$$

$$S_n = E\left[\text{Cov}\left(\mathbf{X}_n, \mathbf{X}_n'|\Theta_n\right)\right],$$

$$Q_{n|n} = \left(I - G_nH_n\right)Q_{n|n-1}. \tag{10.47}$$

b) Movement in parameter space

Case I:

$$\boldsymbol{\beta}_{n+1|n} = \boldsymbol{\beta}_{n|n}, \tag{10.48}$$

$$Q_{n+1|n} = Q_{n|n} + D_{n+1}, \tag{10.49}$$

where

$$D_{n+1} = E\left[\left(\boldsymbol{\beta}(\Theta_{n+1}) - \boldsymbol{\beta}(\Theta_n)\right)\cdot\left(\boldsymbol{\beta}(\Theta_{n+1}) - \boldsymbol{\beta}(\Theta_n)\right)'\right].$$

Case II:

$$\boldsymbol{\beta}_{n+1|n} = A_{n+1}\boldsymbol{\beta}_{n|n} + \mathbf{b}_{n+1}, \tag{10.50}$$

$$Q_{n+1|n} = A_{n+1}Q_{n|n}A_{n+1}' + D_{n+1}, \tag{10.51}$$

where

$$D_{n+1} = E\left[\left(\boldsymbol{\beta}(\Theta_{n+1}) - A_{n+1}\boldsymbol{\beta}(\Theta_n) - \mathbf{b}_n\right) \cdot \left(\boldsymbol{\beta}(\Theta_{n+1}) - A_{n+1}\boldsymbol{\beta}(\Theta_n) - \mathbf{b}_n\right)'\right].$$

Remarks:

- Notice the formal identity of Theorem 10.3 with Theorem 9.8. The row vector $\mathbf{y}_n'$ in Theorem 9.8 will be replaced by the matrix H_n in Theorem 10.3.
- In the literature, the matrix G_n used in (10.46) is referred to as the "Kalman gain".
- The non-evolutionary regression model is included in Case I, because in this case we have $\boldsymbol{\beta}(\Theta_n) = \boldsymbol{\beta}(\Theta)$ for all n (compare third remark after Corollary 9.9). The algorithm of Theorem 10.3 hence lends itself to finding the credibility premium in the multidimensional case (Chapter 7) also if the observation vector is of lower dimension than the estimand vector.

On the proof of Theorem 10.3

We omit a formal proof, because it follows exactly the same series of steps as in Theorem 9.8, except that here we are in the multidimensional setting. □

10.6 Decomposition into an Individual and a Common Component

As in Section 10.4, we treat here the realistic situation that the risk profile of an individual depends on both individual factors and global factors, affecting all the risk profiles of all members of the collective.

As in Section 10.4, let

$$\mathbf{X} = \{\mathbf{X}_1, \mathbf{X}_2, \dots, \mathbf{X}_j, \dots\} \quad \text{with } \mathbf{X}_j = (X_{1j}, X_{2j}, \dots, X_{Nj})'$$

be the time series of the observations of all the risks $i = 1, 2, \dots, N$ of the collective, and let

$$\boldsymbol{\Theta} = \{\Theta_1, \Theta_2, \dots, \Theta_j, \dots\}$$

be the time series of the (multidimensional) risk profiles.

We also have the time series of the (multidimensional) individual premiums

$$\boldsymbol{\mu}(\boldsymbol{\Theta}) = \{\boldsymbol{\mu}(\Theta_1), \boldsymbol{\mu}(\Theta_2), \dots, \boldsymbol{\mu}(\Theta_j), \dots\},$$

where

$$\begin{aligned} \boldsymbol{\mu}(\Theta_j) &= \{\mu_i(\Theta_j) : i = 1, 2, \dots N\}, \\ \mu_i(\Theta_j) &= \text{correct individual premium of the risk} \\ &\qquad i \text{ at timepoint } j. \end{aligned}$$

As in Section 10.4 we assume independence of errors, in particular

$$\mathbf{X}_j = \boldsymbol{\mu}(\Theta_j) + \boldsymbol{\varepsilon}_j,$$

where

$$E[\boldsymbol{\varepsilon}_j] = \mathbf{0},$$
$$\mathrm{Cov}\left(\boldsymbol{\varepsilon}_j, \boldsymbol{\varepsilon}_j^{'}\right) = \sigma^2 I,$$
$$\mathrm{Cov}\left(\boldsymbol{\varepsilon}_j, \boldsymbol{\varepsilon}_k^{'}\right) = \mathbf{0} \quad \text{for } j \neq k.$$

However, for the movement in the parameter space, we now consider another model.

In Section 10.4, in both Example 10.1 and Example 10.2, we merely split each of the innovations $\boldsymbol{\delta}_{n+1}$ into a collective and an individual component. In this section, the innovations are again defined as in the earlier examples, but in addition we take the approach that the individual premium $\boldsymbol{\mu}(\Theta_j)$ is composed of both a global and an individual part. For the correct premium of the ith risk, we then have that

$$\mu_i(\Theta_j) = \psi_j + \nu_{ij}, \tag{10.52}$$

$$\text{where} \quad \psi_j = \text{global component}, \tag{10.53}$$
$$\nu_{ij} = \text{individual component}. \tag{10.54}$$

In order that the global and individual contributions are identifiable, we further assume that

$$\sum_{i=1}^{N} E[\nu_{ij}] = 0. \tag{10.55}$$

Note that (10.55) implies

$$\frac{1}{N}\sum_{i=1}^{N} E[\mu_i(\Theta_j)] = E[\psi_j] = \psi_0.$$

Equations (10.52)–(10.54) in vector and matrix notation give that

$$\boldsymbol{\mu}(\Theta_j) = H\,\boldsymbol{\beta}(\Theta_j), \tag{10.56}$$

$$\text{where} \quad \boldsymbol{\beta}(\Theta_j)' = \left(\psi_j, \boldsymbol{\nu}_j^{'}\right) = \left(\psi_j, \nu_{1j}, \dots, \nu_{Nj}\right), \tag{10.57}$$

$$H = \begin{pmatrix} 1 & 1 & 0 & \dots & 0 \\ \vdots & 0 & \ddots & & \vdots \\ \vdots & \vdots & & \ddots & 0 \\ 1 & 0 & \dots & 0 & 1 \end{pmatrix} = (\mathbf{1}, I)\ , \tag{10.58}$$

where $\mathbf{1}$ is the vector of 1's and I is the identity matrix of dimension $N \times N$. Equation (10.56) is a regression equation, and so the model approach from

Section 10.5 can be used here. Note also that equation (10.45) from Theorem 10.3 may also be written as

$$\boldsymbol{\beta}_{n|n} = \boldsymbol{\beta}_{n|n-1} + G_n \left(\mathbf{X}_n - \boldsymbol{\mu}_{n|n-1} \right)$$

$$\text{where } \mu^{(i)}_{n|n-1} = \psi_{n|n-1} + \nu^{(i)}_{n|n-1}.$$

In the following we reconsider the Examples 10.1 and 10.2 from this point of view of premium decomposition.

Example 10.3 (another form of Example 10.1)

As in Example 10.1 on page 256 it should hold for the movements that

$$\boldsymbol{\mu}(\Theta_{n+1}) = \boldsymbol{\mu}(\Theta_n) + \xi_{n+1}\mathbf{1} + \boldsymbol{\eta}_{n+1}. \tag{10.59}$$

Using the decomposition into global and individual components, we can also write (10.59) as

$$\begin{aligned} \psi_{n+1} &= \psi_n + \xi_{n+1}, \\ \nu_{i,n+1} &= \nu_{i,n} + \eta_{i,n+1}, \end{aligned}$$

where $\left\{\xi_j, \eta_{ij}, \varepsilon_{ij} : j = 1, 2, \ldots;\ i = 1, 2, \ldots, N\right\}$ are independent random variables, which are also independent of the components of $\boldsymbol{\mu}(\Theta_{j-1})$, with first- and second-order moments

$$\begin{aligned} E\left[\xi_j\right] = E\left[\eta_{ij}\right] = E\left[\varepsilon_{ij}\right] &= 0 \quad \text{for all } i \text{ and } j, \\ \operatorname{Var}\left[\varepsilon_{ij}\right] &= \sigma^2 \quad \text{for all } i \text{ and } j, \\ \operatorname{Var}\left[\eta_{ij}\right] &= \tau^2 \quad \text{for all } i \text{ and } j, \\ \operatorname{Var}\left[\xi_j\right] &= \omega^2 \quad \text{for } j = 1, 2, \ldots . \end{aligned}$$

The conditions for the multidimensional credibility regression models from Section 10.5 are satisfied, where in the regression equation (10.42) the regression matrix H_k is replaced by H, as defined in (10.58), and the regression vector $\boldsymbol{\beta}(\Theta_n)$ is given by (10.57). For the matrices S_n and D_n in Theorem 10.3 we have that

$$S_n = \sigma^2 I,$$

$$D_n = \begin{pmatrix} \omega^2 & & & 0 \\ & \tau^2 & & \\ & & \ddots & \\ 0 & & & \tau^2 \end{pmatrix} = \begin{pmatrix} \omega^2 & \mathbf{0}' \\ \mathbf{0} & \tau^2 I \end{pmatrix}.$$

In order to apply Theorem 10.3, it remains to specify the initial values (10.43) and (10.44). If we assume that the process for the global development

$$\{\psi_j : j = 0, 1, 2, \ldots\}$$

begins at the value ψ_0 and that the superimposed individual process

$$\{\nu_{ij} : j = 1, 2, \ldots\}, \quad i = 1, 2, \ldots, N$$

begins at the value $\boldsymbol{\nu}_0$, then we have

$$\boldsymbol{\beta}_{1|0} = \begin{pmatrix} \psi_0 \\ \boldsymbol{\nu}_0 \end{pmatrix}, \tag{10.60}$$

and

$$Q_{1|0} = \begin{pmatrix} \omega^2 & \mathbf{0}' \\ \mathbf{0} & \tau^2 I \end{pmatrix}. \tag{10.61}$$

Remark:

- Assumption (10.61) can be easily generalized to the situation where we have a stochastic starting point. The following recursion formulae still hold and the initial matrix $Q_{1|0}$ still has the special form (10.80), as defined on page 269.

Now, we can just apply the whole machinery of the Kalman Filter as described in Theorem 10.3. However, as already done in Example 10.1, by updating the scalar quantities, the recursive formulae can be written out explicitly using the results of Theorem 10.3.

Let

$$Q_{n|n-1} = \begin{pmatrix} s_{n|n-1} & t_{n|n-1}\,\mathbf{1}' \\ t_{n|n-1}\,\mathbf{1} & q_{n|n-1}\, I + r_{n|n-1}\, E \end{pmatrix}$$

be the loss matrix of $\boldsymbol{\beta}_{n|n-1}$, where $\mathbf{1}$ (respectively E) is a vector of length N (respectively, a matrix of dimension $N \times N$, the elements of which are 1's). This form of the loss matrix is a consequence of the fact that the elements $\eta^{(i)}_{n|n-1}$, $i = 1, 2, \ldots$, are exchangeable.

As we will show, we get the following updating formulae for the components of ψ and ν:

$$\psi_{n|n} = \psi_{n|n-1} + \alpha_n (1 - \gamma_n) \left(\overline{X}_n - \overline{\mu}_{n|n-1} \right) \tag{10.62}$$

$$\begin{aligned} \nu^{(i)}_{n|n} &= \nu^{(i)}_{n|n-1} + \beta_n \left(X_{in} - \mu^{(i)}_{n|n-1} \right) \\ &\quad + \alpha_n (\gamma_n - \beta_n) \left(\overline{X}_n - \overline{\mu}_{n|n-1} \right). \end{aligned} \tag{10.63}$$

The weights α_n, β_n, γ_n can be calculated from the elements of $Q_{n|n-1}$ by applying the following formulae, where, for simplicity, we write s, t, etc. instead of $s_{n|n-1}$, $t_{n|n-1}, \ldots$:

$$\alpha_n = \frac{Np}{Np + q + \sigma^2}, \quad \text{where } p = s + 2t + r, \tag{10.64}$$

$$\beta_n = \frac{q}{q + \sigma^2}, \tag{10.65}$$

$$\gamma_n = \frac{r + t}{(r + t) + (s + t)} = \frac{r + t}{p}. \tag{10.66}$$

The updating of the loss matrix again gives a matrix of the form

$$Q_{n|n} = \begin{pmatrix} s_{n|n} & t_{n|n} \cdot \mathbf{1}' \\ t_{n|n} \cdot \mathbf{1} & q_{n|n}\, I + r_{n|n}\, E \end{pmatrix}. \tag{10.67}$$

The parameters in $Q_{n|n}$ are given by (10.68)–(10.71), where again, for simplicity, the corresponding elements of $Q_{n|n}$ are written with "*", so for example, we write s^* instead of $s_{n|n}$.

$$q^* = (1 - \beta_n)\, q, \tag{10.68}$$

$$s^* = s - \alpha_n (1 - \gamma_n)^2\, p, \tag{10.69}$$

$$t^* = \alpha_n (1 - \gamma_n) \frac{\sigma^2}{N} - s^*, \tag{10.70}$$

$$r^* = \alpha_n (\gamma_n - \beta_n) \frac{\sigma^2}{N} - t^*. \tag{10.71}$$

Finally, for the movement in parameter space we get

$$\psi_{n+1|n} = \psi_{n|n}, \tag{10.72}$$

$$\nu^{(i)}_{n+1|n} = \nu^{(i)}_{n|n}, \tag{10.73}$$

$$Q_{n+1|n} = Q_{n|n} + D_n, \tag{10.74}$$

and by using the special form of the Q-matrices we find

$$Q_{n+1|n} = \begin{pmatrix} s_{n+1|n} & t_{n+1|n} \cdot \mathbf{1}' \\ t_{n+1|n} \cdot \mathbf{1} & q_{n+1|n}\, I + r_{n+1|n}\, E \end{pmatrix}, \tag{10.75}$$

$$\text{where} \quad \begin{aligned} s_{n+1|n} &= s_{n|n} + \omega^2, \\ t_{n+1|n} &= t_{n|n}, \\ q_{n+1|n} &= q_{n|n} + \tau^2, \\ r_{n+1|n} &= r_{n|n}. \end{aligned}$$

Remarks:

- Note that

$$\begin{aligned}\mu_{n|n}^{(i)} &= \psi_{n|n} + \nu_{n|n}^{(i)} \\ &= \mu_{n|n-1}^{(i)} + \beta_n \left(X_{in} - \mu_{n|n-1}^{(i)}\right) \\ &\quad + (1-\beta_n)\,\alpha_n \left(\overline{X}_n - \overline{\mu}_{n|n-1}\right). \end{aligned} \tag{10.76}$$

This is identical to (10.22). But observe that in Example 10.3, in contrast to Example 10.1, we now also have estimators for the global and for the individual components, that we would not "intuitively" have "derived" from (10.22).

- Let $\widetilde{Q}_{n|n}$ be the loss matrix of the vector with components $\mu_{n|n}^{(i)}$ ($i = 1, 2, \dots, N$). Then it holds that

$$\begin{aligned}\widetilde{Q}_{n|n} &= HQ_{n|n}H' \\ &= \widetilde{q}_{n|n}I + \widetilde{p}_{n|n}E, \end{aligned} \tag{10.77}$$

$$\text{where}\quad \widetilde{q}_{n|n} = q_{n|n} = (1-\beta_n)\,q_{n|n-1}, \tag{10.78}$$

$$\begin{aligned}\widetilde{p}_{n|n} &= s_{n|n} + 2\,t_{n|n} + r_{n|n} \\ &= \alpha_n\,(1-\beta_n)\,\frac{\sigma^2}{N} \\ &= (1-\alpha_n)\,(1-\beta_n)^2\,\widetilde{p}_{n|n-1}. \end{aligned} \tag{10.79}$$

(10.77)–(10.79) are identical to (10.25)–(10.27).

- In Example 10.1, we saw that the loss matrices $Q_{n|n-1}$ and $Q_{n|n}$ all have the $D_{p,q}$-form $qI + pE$, $q, p \in R$. Here, in Example 10.3, we see that the loss matrices also have a very special structure. They are of the form

$$\begin{pmatrix} s & t\cdot\mathbf{1}' \\ t\cdot\mathbf{1} & qI + rE \end{pmatrix} \tag{10.80}$$

where s, t, q, r are real numbers. For the initial matrix (10.61) we have $t = 0$ and $r = 0$, but one could just as well use an arbitrary matrix of the form (10.80) as initial value and the whole procedure would still work.

Derivation of (10.62)–(10.71):
We write

$$Q_{n|n-1} := Q = \begin{pmatrix} s & t\cdot\mathbf{1}' \\ t\cdot\mathbf{1} & qI + rE \end{pmatrix}$$

and carry out the calculations as described in Theorem 10.3. It holds that

$$
\begin{aligned}
H_n &= H = (\mathbf{1}\ ,\ I)\,, \\
S_n &= \sigma^2 I, \\
QH' &= \begin{pmatrix} s & t\cdot\mathbf{1}' \\ t\cdot\mathbf{1} & qI+rE \end{pmatrix} \begin{pmatrix} \mathbf{1}' \\ I \end{pmatrix} \\
&= \begin{pmatrix} (s+t)\ \mathbf{1}' \\ q\,I+(r+t)\,E \end{pmatrix}.
\end{aligned}
\tag{10.81}
$$

Because HQH' is the loss matrix of $\boldsymbol{\mu}_{n|n-1}$, it follows from (10.33) that

$$
\begin{aligned}
\left(HQH'+\sigma^2 I\right)^{-1} &= \frac{1}{q+\sigma^2}\left(I-\frac{\alpha}{N}E\right), \\
\alpha &= \frac{Np}{Np+q+\sigma^2}\,, \qquad \text{(cf. (10.35) and (10.24))} \\
p &= \text{quadratic loss of } \mu^{(i)}_{n|n-1} \\
&= s+2t+r.
\end{aligned}
$$

With this we get

$$
\begin{aligned}
G_n &= \frac{1}{q+\sigma^2}\begin{pmatrix} (s+t)\,\mathbf{1}' \\ qI+(r+t)\,E \end{pmatrix}\left(I-\frac{\alpha}{N}E\right) \\
&= \begin{pmatrix} a\,\mathbf{1}' \\ \beta I+bE \end{pmatrix},
\end{aligned}
$$

$$
\begin{aligned}
\text{where}\quad \beta &= \frac{q}{q+\sigma^2}, \quad \text{(cf. (10.34))} \\
a &= \frac{s+t}{q+\sigma^2}\,(1-\alpha) \\
&= \frac{1}{N}\frac{s+t}{p}\alpha \\
&= \frac{1}{N}\alpha\,(1-\gamma)\,, \\
\gamma &= \frac{r+t}{(r+t)+(s+t)} = \frac{r+t}{p}, \\
b &= \frac{r+t}{q+\sigma^2}\,(1-\alpha)-\frac{q}{q+\sigma^2}\frac{\alpha}{N} \\
&= \frac{\alpha}{N}\,(\gamma-\beta)\,.
\end{aligned}
$$

With this we have shown (10.62)–(10.66).

For the updating of the loss matrix $Q_{n|n-1}$ we could also apply Theorem 10.3 and explicitly do the calculations. However, this is a tedious procedure and leads to clumsy, difficult to interpret formulae. For this reason, we approach the problem in another, more direct way. For convenience, we write Q for $Q_{n/n-1}$ (with elements q, s, t, r) and Q^* for $Q_{n/n}$ (with elements q^*, s^*, t^*, r^*). First we show the following relationships, which are interesting in themselves. It holds that

$$(s^* + t^*) = \alpha_n (1 - \gamma_n) \frac{\sigma^2}{N}, \tag{10.82}$$

$$(r^* + t^*) = \alpha_n (\gamma_n - \beta_n) \frac{\sigma^2}{N}, \tag{10.83}$$

$$q^* = \beta_n \sigma^2 = (1 - \beta_n) q. \tag{10.84}$$

Notice, that from equations (10.82) and (10.83) we get

$$\begin{aligned} p^* &= s^* + 2t^* + r^*, \\ &= \alpha_n (1 - \beta_n) \frac{\sigma^2}{N}, \\ &= (1 - \alpha_n) (1 - \beta_n)^2 p, \end{aligned} \tag{10.85}$$

which is identical to formula (10.27).

Derivation of (10.82) and (10.83):
Let

$$\begin{aligned} \mu_{in} &= \psi_n + \nu_{in}, \\ \overline{\mu}_n &= \frac{1}{N} \sum_{i=1}^{N} \mu_{in}. \end{aligned}$$

We have

$$\begin{aligned} s^* + t^* &= E\left[\left(\overline{\mu}_n - \overline{\mu}_{n|n}\right)\left(\psi_n - \psi_{n|n}\right)\right] \\ &= E\left[\overline{\mu}_n \left(\psi_n - \psi_{n|n}\right)\right] \\ &= E\left[\left(\overline{\mu}_n - \overline{X}_n\right)\left(\psi_n - \psi_{n|n}\right)\right] \\ &= E\left[-\overline{\varepsilon}_n \left(\psi_n - \psi_{n|n-1} - \alpha_n (1 - \gamma_n)\left(\overline{X}_n - \overline{\mu}_{n|n-1}\right)\right)\right] \\ &= E\left[\overline{\varepsilon}_n \left(\alpha_n (1 - \gamma_n)\left(\overline{X}_n - \overline{\mu}_n\right)\right)\right] \\ &= \alpha_n (1 - \gamma_n) \frac{\sigma^2}{N}. \end{aligned}$$

For the second and the third equality we have used the fact that $\psi_{n|n}$ is the credibility estimator of ψ_n and therefore $\psi_n - \psi_{n|n}$ is orthogonal to $\overline{\mu}_{n|n}$ and $\overline{X}_n$. Analogously we get

$$\begin{aligned}
r^* + t^* &= E\left[\left(\mu_{kn} - \mu_{n|n}^{(k)}\right)\left(\nu_{in} - \nu_{n|n}^{(i)}\right)\right] \quad (k \neq i) \\
&= E\left[\left(\mu_{kn} - X_{kn}\right)\left(\nu_{in} - \nu_{n|n}^{(i)}\right)\right] \\
&= E\left[-\varepsilon_{kn}\left(\nu_{in} - \nu_{n|n-1}^{(i)} - \beta_n\left(X_{in} - \mu_{n|n-1}^{(i)}\right)\right.\right. \\
&\qquad \left.\left. - \alpha_n\left(\gamma_n - \beta_n\right)\left(\overline{X}_n - \overline{\mu}_{n|n-1}\right)\right)\right] \\
&= E\left[\varepsilon_{kn}\left(\alpha_n\left(\gamma_n - \beta_n\right)\left(\overline{X}_n - \overline{\mu}_n\right)\right)\right] \\
&= \alpha_n\left(\gamma_n - \beta_n\right)\frac{\sigma^2}{N}.
\end{aligned}$$

Thus we have proved (10.82) and (10.83).

Next we show the validity of (10.69). We have that

$$\begin{aligned}
s &= E\left[\left(\psi_{n|n-1} - \psi_n\right)^2\right] \\
&= E\left[\left(\psi_{n|n} - \psi_n\right)^2\right] + E\left[\left(\psi_{n|n} - \psi_{n|n-1}\right)^2\right] \\
&= s^* + E\left[\left(\alpha_n\left(1 - \gamma_n\right)\left(\overline{X_n} - \overline{\mu}_{n|n-1}\right)\right)^2\right] \\
&= s^* + \alpha_n^2\left(1 - \gamma_n\right)^2 E\left[\left(\overline{X_n} - \overline{\mu}_{n|n-1}\right)^2\right].
\end{aligned}$$

$$\begin{aligned}
E\left[\left(\overline{X_n} - \overline{\mu}_{n|n-1}\right)^2\right] &= E\left[\left(\overline{X_n} - \overline{\mu}_n\right)^2\right] + E\left[\left(\overline{\mu}_n - \overline{\mu}_{n|n-1}\right)^2\right] \\
&= \frac{\sigma^2}{N} + \frac{1}{N^2}\left(\mathbf{1}'\left(qI + pE\right)\mathbf{1}\right) \\
&= \frac{1}{N}\left(\sigma^2 + Np + q\right)\right) \\
&= \frac{p}{\alpha_n}.
\end{aligned}$$

From this we get directly that

$$s^* = s - \alpha_n\left(1 - \gamma_n\right)^2 p, \tag{10.86}$$

which is what we had to show.

From equation (10.86) as well as (10.82)–(10.84) it follows that equations (10.70) and (10.71) are true. Finally, equation (10.84) is identical to equation (10.26), the truth of which was shown in example 10.1. This concludes the proof of (10.62)–(10.71). □

Example 10.4 (extended counterpart to Example 10.2)

Analogous to Example 10.2, the difference between Example 10.4 and example 10.3 lies only in the movements in parameter space. All other conditions are identical.

In Example 10.2 (see page 261) the movements in parameter space were modelled by an autoregressive process

$$\boldsymbol{\mu}(\Theta_{n+1}) - \boldsymbol{\mu} = \rho\left(\boldsymbol{\mu}(\Theta_n) - \boldsymbol{\mu}\right) + \boldsymbol{\delta}_{n+1} \tag{10.87}$$

$$\text{where } |\rho| < 1, \quad \boldsymbol{\mu} = E\left[\boldsymbol{\mu}(\Theta_1)\right].$$

In this example we now assume that both the global components and the individual components can be modelled by two different autoregressive processes

$$\psi_{n+1} - \psi_0 = \rho_0\left(\psi_n - \psi_0\right) + \xi_{n+1}, \tag{10.88}$$

$$\nu_{i,n+1} - \nu_{i0} = \rho_1\left(\nu_{in} - \nu_{i0}\right) + \eta_{i,n+1}, \tag{10.89}$$

$$\begin{aligned} \text{where } \psi_0 &= E\left[\psi_n\right], \\ \nu_{i0} &= E\left[\nu_{in}\right], \\ |\rho_0|, |\rho_1| &< 1, \\ \sum_i \nu_{i0} &= 0. \end{aligned}$$

For the components of the $\boldsymbol{\mu}(\Theta_n)$-process we still have

$$\mu(\Theta_n) = \psi_n + \nu_{in},$$

i.e. they are the sum of the processes of the individual and the global components. Notice that the contraction parameter ρ_0 for the global parameter can be different to that for the individual process, ρ_1 . This is not the same as (10.87) (see also the first point in the remarks on page 261). When we consider (10.87) decomposed into individual components and global components, then (10.87) means that both the global and the individual components change with the same contraction parameter. Example 10.4 is therefore a true extension of Example 10.2.

Analogous to Example 10.2, where the explicit formulae for the credibility estimator can be derived, here the credibility formulae can easily be derived from Example 10.3 and Theorem 10.3, (10.50)–(10.51).

From anchoring to updating, nothing changes from example 10.3. Thus, the corresponding formulae can be directly taken from there:

Anchoring: formulae (10.60) and (10.61), (10.90)

Updating: formulae (10.62)–(10.71). (10.91)

Only the equations for the movement in parameter space look different. Instead of (10.72) and (10.73) we now have

$$\psi_{n+1|n} = \rho_0 \, \psi_{n|n} + (1 - \rho_0) \, \psi_0, \tag{10.92}$$

$$\nu_{n+1|n}^{(i)} = \rho_1 \, \nu_{n|n}^{(i)} + (1 - \rho_1) \, \nu_{i0}. \tag{10.93}$$

For the loss matrix, we get from the movement equations and applying Theorem 10.3, (10.50) and (10.51),

$$Q_{n+1|n} = \begin{pmatrix} s_{n+1|n} & t_{n+1|n} \mathbf{1}' \\ t_{n+1|n} \mathbf{1} & q_{n+1|n} I + r_{n+1|n} E \end{pmatrix}. \tag{10.94}$$

$$\text{where} \quad \begin{aligned} s_{n+1|n} &= \rho_0^2 \, s_{n|n} + \omega^2, \\ t_{n+1|n} &= \rho_0 \, \rho_1 \, t_{n|n}, \\ q_{n+1|n} &= \rho_1^2 \, q_{n|n} + \tau^2, \\ r_{n+1|n} &= \rho_1^2 \, r_{n|n}. \end{aligned}$$

Remarks:

- The effect of the movements on the elements of the loss matrix is slightly different from that in Example 10.3 (compare (10.94) with (10.75)).
- Also in the new movement equations, in the step from $Q_{n|n}$ to $Q_{n+1|n}$, the structure (10.80) stays the same. This is the reason why the updating formulae from Example 10.3 can be used without alteration.
- If one is only interested in the sum of individual and global components

  $$\mu_{ni} = \psi_n + \nu_{in},$$

  then the updating formula (10.22) still holds for the credibility estimator. In contrast, the recursion formulae (10.38)–(10.41) from Example 10.2 remain only correct in the case where ρ_1 and ρ_2 are equal.

A

Appendix A: Basic Elements from Probability Theory

A.1 Random Variables, Distribution Functions and Moments

i) **Distribution Function and Density**
Let X be a real-valued random variable defined on a probability space $(\Omega, \mathfrak{A}, P)$.

$F(x) := P(X \leq x)$ defined for every real argument $x \in \mathbb{R}$ is called the *distribution function* of the random variable X.

If there exists a real-valued function $f(x)$ such that

$$F(x) = \int_{-\infty}^{x} f(y)dy$$

for all x, this function is called the *density* of the random variable X.

If the random variable X takes only a countable number of values $\{x_i : i \in A \subseteq \mathbb{N}\}$, then X has a *discrete distribution* with discrete probabilities $p_i := P(X = x_i)$.
Then we have

$$F(x) = \sum_{\{i,\, x_i \leq x\}} p_i.$$

ii) **Expected Value**
For a (Borel measurable) function g we define $E[g(X)]$, the *expected value* of $g(X)$, as follows:
a) If X has a *density* $f(x)$ then

$$E[g(X)] = \int_{-\infty}^{+\infty} g(x)\, f(x)dx.$$

b) If X has a *discrete distribution* then

$$E[g(X)] = \sum_i p_i\, g(x_i).$$

c) In the general case the expected value is defined by the *Stieltjes integral*

$$E[g(X)] = \int g(x)\, dF(x).$$

- The more elementary way to interpret c) uses the *Riemann–Stieltjes integral.* With this interpretation all the above integrals are improper integrals and must converge *absolutely* (this absolute convergence is also required for the case b)).
- The more elegant advanced interpretation uses the *Lebesgue–Stieltjes integral.*

Remark:
The special random variable I_A, which takes the value 1 if A occurs and 0 otherwise, is called *indicator function of the event A*. Observe that $E[I_A] = P(A)$.

iii) **Moments and Central Moments**

moments: $\mu_k(X) := E\left[X^k\right]$,

central moments: $\alpha_k(X) := E[(X - E[X])^k]$.

The most important moments are:
expected value: $\mu_X := E[X]$,

variance: $\sigma_X^2 := \alpha_2(X) = E[(X - \mu_X)^2]$,

skewness: $\gamma(X) := \left(\alpha_3(X)/\sigma_X^3\right)$,

kurtosis: $\kappa(X) := (\alpha_4(X)/\sigma_X^4) - 3$.

Derived quantities:
standard deviation: $\sigma_X := \sqrt{\sigma_X^2} = \sqrt{\text{variance}}$,

coefficient of variation: $\text{CoVa}(X) := \sigma_X/\mu_X$.

A.2 Special Functions

The following functions are sometimes useful:

- **Moment Generating Function**:
  $M_X(z) := E[e^{zX}] = \int\limits_{-\infty}^{+\infty} e^{zx} dF_X(x).$

  The derivatives at zero reproduce the *moments*

$$M_X(0) = 1,$$
$$M_X^{'}(0) = E[X],$$
$$M_X^{(k)}(0) = E[X^k].$$

- **Cumulant Generating Function:**
  $K_X(z) := \ln M_X(z).$

  Its derivatives $K_X^{(k)}(0)$ at zero reproduce the cumulants β_k $(k = 1, 2, \ldots)$.
  It holds, that

$$\beta_1 (X) = \mu_1 (X) = \mu_X,$$
$$\beta_2 (X) = \alpha_2 (X) = \sigma_X^2 ,$$
$$\beta_3 (X) = \alpha_3 (X),$$
$$\text{but } \beta_k \neq \alpha_k (X) \text{ for } k > 3.$$

A.3 Multidimensional Distributions

i) **Multidimensional Distribution Functions**
Let $X_1, X_2, \ldots, X_n$ be random variables defined on the probability space $(\Omega, \mathfrak{A}, P)$.

$F(x_1, x_2, \ldots, x_n) := P(X_1 \leq x_1, X_2 \leq x_2, \ldots, X_n \leq x_n)$ defined for every real n-tuple $(x_1, x_2, \ldots, x_n) \in \mathbb{R}^n$ is called the *n-dimensional distribution function* of the random variables $X_1, X_2, \ldots, X_n$.

The *marginal distribution functions* of the random variables X_k are $P(X_k \leq x_k) = F(\infty, \ldots, \infty, x_k, \infty, \ldots, \infty) = F(x_k)$.

Remark on the notation:
To abbreviate the notation we often let the argument of the function tell us what distribution function we are looking at. For instance, we denote by $F(x_1, x_2, \ldots, x_n)$ the *n-dimensional distribution function* of $X_1, X_2, \ldots, X_n$, by $F(x_k)$ the marginal distribution function of X_k, by $F(x_l)$ the marginal distribution function of X_l. The same notation philosophy is also used for other functions (densities, conditional distribution functions, etc.).

ii) **Independence**
The random variables $X_1, X_2, \ldots, X_n$ are *independent* if the n-dimensional

distribution function factorizes, i.e. if

$$F(x_1, x_2, \dots, x_n) = \prod_{k=1}^{n} F(x_k) .$$

For independent random variables we have

$$E\left[\prod_{k=1}^{n} g_k(X_k)\right] = \prod_{k=1}^{n} E[g_k(X_k)]$$

for any (integrable) functions g_k.

iii) **Covariance and Correlation Coefficient**
The covariance between two random variables X and Y is defined by

$$\mathrm{Cov}(X, Y) := E[(X - \mu_X)(Y - \mu_Y)] .$$

If X and Y are *independent* then $\mathrm{Cov}(X, Y) = 0$. However, the converse is *not* true.

From the covariance one obtains the *correlation coefficient*

$$\rho(X, Y) := \frac{\mathrm{Cov}(X, Y)}{\sigma_X \cdot \sigma_Y} .$$

It always holds that $-1 \leqq \rho(X, Y) \leqq +1$. The border cases -1 and $+1$ are obtained if $X = aY + b$.

A.4 Conditional Probability and Conditional Expectation

i) **Conditional Probability**
Let $(\Omega, \mathfrak{A}, P)$ be a probability space.

For $A, B \in \mathfrak{A}$ and $P(B) > 0$, the conditional probability of A given B is defined by

$$P(A|B) = \frac{P(A \cap B)}{P[B]} .$$

Theorem A.1 (Theorem of total probability).
Let B_i, $i = 1, 2, \dots, I$ be a partition of Ω, i.e. $\bigcup_{i=1}^{I} B_i = \Omega$, $B_i \cap B_k = \phi$ for $i \neq k$. Then

$$P(A) = \sum_{i=1}^{I} P(A|B_i) P(B_i) .$$

Theorem A.2 (Bayes Theorem).
Let B_i, $i = 1, 2, \ldots, I$, be a partition of Ω and $P(A) > 0$, then it holds that

$$P(B_i | A) = \frac{P(A | B_i) P(B_i)}{\sum_{i=1}^{I} P(A | B_i) P(B_i)}.$$

Remarks:

- Let X and Y be two random variables with two-dimensional density $f(x, y)$ and marginal densities $f(x)$ and $f(y)$. Then the analogue to the conditional probability is the conditional density of X given $Y = y$, which is

$$f(x | y) = \frac{f(x, y)}{f(y)}.$$

- We further have

$$f(x) = \int_{-\infty}^{+\infty} f(x | y) f(y) \, dy$$

and

$$f(x | y) = \frac{f(x, y)}{f(y)} = \frac{f(y | x) f(x)}{\int_{-\infty}^{+\infty} f(y | x) f(x) \, dx}.$$

ii) **General Definition of Conditional Expectation and Conditional Probability**

a) The conditional expectation of X given Y is a random variable denoted by $E[X | Y]$, which takes for every realization $Y = y$ the value $E[X | Y = y]$. The y-valued function $E[X | Y = y]$ is such that

$$E[X | Y = y] dF(y) = \int_{x = -\infty}^{+\infty} x \, dF(x, y).$$

This notation is useful, but to understand its meaning one has to integrate both sides over an arbitrary (measurable) y-set.

b) If one chooses the indicator function I_A of the event A instead of X this general definition also yields *conditional probabilities*

$$P[A | Y = y] = E[I_A | Y = y].$$

One sees that in the case of the existence of densities $f(y)$ and $f(x, y)$ we get

$$E[X|Y=y] = \int_{x=-\infty}^{+\infty} \frac{x\,f(x,y)}{f(y)} dx\,,$$
$$= \int_{-\infty}^{+\infty} x\,f(x|y)\,dx,$$

where $f(x|y) = \frac{f(x,y)}{f(y)}$ is the conditional density of X given $Y=y$.

An analogous construction for conditional expectations is obtained if X and Y have discrete probability distributions.

iii) **Properties of Conditional Expectation**
The following properties of conditional expectation are important and often used in the book:

- $E[E[X|Y]] = E[X]$,
- $E[E[X|Y,Z]|Y] = E[X|Y]$,
- If $X = f(Y)$ (more general, if X is Y-measurable) then $E[X|Y] = X$,
- If X and Y are independent then $E[X|Y] = E[X]$.
- Decomposition of covariance:

$$\text{Cov}(X,Y) = E\left[\text{Cov}(X,Y|Z)\right] + \text{Cov}(E[X|Z], E[Y|Z]),$$

which leads for $Y = X$ to

$$\text{Var}(X) = E[\text{Var}(X|Z)] + \text{Var}(E[X|Z])\,.$$

A.5 Two Useful Results

The following two results are often helpful and used in the book.

Theorem A.3. *Let $X_1, X_2, \ldots, X_I$ be independent random variables with $E[X_i] = \mu$ and $Var\,(X_i) = \sigma_i^2$ for $i = 1, 2, \ldots, I$.*
Then the minimum variance unbiased linear estimator of μ is given by

$$\hat{\mu} = \left(\sum\nolimits_{i=1}^{I} \sigma_i^{-2}\right)^{-1} \sum\nolimits_{i=1}^{I} \sigma_i^{-2} X_i,$$

i.e. a weighted mean with the inverse of the variance as weights.

Theorem A.4. *Let $X_1, X_2, \ldots, X_I$ be conditionally independent random variables given Θ with $E[X_i|\Theta] = \mu(\Theta)$ and $E\left[(X_i - \mu(\Theta))^2\right] = \tau_i^2$ for $i = 1, 2, \ldots, I$.*
Then the best collectively unbiased linear estimator of $\mu(\Theta)$ is given by

$$\widehat{\mu(\Theta)} = \left(\sum\nolimits_{i=1}^{I} \tau_i^{-2}\right)^{-1} \sum\nolimits_{i=1}^{I} \tau_i^{-2} X_i.$$

Remark:

- By "best" we mean the estimator with minimum quadratic loss and by "collectively unbiased", that $E\left[\widehat{\mu(\Theta)}\right] = E\left[\mu(\Theta)\right]$ for all possible probability measures.

B

Appendix B: The Hilbert Space $\mathcal{L}^2$

In this appendix we summarize – as much as needed for this text – the theory of the Hilbert space $\mathcal{L}^2$ of the real-valued square integrable random variables. All results are given without proofs. Readers interested in the proofs and more details are referred e.g. to Kolmogorov and Fomin [KF70] or – in the credibility context – to F. De Vylder [DV77].

A Hilbert space is a closed (infinite-dimensional) linear space endowed with a scalar product. The basic idea of using such spaces is to relate abstract probabilistic concepts to geometric intuition. In the following we define the Hilbert space $\mathcal{L}^2$.

Definition B.1. *We define by $\mathcal{L}^2$ the space of real-valued random variables with finite first- and second-order moments, i.e.*

$$\mathcal{L}^2 := \{X : X = \textit{real-valued r.v. with } E[X^2] < \infty\}.$$

Definition B.2. *The inner product of two elements X and Y in $\mathcal{L}^2$ is defined by*

$$<X,Y> \; := E\left[X \cdot Y\right].$$

Based on the inner product we can define the norm, the distance and the orthogonality relation between two elements in $\mathcal{L}^2$.

Definition B.3.

i) Norm of an element in $\mathcal{L}^2$:

$$\|X\| := E\left[X^2\right]^{1/2};$$

Definition B.4. *Distance $d(X,Y)$ between two elements in $\mathcal{L}^2$:*

$$d(X,Y) := \|X - Y\|.$$

Remark:

- Two random variables X and $Y \in \mathcal{L}^2$ with $d(X,Y) = 0$, i.e. two random variables, which are equal almost everywhere (i.e. up to a set with probability 0) are considered as the same element in $\mathcal{L}^2$.

Definition B.5. *Two elements X and Y in $\mathcal{L}^2$ are* ***orthogonal*** *$(X \perp Y)$, if $<X, Y> = 0$.*

Remarks:

- $\mathcal{L}^2$ is a linear space, i.e.
  if $X_1, X_2, \ldots, X_n \in \mathcal{L}^2$ then each finite linear combination $X = \sum_{i=1}^n a_i X_i \in \mathcal{L}^2$ $(a_i \in \mathbb{R})$.
- $\mathcal{L}^2$ is generally infinite dimensional (for infinite probability spaces).
- $\mathcal{L}^2$ is a closed space, i.e. every Cauchy sequence $\{X_1, X_2, \ldots\}$ with elements $X_i \in \mathcal{L}^2$ converges to an element in $\mathcal{L}^2$. $\{X_1, X_2, \ldots\}$ is a Cauchy sequence, if for every $\varepsilon > 0$ there exists a n_ε such that $d(X_n, X_m) < \varepsilon$ for $n, m \geqslant n_\varepsilon$.
- As already mentioned, the advantage of looking at $\mathcal{L}^2$ as a Hilbert space is that we can transfer our knowledge and intuition from linear spaces (for instance the three-dimensional Euclidian space) to the space $\mathcal{L}^2$, which makes it much easier to understand and to visualize things like orthogonality relations, projections on subspaces, linearity and iterativity of the projection operator.

Definition B.6. *$\mathbb{P} \subset \mathcal{L}^2$ is a* ***subspace*** *of $\mathcal{L}^2$, if $\mathbb{P}$ is non-empty and closed with respect to finite linear combinations.*

Remark:

- For every subspace $\mathbb{P}$ it holds that the "degenerate" random variable $0 \in \mathbb{P}$.

Definition B.7. *$\mathbb{Q} \subset \mathcal{L}^2$ is an* ***affine or translated subspace*** *of $\mathcal{L}^2$ if*

$$\mathbb{Q} = Z + \mathbb{P} := \{X : X = Z + Y, Y \in \mathbb{P}\},$$

where $Z \in \mathcal{L}^2$ and $\mathbb{P} \subset \mathcal{L}^2$ is a subspace of $\mathcal{L}^2$.

Remarks:

- $\mathbb{Q} \subset \mathcal{L}^2$ is an affine subspace of $\mathcal{L}^2$, if $\mathbb{Q}$ is non-empty and closed with respect to all *normed linear combinations* (if $Z_1, \ldots, Z_n \in \mathbb{Q}$, then $X = \sum_{i=1}^n a_i Z \in \mathbb{Q}$ for all $a_i \in \mathbb{R}$ with $\sum_{i=1}^n a_i = 1$).
- If $\mathbb{Q} \subset \mathcal{L}^2$ is an affine subspace of $\mathcal{L}^2$ and $Z \in \mathbb{Q}$, then $\mathbb{P} = \mathbb{Q} - Z := \{X : X = Y - Z,\ Y \in \mathbb{Q}\}$ is a subspace.

Special subspaces and affine subspaces:
Consider the random vector $\mathbf{X} = (X_1, X_2, \ldots, X_n)'$ with components $X_i \in \mathcal{L}^2$.

- $G(\mathbf{X}) := \{Z : Z = g(\mathbf{X}),\ g =$real-valued function and $g(\mathbf{X})$ square integrable$\}$.
  $G(\mathbf{X})$ is a closed subspace.
- $L(\mathbf{X}, 1) := \left\{ Z : Z = a_0 + \sum_{j=1}^{n} a_j X_j, \quad a_0, a_1, \ldots \in \mathbb{R} \right\}$.
  $L(\mathbf{X}, 1)$ is a closed subspace.
- Under the assumptions of Theorem A.4 in Appendix A we want to estimate $\mu(\Theta)$. The *collectively unbiased linear estimators* of $\mu(\Theta)$ form the closed affine subspace
  $L_e(\mathbf{X}) := \{Z : Z = \sum_{j=1}^{n} a_j X_j, \quad a_1, a_2, \ldots \in \mathbb{R},\ E[Z] = E[\mu(\Theta)]\}$.

Definition B.8. *For a closed subspace (or a closed affine subspace) $M \subset \mathcal{L}^2$ we define the orthogonal projection of $Y \in \mathcal{L}^2$ on M as follows: $Y^* \in M$ is the* ***orthogonal projection of*** *Y* ***on*** *M ($Y^* = \mathrm{Pro}(Y|\, M)$) if $Y - Y^* \perp M$, i.e. $Y - Y^* \perp Z_1 - Z_2$ for all $Z_1, Z_2 \in M$.*

Remark:

- If M is a subspace, then the condition $Y - Y^* \perp M$ can also be written as

$$<Y - Y^*, Z> = 0 \text{ for all } Z \in M.$$

Theorem B.9. *Y^* always exists and is unique.*

Theorem B.10. *The following statements are equivalent:*

i) $Y^* = \mathrm{Pro}\,(Y|\, M)$.
ii) $Y^* \in M$ *and* $<Y - Y^*, Z - Y^*> = 0$ *for all* $Z \in M$.
iii) $Y^* \in M$ *and* $\|Y - Y^*\| \leq \|Y - Z\|$ *for all* $Z \in M$.

Theorem B.11 (linearity of projections).

a) The following linearity property holds true for the projection on a closed linear subspace $\mathbb{P}$:

$$\mathrm{Pro}\,(aX + bY|\, \mathbb{P}) = a\,\mathrm{Pro}\,(X|\, \mathbb{P}) + b\,\mathrm{Pro}\,(Y|\, \mathbb{P})\,.$$

b) The following "normed linearity property" holds true for the projection on an affine linear subspace $\mathbb{Q}$:

$$\mathrm{Pro}\left((a+b)^{-1} (aX + bY) \Big|\, \mathbb{Q} \right) = (a+b)^{-1} \left(a\,\mathrm{Pro}\,(X|\, \mathbb{Q}) + b\,\mathrm{Pro}\,(Y|\, \mathbb{Q})\right).$$

Theorem B.12 (iterativity of projections). *Let M and M' be closed subspaces (or closed affine subspaces) of $\mathcal{L}^2$ with $M \subset M'$, then we have*

$$\mathrm{Pro}\,(Y\,|M\,) = \mathrm{Pro}\,(\mathrm{Pro}\,(Y\,|M'\,)\,|M\,)$$

and

$$\|Y - \mathrm{Pro}\,(Y\,|M\,)\|^2 = \|Y - \mathrm{Pro}\,(Y \mid M')\|^2 + \|\mathrm{Pro}\,(Y\,|M'\,) - \mathrm{Pro}\,(Y\,|M\,)\|^2\ .$$

Remark:

- Note that geometrically, the last equation corresponds to the Theorem of Pythagoras.

C

Appendix C: Solutions to the Exercises

C.1 Exercises to Chapter 2

Exercise 2.1

a)

$\vartheta =$	good	medium	bad	P^{coll}
p^{ind}	300	500	1 000	
$u(\vartheta)$	65%	30%	5%	395

b) With the Bayes-theorem we can calculate the posterior probabilities $u(\vartheta\,|x)$ and from that $P^{\text{Bayes}} = E\left[P^{\text{ind}}\middle|\, x\right]$.

x	$\vartheta =$	good	medium	bad	P^{Bayes}
	$P^{\text{ind}} =$	300	500	1 000	
0	$u(\vartheta\,\|x) =$	65.64%	29.67%	4.69%	392.14
1 000	$u(\vartheta\,\|x) =$	49.37%	37.97%	12.66%	464.56

c) The observation $\mathbf{X}$ is now a two-dimensional vector with possible outcomes $\mathbf{x}_1 = (0,0)$, $\mathbf{x}_2 = (10\,000, 0)$, $\mathbf{x}_3 = (10,\ 10\,000)$, $\mathbf{x}_4 = (10\,000,\ 10\,000)$.

With the Bayes-theorem, we can again calculate the posterior probability distributions $u(\vartheta\,|\mathbf{x})$ and P^{Bayes}.

$\mathbf{x}$	$\vartheta =$	good	medium	bad	P^{Bayes}
	$p^{\text{ind}} =$	300	500	1 000	
$\mathbf{x}_1$	$u(\vartheta\,\|\mathbf{x}) =$	66.27%	29.34%	4.39%	389.40
$\mathbf{x}_2$	$u(\vartheta\,\|\mathbf{x}) =$	50.22%	37.83%	11.95%	459.30
$\mathbf{x}_3$	$u(\vartheta\,\|\mathbf{x}) =$	50.22%	37.83%	11.95%	459.30
$\mathbf{x}_4$	$u(\vartheta\,\|\mathbf{x}) =$	31.88%	40.87%	27.25%	572.48

Exercise 2.2

a) Moment method estimator of the parameters:

$$\begin{aligned} \widehat{\mu}_N &= 0.089, \quad \widehat{\sigma}_N^2 = 0.0976 \\ \Rightarrow \quad \widehat{\gamma} &= 0.878, \quad \widehat{\beta} = 9.911. \end{aligned}$$

Fit of the negative binomial distribution:

number of policies with k claims:						
$k =$	0	1	2	3	4	>5
observed:	143 991	11 588	1 005	85	6	2
fitted:	143 990	11 593	998	88	8	1

b)

$$\begin{aligned} \operatorname{CoVa}(\Theta) &= \frac{\sqrt{\operatorname{Var}[\Theta]}}{E[\Theta]} = \gamma^{-1}, \\ \Rightarrow \widehat{\operatorname{CoVa}(\Theta)} &= \widehat{\gamma}^{-1} = 1.07. \end{aligned}$$

c) Let N_3 be the claim number within the last three years of a randomly drawn policy. N_3 is conditionally, given Θ, Poisson distributed with parameter $\Theta_3 = 3\Theta$. Hence N_3 has a negative binomial distribution with estimated parameters $\widehat{\gamma}_3 = \widehat{\gamma}$ and $\widehat{\beta}_3 = \widehat{\beta}/3$.

$$\Rightarrow P(N_3 = 0) = \left(\frac{\widehat{\beta}}{\widehat{\beta}+3}\right)^{\widehat{\gamma}} = 0.793.$$

$\Rightarrow$ impact on premium: estimated reduction $= 0.15 \cdot 0.793 = 11.9\%$.

d) For $N_3 = 0$ we have

$$\begin{aligned} \widehat{\Theta^{\text{Bayes}}} &= \frac{\widehat{\beta}}{\widehat{\beta}+3} \widehat{\Theta^{\text{coll}}}, \\ \frac{\widehat{\Theta^{\text{Bayes}}}}{\widehat{\Theta^{\text{coll}}}} &= \frac{\widehat{\beta}}{\widehat{\beta}+3} = 76.8\%. \end{aligned}$$

$\Rightarrow$ The discount rate obtained with the empirical Bayes estimator would be 23.2%. The no-claims discount of 15% is below that and on the conservative side, which is reasonable.

Exercise 2.3

Θ_i is Gamma distributed with parameters γ, $\beta = \gamma$.
$\mathrm{CoVa}(\Theta_i) = 0.2 \quad \Rightarrow \quad \gamma = 25.$

$$\Theta^{\mathrm{Bayes}} = \frac{\gamma + N}{\gamma + \mu_N} = \frac{25 + 15}{25 + 25} = 0.8.$$

Exercise 2.4

We denote by N_n, $n = 1, 2$, the claim number of the total fleet in n years.
From $\mathrm{CoVa}(\Theta_i) = 0.2 \Rightarrow \gamma = 25.$

a)

Vehicle type	# of risks	a priori frequency in ‰	a priori expected claim numbers in one year
lorry	30	300	9.00
delivery van	30	100	3.00
passenger car	10	75	0.75
Total	70		12.75

$$\Theta^{\mathrm{Bayes}} = \frac{25 + N_1}{25 + 12.75}.$$

b)

$$\Theta^{\mathrm{Bayes}} = \begin{cases} 1.09 \text{ if } N_1 = 16, \\ 0.89 \text{ if } N_2 = 20. \end{cases}$$

$$\text{experience correction term} = \begin{cases} +9\% \text{ if } N_1 = 16, \\ -11\% \text{ if } N_2 = 20. \end{cases}$$

c) A priori distribution of Θ: $\Gamma(\gamma, \gamma)$.
A posteriori distribution of Θ given $\mathrm{N}_2 = 20$: $\Gamma(25 + 20, 25 + 25.5)$.

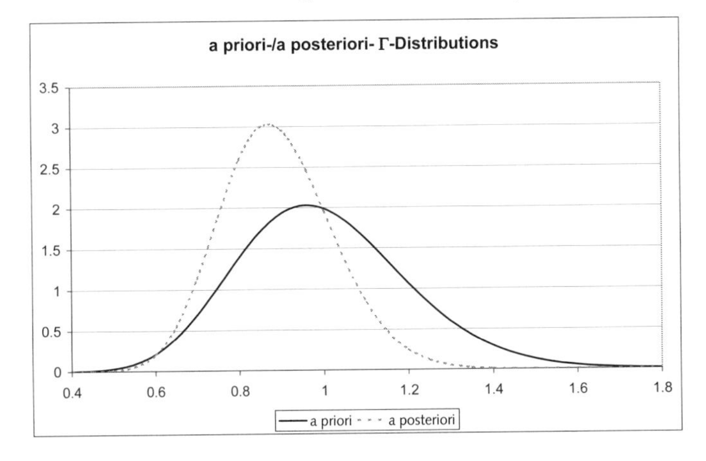

Exercise 2.5

a) $\Theta \sim \text{Beta}(a,b)$.

$$E[\Theta] = \frac{a}{a+b} = 0.01,$$

$$\text{CoVa}(\Theta) = \sqrt{\frac{b}{a(1+a+b)}} = 0.5,$$

$$a = 3.95 \quad , \quad b = 391.05.$$

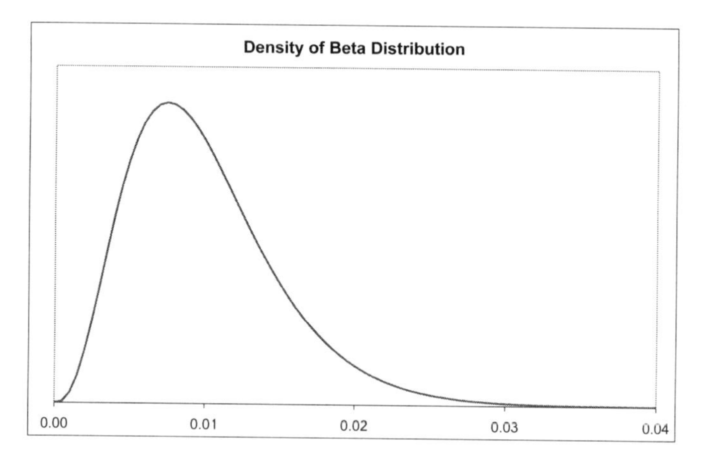

b) 13.6 ‰.

c)

$$\Theta_i^{\text{Bayes}} = \frac{3.95 + N}{3.95 + 391.05 + w_i},$$

$$\text{Decision rule} = \begin{cases} +20\% \text{ if } \Theta_i^{\text{Bayes}} > 13‰, \\ -20\% \text{ if } \Theta_i^{\text{Bayes}} < 7‰. \end{cases}$$

i) $w_i = 500$

$$\text{Decision rule} = \begin{cases} +20\% \text{ if } N \geqslant 8, \\ -20\% \text{ if } N \leqslant 2. \end{cases}$$

ii) $w_i = 2\,000$

$$\text{Decision rule} = \begin{cases} +20\% \text{ if } N \geqslant 28, \\ -20\% \text{ if } N \leqslant 12. \end{cases}$$

Exercise 2.6

a)

$$\widehat{\vartheta}_{MLE} = 1.70,$$
$$\Theta^{\text{Bayes}} = 2.20.$$

b) Denote by n_{2m} the expected number of claims exceeding 2 Million CHF, by $\mu_{2m} = E\left[X - 2\text{ Million} \mid X > 2\text{ Million}\right]$ the expected value of these claims and by $P = n_{2m} \cdot \mu_{2m}$ the pure risk premium. With the different estimators of the Pareto parameter we obtain:

Pareto parameter	n_{2m}	μ_{2m}	P
$\Theta_i^{\text{Bayes}} = 2.2$	0.76	1.66 mio	1.26 mio
$\widehat{\vartheta}_{MLE} = 1.7$	1.08	2.87 mio	3.10 mio
$\Theta^{\text{coll}} = 2.5$	0.62	1.33 mio	0.82 mio

Exercise 2.7

a) Family of the normal distributions.

b) $Y_i^{(\nu)} \sim \text{Lognormal}\left(\Theta_i, \sigma^2\right)$.
$\Theta_i \sim \text{Normal}\left(\mu, \tau^2\right)$.

$$\widehat{\vartheta}_{MLE} = \frac{1}{n}\sum_{y=1}^{n} \ln\left(Y_i\right),$$
$$\Theta_i^{\text{Bayes}} = \alpha\, \widehat{\vartheta}_{MLE} + (1-\alpha)\,\mu,$$
$$\text{where } \alpha = \frac{n}{n + \frac{\sigma^2}{\tau^2}}.$$

From $\text{CoVa}\left(Y_i^{(\nu)} | \Theta_i\right) = \sqrt{e^{\sigma^2} - 1} = 4 \Rightarrow \sigma^2 = \ln(17) = 2.8332$.
From $\mu = 7.3$ and $\text{CoVa}\left(\Theta_i\right) = 10\% \Rightarrow \tau^2 = 0.5329$.

c) The estimated expected values of the claim amount obtained with the different estimators of Θ_i are:

$$\widehat{\mu}_Y^{(1)} := e^{\widehat{\vartheta}_{MLE} + \frac{\sigma^2}{2}},\ \widehat{\mu}_Y^{(2)} := e^{\Theta^{\text{Bayes}} + \frac{\sigma^2}{2}},\ \widehat{\mu}_Y^{(3)} := e^{\Theta^{\text{coll}} + \frac{\sigma^2}{2}}.$$

Results:

i	$\widehat{\vartheta}_{MLE}$	Θ^{Bayes}	Θ^{coll}	$\widehat{\mu}_Y^{(1)}$	$\widehat{\mu}_Y^{(2)}$	$\widehat{\mu}_Y^{(3)}$	observed mean
1	6.85	7.01	7.30	3 906	4 561	6 103	15 396
2	7.45	7.37	7.30	7 118	6 576	6 103	3 347

Comments:
The estimates $\widehat{\mu}_Y^{(2)}$ (and $\widehat{\mu}_Y^{(1)}$) are bigger for the second class even if the observed mean of the second class is much lower than the one of the first class. The point is that the observed mean of the first class is highly influenced by one outlier observation, and that such a single outlier is weighed down by the maximum-likelihood estimator.

Exercise 2.8

a) Family of the Beta distributions.
b)

$$\Theta^{\text{Bayes}} = \frac{a+\gamma n}{a+b+\gamma n+X_\bullet},$$
$$\text{where } X_\bullet = \sum_{j=1}^{n} X_j \, .$$

Exercise 2.9

$$\begin{aligned} X_{n+1}^{\text{Bayes}} &= G(X_1,\dots,X_n) = g(\mathbf{X}), \\ &\text{where} \\ g(\mathbf{X}) &= \arg\min_h \mathrm{E}\left[(h(\mathbf{X}) - X_{n+1})^2\right]. \end{aligned}$$

$$\begin{aligned} E\left[(h(\mathbf{X}) - X_{n+1})^2\right] &= E\left[E\left[(h(\mathbf{X}) - X_{n+1})^2 \middle| \Theta\right]\right] \\ &= E\left[E\left[(h(\mathbf{X}) - \mu(\Theta) + \mu(\Theta) - X_{n+1})^2 \middle| \Theta\right]\right] \\ &= E\left[(h(\mathbf{X}) - \mu(\Theta))^2\right] + E\left[(X_{n+1} - \mu(\Theta))^2\right]. \end{aligned}$$

Hence

$$\begin{aligned} X_{n+1}^{\text{Bayes}} &= \arg\min_h E\left[(h(\mathbf{X}) - \mu(\Theta))^2\right], \\ \Rightarrow X_{n+1}^{\text{Bayes}} &= \mu(\Theta)^{\text{Bayes}}. \end{aligned}$$

C.2 Exercises to Chapter 3

Exercise 3.1

a) The linearity property of the credibility estimator follows directly from the linearity property of the projection-operator on a closed linear subspace.

b) We define

$$
\begin{aligned}
\mu_k &:= E\left[\mu_k(\Theta)\right], \quad k = 1, 2, \ldots, p; \\
\mu &:= \sum_{k=1}^{p} a_k \mu_k; \\
L_e^{(k)}(\mathbf{X}) &:= \left\{ \widehat{\mu(\Theta)} : \widehat{\mu(\Theta)} = \sum_{j=1}^{n} b_i X_j, E\left[\widehat{\mu(\Theta)}\right] \equiv \mu_k \right\}; \\
L_e(\mathbf{X}) &:= \left\{ \widehat{\mu(\Theta)} : \widehat{\mu(\Theta)} = \sum_{j=1}^{n} b_i X_j, E\left[\widehat{\mu(\Theta)}\right] \equiv \mu \right\}; \\
L_0(\mathbf{X}) &:= \left\{ \widehat{\mu(\Theta)} : \widehat{\mu(\Theta)} = \sum_{j=1}^{n} b_i X_j, E\left[\widehat{\mu(\Theta)}\right] \equiv 0 \right\};
\end{aligned}
$$

"$\equiv$" means that the equality has to be fulfilled, whatever the probability measure $dP(\mathbf{X}, \vartheta)$ resp. whatever the vector $\boldsymbol{\mu} = (\mu_1, \ldots, \mu_p)'$ might be.

We have to show that

$$
\mu(\Theta) - \widehat{\widehat{\mu(\Theta)}}^{\text{hom}} \perp \widehat{\mu(\Theta)} - \widehat{\widehat{\mu(\Theta)}}^{\text{hom}} \qquad \text{for all } \widehat{\mu(\Theta)} \in L_l(\mathbf{X}).
$$

First we note that

$$
\begin{aligned}
L_e^{(k)}(\mathbf{X}) &= \widehat{\widehat{\mu_k(\Theta)}}^{\text{hom}} + L_0(\mathbf{X}) \\
&= \left\{ \widehat{\mu_k(\Theta)} : \widehat{\mu_k(\Theta)} = \widehat{\widehat{\mu_k(\Theta)}}^{\text{hom}} + Y, Y \in L_0(\mathbf{X}) \right\}, \\
L_e(\mathbf{X}) &= \left\{ \widehat{\mu(\Theta)} : \widehat{\mu(\Theta)} = \sum_{k=1}^{P} a_k \widehat{\mu_k(\Theta)}, \widehat{\mu_k(\Theta)} \in L_e^{(k)}(\mathbf{X}) \right\},
\end{aligned}
$$

$\Rightarrow$ for all $\widehat{\mu(\Theta)} \in L_e(\mathbf{X})$ it holds that

$$
\mu_k(\Theta) - \widehat{\widehat{\mu_k(\Theta)}}^{\text{hom}} \perp \widehat{\mu(\Theta)} - \widehat{\widehat{\mu_e(\Theta)}}^{\text{hom}} \qquad \text{for all } k, l = 1, 2, \ldots, p.
$$

$\Rightarrow$ for all $\widehat{\mu(\Theta)} \in L_e(\mathbf{X})$ it holds that

$$\mu(\Theta) - \widehat{\widehat{\mu_k(\Theta)}}^{\text{hom}} = \sum_{k=1}^{p} a_k \left(\mu_k(\Theta) - \widehat{\widehat{\mu_k(\Theta)}}^{\text{hom}} \right)$$

is orthogonal to

$$\widehat{\mu(\Theta)} - \widehat{\widehat{\mu_k(\Theta)}}^{\text{hom}} = \sum_{k=1}^{p} a_k \left(\mu_k(\Theta) - \widehat{\widehat{\mu_k(\Theta)}}^{\text{hom}} \right).$$

Exercise 3.2

Let $\widehat{\widehat{\mu(\Theta)}}$ be the credibility estimator of $\mu(\Theta)$ based on $\mathbf{X}_1$. If $\mathbf{X}_2$ is independent of $\mathbf{X}_1$ and $\mu(\Theta)$, we then have

$$\mathrm{Cov}\left(\mu(\Theta), X_{2j}\right) = \mathrm{Cov}\left(\widehat{\widehat{\mu(\Theta)}}, X_{2j}\right) = 0$$

for every component X_{2j} of the vector $\mathbf{X}_2$. Hence the normal equations are also fulfilled for $\mathbf{X}_2$, i.e. $\widehat{\widehat{\mu(\Theta)}}$ is the credibility estimator based on $\mathbf{X} = (\mathbf{X}_1', \mathbf{X}_2')'$.

Exercise 3.3

a)

$$\begin{aligned} \widehat{\widehat{X}}_{n+1} &= \mathrm{Pro}\left(X_{n+1} L \,|(1, \mathbf{X})\right), \\ &= \mathrm{Pro}\left(\mathrm{Pro}\left(X_{n+1} \,| L\left(1, \mu(\Theta), \mathbf{X}\right)\right) | L\left(1, \mathbf{X}\right)\right). \end{aligned}$$

The inner projection is equal to $\mu(\Theta)$, which can easily be seen by checking, that it fulfils the normal equations.
Hence

$$\widehat{\widehat{X}}_{n+1} = \mathrm{Pro}\left(\mu(\Theta) \,| L\left(\mathbf{X}, \mathbf{1}\right)\right) = \widehat{\widehat{\mu(\Theta)}}.$$

b)

$$\begin{aligned} E\left[\left(\widehat{\widehat{X}}_{n+1} - X_{n+1}\right)^2\right] &= E\left[\left(X_{n+1} - \mu(\Theta)\right)^2\right] + E\left[\left(\mu(\Theta) - \widehat{\widehat{\mu(\Theta)}}\right)^2\right] \\ &= \sigma^2 + \alpha \frac{\sigma^2}{n}. \end{aligned}$$

Exercise 3.4
Structural parameters: $\mu_0 = 395$, $\sigma^2 = 3\,766\,500$, $\tau^2 = 27\,475$.

a) $n = 1, \alpha = 0.72\%$.

$$\widehat{\widehat{\mu(\Theta)}} = \begin{cases} 392.14 \text{ if } X = 0, \\ 464.56 \text{ if } X = 10\,000. \end{cases}$$

b) $n = 2, \alpha = 1.44\%$.

$$\widehat{\widehat{\mu(\Theta)}} = \begin{cases} 382.32 \text{ if } \overline{X} = 0 \\ 461.22 \text{ if } \overline{X} = 5\,000 \\ 533.11 \text{ if } \overline{X} = 10\,000. \end{cases}$$

c) For $n = 1$, $P^{\text{Cred}} = P^{\text{Bayes}}$, but for $n = 2$, this equality no longer holds.

Reason:
After one year's claim experience, P^{Bayes} is by definition a function of the two possible outcomes of X, i.d.

$$P^{\text{Bayes}}(X) = \begin{cases} a_0 \text{ if } X = 0, \\ a_1 \text{ if } X = 10\,000. \end{cases}$$

One can always draw a straight line through two points in the plane, hence $P^{\text{Bayes}}(X)$ can always be written as a linear function of X. Since P^{Cred} is the best linear approximation to P^{Bayes}, P^{Cred} coincides with P^{Bayes}.

After two years' claim experience, P^{Bayes} is by definition a function of the four possible outcomes of the two-dimensional observation vector (X_1, X_2), resp. a function of the three possible outcomes of $\overline{X} = (X_1 + X_2)/2$, i.e.

$$P^{\text{Bayes}}(\overline{X}) = \begin{cases} a_0 \text{ if } X = 0, \\ a_1 \text{ if } X = 5\,000, \\ a_2 \text{ if } X = 10\,000. \end{cases}$$

In general, one cannot draw a straight line through three points in the plane. Therefore, P^{Cred} is no longer equal to P^{Bayes}, but it is still the best (with respect to quadratic loss) linear approximation to P^{Bayes}.

Exercise 3.5
$n =$ number of claims (exposure),
$N =$ number of large claims.
Given Θ (and n), N has a Binomial distribution with parameters n and Θ.
For $X = \frac{N}{n}$, it holds that

$$\mu(\Theta) = E[X|\Theta] = \Theta,$$
$$\sigma^2(\Theta) = \text{Var}[X|\Theta] = \frac{1}{n}\Theta(1-\Theta).$$

Structural parameters:

$$\mu = E[\Theta] = 2\%,$$
$$\tau^2 = \text{Var}[\Theta] = 0.01\%,$$
$$\sigma^2 = E\left[\sigma^2(\Theta)\right] = \frac{1}{n}\left(\mu - \tau^2 - \mu^2\right) = \frac{1}{n}(0.0196),$$
$$n = 120, \quad X = \frac{6}{120} = 5\%,$$
$$\alpha = \frac{1}{1+\frac{\sigma^2}{\tau^2}} = 38.0\%, \quad \widehat{\widehat{\Theta}} = \alpha X + (1-\alpha)\mu = 3.14\%.$$

Alternative solution:
We approximate the Binomial distribution with the Poisson distribution, i.e. we assume that N is conditionally Poisson-distributed with Poisson parameter $\lambda = m\Theta$. Then

$$\sigma^2(\Theta) = \text{Var}[X|\Theta] = \frac{1}{n}\Theta,$$
$$\sigma^2 = E\left[\sigma^2(\Theta)\right] = \frac{1}{n}0.02,$$
$$\alpha = \frac{1}{1+\frac{\sigma^2}{\tau^2}} = 37.5\%, \quad \widehat{\widehat{\Theta}} = \alpha X + (1-\alpha)\mu = 3.13\%.$$

C.3 Exercises to Chapter 4

Exercise 4.1

a) $\kappa = 3\,000$ mio, $\widehat{\mu}_0 = 0.80$

Risk group	1	2	3	4	5	Total
α_i	59%	77%	59%	86%	41%	
$\widehat{\widehat{\mu(\Theta_i)}}^{\text{hom}}$	0.93	0.42	1.06	0.89	0.72	0.78
$w_i \cdot \widehat{\widehat{\mu(\Theta_i)}}^{\text{hom}}$	4 038	4 239	4 628	16 268	1 517	30 690

b) $\kappa = 6\,000$ mio.

Risk group	1	2	3	4	5	Total
α_i	42%	63%	42%	75%	26%	
$\widehat{\widehat{\mu(\Theta_i)}}^{\text{hom}}$	0.89	0.48	0.98	0.87	0.74	0.78
$w_i \cdot \widehat{\widehat{\mu(\Theta_i)}}^{\text{hom}}$	3 860	4 938	4 280	16 044	1 568	30 690

c) $\widehat{\sigma}^2 = 261$ mio, $\widehat{\tau}^2 = 0.102$ mio, $\widehat{\kappa} = 2\,558$ mio.

Risk group	1	2	3	4	5	Total
α_i	63%	80%	63%	88%	45%	
$\widehat{\widehat{\mu(\Theta_i)}}^{\text{hom}}$	0.94	0.40	1.08	0.89	0.71	0.78
$w_i \cdot \widehat{\widehat{\mu(\Theta_i)}}^{\text{hom}}$	4 075	4 106	4 703	16 304	1 502	30 690

d) Obvious from the above tables.

e) $\widehat{\sigma}^2 = 356$ mio, $\widehat{\tau}^2 = 0.066$ mio, $\widehat{\kappa} = 5\,404$ mio.

Risk group	1	2	3	4	5	Total
μ_i	0.60	0.68	1.28	0.86	0.43	0.78
$\widehat{\widehat{\mu(\Theta_i)}}^{\text{hom}}$	0.75	0.45	1.26	0.89	0.46	0.78
$w_i \cdot \widehat{\widehat{\mu(\Theta_i)}}^{\text{hom}}$	3 261	4 635	5 493	16 340	962	30 690

Exercise 4.2

a) $\widehat{\tau}^2 = 0 \Longrightarrow \widehat{\widehat{\mu(\Theta_i)}}^{\text{hom}} = 482$ for all $i = 1, 2, \ldots, 10$.

b) $\widehat{\sigma}^2 = 6\,021\,772$, $\widehat{\tau}^2 = 293$, $\widehat{\kappa} = 20\,577$, $\widehat{\mu} = 189$.
Resulting credibility estimates: see table "Results of Exercise 4.2b)".

region	# of years at risk (yr)	normal claims cred. results weight α_i	est. of $\mu(\Theta_i)$	relative $\mu(\Theta_i)$/Total
1	41 706	67%	183	0.95
2	9 301	31%	194	1.01
3	33 678	62%	210	1.09
4	3 082	13%	188	0.98
5	27 040	57%	169	0.88
6	44 188	68%	182	0.95
7	10 049	33%	178	0.93
8	30 591	60%	204	1.06
9	2 275	10%	185	0.96
10	109 831	84%	197	1.03
Total	311 741		192	1.00

Table: *Results of Exercise 4.2b)*

c) $\widehat{\tau}^2 = 0 \Longrightarrow \widehat{\widehat{\mu(\Theta_i)}}^{\text{hom}} = 290$ for $i = 1, 2, \ldots, 10$.

d) Summing up the results from b) and c) yields the estimates listed in the table "Results of Exercise 4.2 d)".

region	# of years at risk (yr)	all claims est. of $\mu(\Theta_i)$	relative $\mu(\Theta_i)$/Total
1	41 706	473	0.98
2	9 301	484	1.00
3	33 678	500	1.04
4	3 082	478	0.99
5	27 040	458	0.95
6	44 188	472	0.98
7	10 049	468	0.97
8	30 591	494	1.03
9	2 275	474	0.98
10	109 831	487	1.01
Total	311 741	482	1.00

Table: *Results of Exercise 4.2d)*

Exercise 4.3

region	# of years at risk	claim frequency in ‰							
		normal claims				large claims			
		obs.	cred. results			obs.	cred. results		
			weight α_i	absol. $\lambda(\Theta_i)$	relative $\lambda(\Theta_i)$/Tot.		weight α_i	absol. $\lambda(\Theta_i)$	relative $\lambda(\Theta_i)$/Tot.
1	41 706	62.2	98.7%	62.3	0.84	0.50	18.1%	0.68	0.94
2	9 301	82.0	94.5%	81.5	1.10	0.86	4.7%	0.73	1.01
3	33 678	98.0	98.4%	97.6	1.32	0.77	15.1%	0.73	1.01
4	3 082	75.6	85.0%	75.1	1.01	0.32	1.6%	0.71	0.99
5	27 040	61.8	98.0%	62.0	0.84	0.55	12.5%	0.70	0.97
6	44 188	70.1	98.8%	70.2	0.95	0.75	18.9%	0.72	1.00
7	10 049	62.8	94.9%	63.3	0.85	1.19	5.0%	0.74	1.03
8	30 591	79.4	98.3%	79.2	1.07	0.82	13.9%	0.73	1.02
9	2 275	51.0	80.7%	55.1	0.74	0.00	1.2%	0.71	0.98
10	109 831	75.1	99.5%	75.1	1.01	0.76	36.7%	0.74	1.02
Total	311 741	74.0		74.0	1.00	0.72		0.72	1.00

Exercise 4.4

a) Estimates of structural parameters:
$\widehat{\mu}_0 = 2\,573$,
$\widehat{\sigma}^2 = 1.812\text{E}7$, $\widehat{\tau}^2 = 2.794\text{E}4$, $\widehat{\kappa} = \frac{\widehat{\sigma}^2}{\widehat{\tau}^2} = 648.$

Credibility results:

region	normal claims			
	claim number	average claim amount in CHF		
		obs.	cred. results	
		$X_i^{(n)}$	α_i	$\mu(\Theta_i)$
1	2 595	2 889	80%	2 826
2	763	2 490	54%	2 528
3	3 302	2 273	84%	2 323
4	233	2 403	26%	2 528
5	1 670	2 480	72%	2 506
6	3 099	2 558	83%	2 561
7	631	2 471	49%	2 523
8	2 428	2 705	79%	2 677
9	116	2 836	15%	2 613
10	8 246	2 651	93%	2 645
Total	23 083	2 593		2 593

b) $\widehat{\tau}^2 = 0 \Longrightarrow \widehat{\widehat{\mu(\Theta_i)}}^{\text{hom}} = 401\,000 \quad \text{for } i = 1, 2, \ldots, 10.$

Exercise 4.5

region	years at risk at risk	pure risk premium			pure risk premium relative		
			with results of			with results of	
		obs.	ex. 4.3 and ex. 4.4	ex. 4.2	obs.	ex. 4.3 and ex. 4.4	ex. 4.2
1	41 706	422	449	473	0.88	0.93	0.98
2	9 301	626	499	484	1.30	1.04	1.00
3	33 678	446	520	500	0.93	1.08	1.04
4	3 082	246	477	478	0.51	0.99	0.99
5	27 040	413	435	458	0.86	0.90	0.95
6	44 188	566	470	472	1.18	0.98	0.98
7	10 049	569	457	468	1.18	0.95	0.97
8	30 591	473	507	494	0.98	1.05	1.03
9	2 275	145	418	474	0.30	0.87	0.98
10	109 831	494	494	487	1.03	1.03	1.01
Total	311 741	482	482	482	1.00	1.00	1.00

Short comment:
By comparing the figures in the two last columns it is clearly seen that the results obtained by estimating the frequency and the average claim amount differentiate much better between regions than the results obtained by estimating directly the pure risk premium based on the observed total claims cost per year at risk.

Exercise 4.6

a) Estimates of structural parameters:
$\widehat{\mu}_0 = 960$,
$\widehat{\sigma}^2 = 2.287\text{E}8$, $\widehat{\tau}^2 = 1.793\text{E}5$, $\widehat{\kappa} = \frac{\widehat{\sigma}^2}{\widehat{\tau}^2} = 1\,275$.
Estimate of the heterogeneity parameter $\text{CoVa}\left(\mu\left(\Theta_i\right)\right) : \widehat{\tau}/\widehat{\mu}_0 = 44\%$.(a rather heterogeneous portfolio).
Resulting credibility estimates: see table "Results of Exercise 4.6a)".

risk class	years at risk	observations X_i	X_i rel.	cred. estim. of α_i	$\mu(\Theta_i)$	$\mu(\Theta_i)$ rel.
A1	17 309	543	0.62	93%	572	0.65
A2	41 446	600	0.69	97%	611	0.70
A3	2 956	238	0.27	70%	456	0.52
A4	4 691	958	1.10	79%	958	1.10
A5	19 336	1 151	1.32	94%	1 139	1.30
B1	260	646	0.74	17%	907	1.04
B2	4 672	1 286	1.47	79%	1 216	1.39
B3	326	310	0.35	20%	828	0.95
B4	1 216	1 884	2.15	49%	1 411	1.61
B5	15 406	1 549	1.77	92%	1 504	1.72
Total	107 618	875	1.00		875	1.00

Table: *Results of Exercise 4.6a)*

b) Estimates of structural parameters:
Claim frequency:
$\widehat{\lambda}_0 = 0.1163$,
$\widehat{\sigma}_F^2 = \widehat{\lambda}_0 = 0.1163$, $\widehat{\tau}_F^2 = 0.0020$, $\widehat{\kappa}_F = \frac{\widehat{\sigma}^2}{\widehat{\tau}^2} = 58$.
Estimate of the heterogeneity parameter $\text{CoVa}\left(\lambda\left(\Theta_i\right)\right) : \widehat{\tau}_F/\widehat{\lambda}_0 = 38\%$.

Claim severity:
$\widehat{\tau}_Y^2 = 0$.

Resulting credibility estimates: see table "Results of Exercise 4.6b)".

c) Estimates of structural parameters:

Normal claims:	frequency	$\widehat{\lambda}_0 = 0.1147$, $\widehat{\kappa}_F = 59$,
	severity	$\widehat{\mu}_0 = 3\,953$, $\widehat{\kappa}_Y = 26$.
Large claims:	frequency	$\widehat{\lambda}_0 = 0.0014$, $\widehat{\kappa}_F = 3\,723$,
	severity (in 1000)	$\widehat{\mu}_0 = 339$, $\widehat{\kappa}_Y = \infty$ $(\widehat{\tau}_Y^2 = 0)$.

Credibility results: see tables "Results of Exercise 4.6c), normal claims" and "Results of Exercise 4.6c), large claims"

Of course, the pure risk premium for the different risk classes is obtained by summing up the pure risk premiums of the normal and large claims.

	claim frequency in %o				average claim amount				risk pr.
risk	years at	obs.	cred. estim.		# of	obs.	cred. estim.		estim. of
class	risk	F_i	α_i	$\mu_F(\Theta_i)$	claims	Y_i	α_i	$\mu_Y(\Theta_i)$	$\mu(\Theta_i)$
A1	17 309	61	99.7%	61	1 058	8 885	0%	8 554	524
A2	41 446	76	99.9%	76	3 146	7 902	0%	8 554	650
A3	2 956	81	98.1%	81	238	2 959	0%	8 554	695
A4	4 691	93	98.8%	93	434	10 355	0%	8 554	794
A5	19 336	128	99.7%	128	2 481	8 971	0%	8 554	1 097
B1	260	135	81.8%	131	35	4 801	0%	8 554	1 123
B2	4 672	177	98.8%	176	825	7 281	0%	8 554	1 504
B3	326	98	84.9%	101	32	3 163	0%	8 554	863
B4	1 216	148	95.4%	147	180	12 725	0%	8 554	1 254
B5	15 406	167	99.6%	167	2 577	9 259	0%	8 554	1 429
Total	107 618	102		102	11 006	8 554		8 554	875

Table: *Results of Exercise 4.6b)*

	claim frequency in %o				average claim amount				risk pr.
risk	years at	obs.	cred. estim.		# of	obs.	cred. estim.		estim. of
class	risk	F_i	α_i	$\mu_F(\Theta_i)$	claims	Y_i	α_i	$\mu_Y(\Theta_i)$	$\mu(\Theta_i)$
A1	17 309	60	99.7%	60	1 043	3 179	97.5%	3 198	193
A2	41 446	75	99.9%	75	3 106	3 481	99.2%	3 485	261
A3	2 956	81	98.1%	81	238	2 959	90.0%	3 058	248
A4	4 691	91	98.8%	92	429	4 143	94.2%	4 132	379
A5	19 336	126	99.7%	126	2 446	4 804	98.9%	4 795	606
B1	260	135	81.6%	131	35	4 801	57.1%	4 437	581
B2	4 672	174	98.8%	174	815	3 518	96.9%	3 532	613
B3	326	98	84.7%	101	32	3 163	54.9%	3 519	354
B4	1 216	143	95.4%	142	174	3 786	86.9%	3 808	540
B5	15 406	165	99.6%	165	2 547	5 579	99.0%	5 562	919
Total	107 618	101		101	10 865	4 267		4 267	431

Table: *Results of Exercise 4.6 c) normal claims*

risk class	claim frequency in %o: years at risk	obs. F_i	cred. estim. α_i	$\mu_F(\Theta_i)$	average claim amount in 1'000: # of claims	obs. Y_i	cred. estim. α_i	$\mu_Y(\Theta_i)$	risk pr. estim. of $\mu(\Theta_i)$
A1	17 309	0.87	82.3%	0.97	15	406	0%	339	328
A2	41 446	0.97	91.8%	1.00	40	351	0%	339	340
A3	2 956	0.00	44.3%	0.80	0	-	0%	339	272
A4	4 691	1.07	55.8%	1.23	5	543	0%	339	417
A5	19 336	1.81	83.9%	1.75	35	300	0%	339	593
B1	260	0.00	6.5%	1.35	0	-	0%	339	456
B2	4 672	2.14	55.7%	1.83	10	314	0%	339	620
B3	326	0.00	8.1%	1.33	0	-	0%	339	449
B4	1 216	4.93	24.6%	2.30	6	272	0%	339	780
B5	15 406	1.95	80.5%	1.85	30	322	0%	339	626
Total	107 618	1.31		1.31	141	339		339	444

Table: *Results of Exercise 4.6 c) large claims*

d) The following table and graph summarize the estimates of the pure risk premiums $\mu(\Theta_i)$ obtained in a), b) and c)

risk class	years at risk	observations X_i	X_i rel.	estimate of $\mu(\Theta_i)$ a)	b)	c)	estimate of $\mu(\Theta_i)$ rel. a)	b)	c)
A1	17 309	543	0.62	572	524	521	0.65	0.60	0.60
A2	41 446	600	0.69	611	650	602	0.70	0.74	0.69
A3	2 956	238	0.27	456	695	520	0.52	0.79	0.59
A4	4 691	958	1.10	958	794	796	1.10	0.91	0.91
A5	19 336	1 151	1.32	1 139	1 097	1 200	1.30	1.25	1.37
B1	260	646	0.74	907	1 123	1 037	1.04	1.28	1.19
B2	4 672	1 286	1.47	1 216	1 504	1 234	1.39	1.72	1.41
B3	326	310	0.35	828	863	803	0.95	0.99	0.92
B4	1 216	1 884	2.15	1 411	1 254	1 320	1.61	1.43	1.51
B5	15 406	1 549	1.77	1 504	1 429	1 545	1.72	1.63	1.77
Total	107 618	875	1.00	875	875	875	1.00	1.00	1.00

Table: *Results of Exercise 4.6 d)*

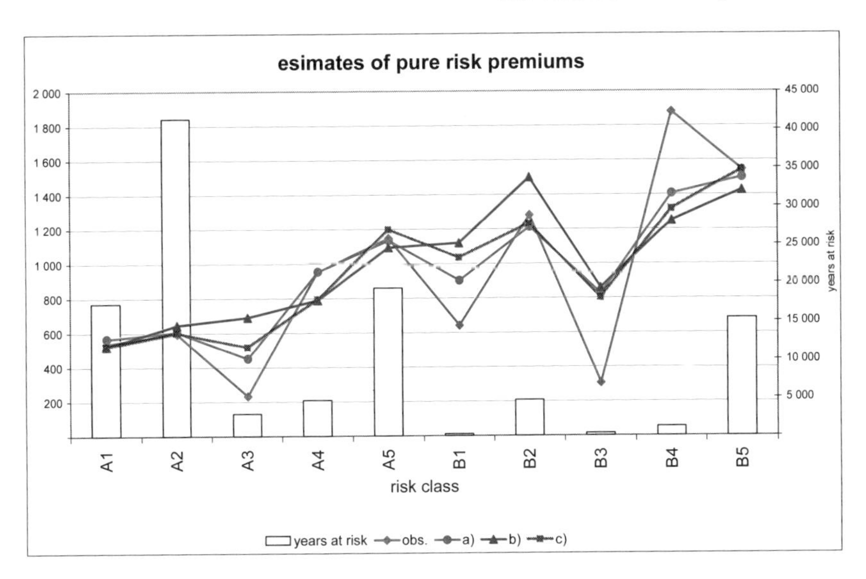

Exercise 4.7
Estimates of sum of squares and of κ:
$\widehat{SS}_{tot} = 9.017\text{E}12$, $SS_b = 6.958\text{E}12$, $\widehat{SS}_w = \widehat{SS}_{tot} - SS_b = 9.064\text{E}12$,
$\widehat{\kappa} = 1\,744$.
Resulting credibility estimates: see table "Results of Exercise 4.7" below.

Exercise 4.8
Estimates of structural parameters:
Claim frequency:
$\widehat{\lambda}_0 = \widehat{\sigma}^2 = 1.0541$, $\widehat{\tau}^2 = 0.0265$, $\widehat{\kappa} = \frac{\widehat{\sigma}^2}{\widehat{\tau}^2} = 40$.
Resulting credibility estimates: see table "Results of Exercise 4.8".

Exercise 4.10
Estimates of structural parameters:
$\widehat{\vartheta}_0 = 1.3596$, $\widehat{\tau}^2 = 0.0229$, $\widehat{\kappa} = \left(\widehat{\vartheta}_0\right)^2 / \widehat{\tau}^2 = 81$.
Resulting credibility estimates: see table "Results of Exercise 4.9".

region	# of claims	average claim amount				
		observed		Cred.-weight	Cred.est. of	
		X_i	X_i rel	α_i	$\mu(\Theta_i)$	$\mu(\Theta_i)$ rel
1	5 424	4 593	0.98	76%	4 616	0.99
2	160	4 514	0.96	8%	4 671	1.00
3	885	4 216	0.90	34%	4 527	0.97
4	10 049	4 303	0.92	85%	4 360	0.93
5	2 469	4 350	0.93	59%	4 489	0.96
6	1 175	4 598	0.98	40%	4 650	0.99
7	2 496	4 897	1.05	59%	4 810	1.03
8	7 514	4 652	0.99	81%	4 659	0.99
9	482	5 030	1.07	22%	4 760	1.02
10	2 892	4 693	1.00	62%	4 690	1.00
11	839	5 097	1.09	32%	4 819	1.03
12	3 736	4 226	0.90	68%	4 372	0.93
13	2 832	4 879	1.04	62%	4 805	1.03
14	548	4 603	0.98	24%	4 666	1.00
15	406	4 555	0.97	19%	4 661	0.99
16	6 691	4 658	0.99	79%	4 664	1.00
17	1 187	4 734	1.01	40%	4 705	1.00
18	2 105	4 463	0.95	55%	4 564	0.97
19	1 533	4 693	1.00	47%	4 689	1.00
20	2 769	4 909	1.05	61%	4 822	1.03
21	5 602	5 078	1.08	76%	4 985	1.06
22	283	4 443	0.95	14%	4 652	0.99
23	9 737	5 112	1.09	85%	5 047	1.08
24	2 839	5 187	1.11	62%	4 996	1.07
25	1 409	4 513	0.96	45%	4 608	0.98
25	20 267	4 630	0.99	92%	4 634	0.99
26	6 977	4 653	0.99	80%	4 660	0.99
Total	103 303	4 685	1.00		4 685	1.00

Table: *Results of Exercise 4.7*

fleet	# of year at risks	# of claims obs.	a priori exp.	standardized claim frequency obs.	cred. weight	cred. est. of Θ_i
1	103	18	23	0.78	37%	0.95
2	257	27	16	1.69	29%	1.24
3	25	2	4	0.50	9%	1.00
4	84	20	17	1.18	30%	1.09
5	523	109	120	0.91	75%	0.94
6	2 147	518	543	0.95	93%	0.96
7	72	11	10	1.10	20%	1.06
8	93	16	13	1.23	25%	1.10
9	154	116	129	0.90	76%	0.94
10	391	156	118	1.32	75%	1.25
Total	3 849	993	993	1.00		

Table: *Results of Exercise 4.8*

group	1	2	3	4	5	6
# of claims	34	22	27	8	12	24
MLE* of Θ_i	1.77	1.58	1.30	1.00	0.87	1.04
cred. weight α_i	28.2%	19.7%	23.5%	6.9%	10.9%	21.2%
cred. est. of Θ_i	1.48	1.40	1.35	1.33	1.31	1.29

Table: *Results of Exercise 4.10*

C.4 Exercises to Chapter 5

Exercise 5.1

a) The following table shows the values of

$$\alpha_Z = \frac{P}{P + \kappa_Z},$$
$$R(m) = \rho_{XY}^2 \, \alpha_Z,$$

if we replace the structural parameters in these formulae with the estimated values given in the table of Exercise 5.1. From the graphs of $R(m)$ we can see, that this function is rather flat and that we can choose the truncation point within a large area without losing much accuracy. Indeed, the main aim of credibility with truncation is to meet the requirement that large premium jumps because of one or a few large claims should be avoided.

trunc.point	P = 60 000		P = 300 000		P = 1 000 000	
m	α_Z	R(m)	α_Z	R(m)	α_Z	R(m)
∞	25.7%	0.257	63.4%	0.634	85.2%	0.852
100 000	27.0%	0.267	64.9%	0.642	86.0%	0.851
95 000	27.2%	0.269	65.2%	0.644	86.2%	0.851
90 000	27.5%	0.271	65.5%	0.645	86.3%	0.851
85 000	27.8%	0.274	65.9%	0.648	86.5%	0.851
80 000	28.2%	0.276	66.2%	0.650	86.7%	0.851
75 000	28.5%	0.279	66.6%	0.652	86.9%	0.851
70 000	28.9%	0.282	67.0%	0.654	87.1%	0.851
65 000	29.3%	0.286	67.5%	0.658	87.4%	0.852
60 000	29.8%	0.290	68.0%	0.661	87.6%	0.852
55 000	30.5%	0.296	68.7%	0.666	88.0%	0.853
50 000	31.3%	0.303	69.5%	0.673	88.4%	0.856
45 000	32.6%	0.315	70.7%	0.684	89.0%	0.860
40 000	34.0%	0.328	72.0%	0.694	89.6%	0.863
35 000	35.5%	0.338	73.4%	0.699	90.2%	0.860
30 000	36.4%	0.348	74.1%	0.710	90.5%	0.867
25 000	37.6%	0.359	75.1%	0.717	90.9%	0.869
20 000	38.9%	0.363	76.1%	0.710	91.4%	0.853
15 000	41.4%	0.374	77.9%	0.704	92.2%	0.832
10 000	44.8%	0.373	80.2%	0.669	93.1%	0.777
5 000	50.0%	0.359	83.4%	0.598	94.3%	0.677

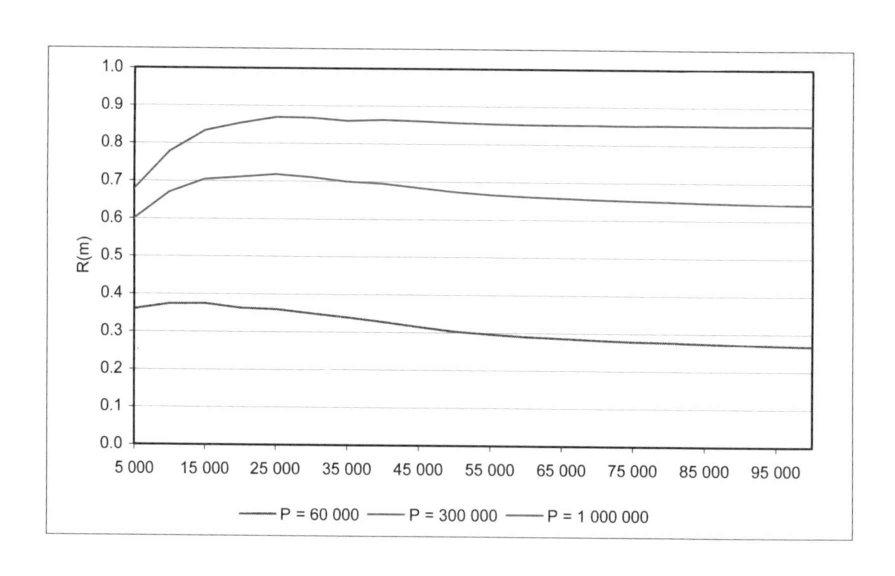

b) The formulae for the correction factor are

$$\widehat{\widehat{\mu(\Theta_i)}} = \begin{cases} 1 + \frac{P}{P+84\,924} \cdot (Z_{15\,000} - 0.81)\,, & \text{if } 20\,000 \le P < 100\,000, \\ 1 + \frac{P}{P+105\,033} \cdot (Z_{30\,000} - 0.90), & \text{if } 100\,000 \le P < 1\,000\,000, \\ 1 + \frac{P}{P+173\,531} \ (X - 1), & \text{if } 1\,000\,000 \le P. \end{cases}$$

For the five policies we get:

	P	total claim amount original	total claim amount with trunc.	loss ratio original	loss ratio with trunc.	estimate of $\mu_X(\Theta_i)$
policy 1	20 000			0%	0%	0.82
policy 2	20 000	36 000	36 000	180%	180%	1.22
policy 3	20 000	36 000	19 000	180%	95%	1.03
policy 4	300 000	300 000	300 000	100%	100%	1.08
policy 5	300 000	360 000	220 000	120%	73%	0.87

Remarks:
The claim free policy 1 gets a discount of 18% compared to the basic premium rate. Policies 2 and 3 both have a very unfavourable claims experience: the observed loss ratio is 180% for both. However, with policy 3, this high loss ratio is mainly caused by one large claim. Therefore, the surcharge for policy 3 is only 3%, whereas it is 22% for policy 2. The observed loss ratio of policy 4 coincides with the a priori expected value. Nevertheless it gets a surcharge of 8%. The reason is that policy 4 had a loss ratio of 100% despite the fact that it was a period without any larger claim (above 30 000). Therefore the loss ratio of 100% is considered an unfavourable observation. The opposite holds true for policy 5. Its observed loss ratio is even higher, namely 120%. But it is caused by two larger claims. Otherwise the claims experience was very good. Therefore it gets a discount of 13%.

Exercise 5.2

The table “Results of Exercise 5.2” below shows the credibility estimates with and without truncating the individual claims:

region	# of claims	observed					cred. estimators	
		original	truncated				semilinear	Bü-Str
		X_i	Y_i	$\alpha_X^{(i)}$	$\alpha_Y^{(i)}$	$\tau_{XY}/\tau_Y^2\alpha_Y^{(i)}$	$\mu_X^{(f)}(\theta_i)$	$\mu_X^{(BS)}(\theta_i)$
1	1 785	11 224	4 760	59%	61%	150%	9 182	10 092
2	37	6 348	4 887	3%	3%	8%	8 536	8 415
3	280	13 273	4 932	18%	20%	48%	8 904	10 818
4	3 880	11 754	4 631	76%	77%	190%	8 781	10 955
5	846	6 559	4 275	40%	42%	105%	7 750	7 703
6	464	10 045	5 071	27%	29%	71%	9 340	8 901
7	136	5 968	4 174	10%	11%	26%	8 232	8 230
8	853	5 407	3 974	41%	43%	105%	6 987	7 232
9	2 151	5 483	4 135	63%	65%	161%	6 820	6 583
10	155	5 133	4 139	11%	12%	29%	8 177	8 108
11	653	6 815	4 504	34%	36%	89%	8 347	7 906
12	252	12 161	4 877	17%	18%	44%	8 810	9 262
13	1 228	8 670	4 849	50%	52%	127%	9 348	8 572
14	865	6 780	4 482	41%	43%	106%	8 267	7 783
15	145	6 273	4 949	10%	11%	28%	8 732	8 248
16	127	6 694	4 070	9%	10%	25%	8 185	8 312
17	2 151	8 757	4 624	63%	65%	161%	8 707	8 654
18	449	6 535	4 514	26%	28%	69%	8 393	7 963
19	784	7 068	4 191	39%	41%	100%	7 580	7 934
20	438	11 453	5 243	26%	28%	68%	9 586	10 287
21	1 011	9 010	5 224	45%	47%	115%	10 305	8 715
22	1 692	7 666	4 671	58%	60%	147%	8 853	8 010
23	82	8 856	3 710	6%	7%	16%	8 139	8 500
24	2 813	7 967	4 094	69%	71%	175%	6 500	6 739
25	729	9 213	4 523	37%	39%	96%	8 382	8 748
26	329	6 727	5 157	21%	22%	55%	9 258	8 112
27	5 743	8 279	4 686	82%	83%	205%	9 077	8 314
28	3 002	8 149	4 491	71%	72%	178%	8 163	8 245
Total	33 080	8 486	4 555				8 421	8 486

Table: *Results of Exercise 5.2*

Exercise 5.3

a) Structural parameters:
$\widehat{\mu}_X = 5\,839$,
$\widehat{\sigma}_X^2 = 2.843E+09$, $\widehat{\tau}_X^2 = 1.165E+05$, $\widehat{\kappa}_X = \frac{\widehat{\sigma}_X^2}{\widehat{\tau}_X^2} = 24\,407$.

The following table shows the resulting estimates of the credibility weights $\alpha_X^{(i)}$ and of $\mu_X(\Theta_i)$.

risk group	# of claims	observations X_i	cred. estimates $\alpha_X^{(i)}$	$\mu_X(\Theta_i)$
1	92	3 478	0.4%	5 830
2	542	5 204	2.2%	5 825
3	57	2 092	0.2%	5 830
4	317	4 437	1.3%	5 821
5	902	4 676	3.6%	5 797
6	137	6 469	0.6%	5 842
7	48	12 285	0.2%	5 851
8	528	3 717	2.1%	5 794
9	183	14 062	0.7%	5 900
10	365	9 799	1.5%	5 897
Total	3 171	5 825		5 825

Note that due to the small credibility weights, the credibility estimates of $\mu_X(\Theta_i)$ are very close to each other.

b) Structural parameters:
$\widehat{\mu}_Y = 6.66$,
$\widehat{\sigma}_Y^2 = 3.8149$, $\widehat{\tau}_Y^2 = 0.0527$, $\widehat{\kappa}_Y = \frac{\widehat{\sigma}_Y^2}{\widehat{\tau}_Y^2} = 72.$

With these parameters we get the following estimates for $\alpha_Y^{(i)}$, $\mu_Y(\Theta_i)$ and $\mu_X(\Theta_i)$.

risk group	# of claims	logarithms Y_i	cred. estimates $\alpha_Y^{(i)}$	$\mu_Y(\Theta_i)$	estimate of $\mu_X(\Theta_i)$
1	92	6.47	56%	6.56	4 737
2	542	6.78	88%	6.77	5 848
3	57	6.45	44%	6.57	4 806
4	317	6.73	81%	6.72	5 571
5	902	6.59	93%	6.60	4 930
6	137	6.60	65%	6.62	5 064
7	48	7.30	40%	6.92	6 805
8	528	6.40	88%	6.43	4 186
9	183	7.25	72%	7.08	8 035
10	365	6.33	83%	6.39	3 995
Total	3 171	6.62			5 125

c)

risk group	# of claims	average claim size (absolute) obs.	estimated a)	estimated b)	average claim size (relative) obs.	estimated a)	estimated b)
1	92	3 478	5 830	4 737	0.597	1.001	0.924
2	542	5 204	5 825	5 848	0.893	1.000	1.141
3	57	2 092	5 830	4 806	0.359	1.001	0.938
4	317	4 437	5 821	5 571	0.762	0.999	1.087
5	902	4 676	5 797	4 930	0.803	0.995	0.962
6	137	6 469	5 842	5 064	1.110	1.003	0.988
7	48	12 285	5 851	6 805	2.109	1.004	1.328
8	528	3 717	5 794	4 186	0.638	0.995	0.817
9	183	14 062	5 900	8 035	2.414	1.013	1.568
10	365	9 799	5 897	3 995	1.682	1.012	0.780
Total	3 171	5 825	5 825	5 125	1.000	1.000	1.000

Short comments:
Method a) (Bühlmann–Straub) yields nearly no differences between risks. This is of course due to the large claims having small credibility weights. The results of method b) depend very much on the distributional assumptions. By taking the logarithms, the influence of large claims is weighed down. The associated credibility weights attributed to the "transformed" claims are much higher, and therefore the results differ much more between risks. However, contrary to the homogeneous estimator in the Bühlmann–Straub model, the resulting average claim amount over the whole portfolio does not coincide any more with the observed one. Indeed, in this case it is much lower, namely 5 125 compared to an overall observed average of 5 825.

Exercise 5.4

a) The estimate of the structural parameter τ^2 is zero, i.e. $\widehat{\widehat{\mu(\Theta_i)}}^{emp} = 6\,443$ for all i.
Comment:
Possible underlying differences between the risk groups cannot be recognized because of the noise caused by the large claims.

b) For the large claims (bodily injury) the estimate of τ^2 is again zero. This is not surprising and means, that the expected value of the large claims is the same for all risk groups. However, the large claims do nevertheless differentiate between risk groups, because the probability that a claim is a large claim is not the same for all risk groups. The table below shows the observations and the resulting estimates.

risk group	observations n_i	$n_i^{(n)}$	$n_i^{(l)}$	$n_i^{(l)}/n_i$	X_i	$X_i^{(n)}$	$X_i^{(l)}$ in 1 000	credibility estimates of $\mu^{(n)}(\theta_i)$	$\mu^{(l)}(\theta_i)$ in 1 000	Π_i	estimate of $\mu(\theta_i)$
1	2 616	2 595	21	0.80%	6'727	2'889	481	2 831	401	0.949%	6 613
2	771	763	8	1.04%	7'548	2'490	490	2 526	401	0.966%	6 380
3	3 328	3 302	26	0.78%	4'513	2'273	289	2 318	401	0.942%	6 080
4	234	233	1	0.43%	3'239	2'403	198	2 524	401	0.959%	6 350
5	1 685	1 670	15	0.89%	6'624	2'480	468	2 504	401	0.959%	6 331
6	3 132	3 099	33	1.06%	7'989	2'558	518	2 560	401	0.974%	6 446
7	643	631	12	1.87%	8'901	2'471	347	2 520	401	0.987%	6 456
8	2 453	2 428	25	1.02%	5'898	2'705	316	2 679	401	0.969%	6 544
9	116	116		0.00%	2'836	2'836	0	2 616	401	0.960%	6 444
10	8 330	8 246	84	1.01%	6'517	2'651	386	2 646	401	0.975%	6 535
Total	23 308	23 083	225	0.97%	6'443	2'593	401	2 593	401	0.967%	6 443

C.5 Exercises to Chapter 6

Exercise 6.1

a) Structural parameters:

$\widehat{\mu} = 0.962, \quad \widehat{\sigma}^2 = 32.76,$
$\widehat{\tau}_1^2 = 0.0172, \ \widehat{\kappa}_1 = \widehat{\sigma}^2/\widehat{\tau}_1^2 = 1\,900,$
$\widehat{\tau}_2^2 = 0.00897, \widehat{\kappa}_2 = \widehat{\tau}_1^2/\widehat{\tau}_2^2 = 1.992.$

Credibility estimates:

risk h	cred. estim. on level 2: cred. weight	μ_0	$B_h^{(2)}$	Cred. estim.	cred. estim. on level 1, own company: cred. weight	$B_1^{(1)}$	cred. estim.	other companies: Cred. weight	$B_2^{(1)}$	cred. estim.
1	45%	0.96	0.78	0.88	67%	0.81	0.83	88%	0.75	0.76
2	34%	0.96	0.89	0.94	38%	0.94	0.94	62%	0.86	0.89
3	41%	0.96	1.00	0.98	60%	0.93	0.95	71%	1.06	1.04
4	35%	0.96	0.98	0.97	37%	1.14	1.03	65%	0.89	0.92
5	35%	0.96	0.95	0.96	22%	0.73	0.91	83%	1.01	1.00
6	37%	0.96	0.93	0.95	41%	1.04	0.99	70%	0.86	0.89
7	37%	0.96	1.30	1.09	41%	1.48	1.25	70%	1.19	1.16
8	32%	0.96	0.90	0.94	33%	1.10	0.99	59%	0.79	0.85

b) Structural parameters:

$\widehat{\mu} = 0.962, \quad \widehat{\sigma}^2 = 32.76,$
$\widehat{\tau}_1^2 = 0.0239, \widehat{\kappa}_1 = \widehat{\sigma}^2/\widehat{\tau}_1^2 = 1\,368,$
$\widehat{\tau}_2^2 = 0.$

cred. estim. on level 2					cred. estim. on level 1						
comp. h	cred. weight	μ_0	$B_h^{(2)}$	cred. estim.	risk i	own company cred. weight	$B_1^{(1)}$	cred. estim.	other companies cred. weight	$B_2^{(1)}$	cred. estim.
1 (own)	0%	0.96	1.02	0.96	1	74%	0.81	0.85	91%	0.75	0.77
2 (others)	0%	0.96	0.93	0.96	2	46%	0.94	0.95	69%	0.86	0.89
					3	68%	0.93	0.94	77%	1.06	1.04
					4	45%	1.14	1.04	72%	0.89	0.91
					5	28%	0.73	0.90	87%	1.01	1.00
					6	49%	1.04	1.00	77%	0.86	0.88
					7	49%	1.48	1.22	77%	1.19	1.14
					8	40%	1.10	1.02	67%	0.79	0.85

Remark:
Since $\widehat{\tau}_2^2 = 0$, the credibility weights on the second level are zero and we have to estimate μ_0 by

$$\widehat{\mu}_0 = \sum_h \frac{w_h^{(2)}}{w_\bullet^{(2)}} B_h^{(2)}.$$

Of course, in this case one could forget about the second hierarchical level, and the above results coincide with the ones obtained with the Bühlmann-Straub model.

c) In model a), each of the eight risk groups is characterized by its profile Φ_h, and the company's own portfolio in each of these risk groups is characterized by further characteristics. In one risk group, the quality of the individual risks of the company might be better, and in another risk group worse than the quality of the risks with the other companies. In model b) however, the company itself is characterized by its profile and might attract in general over all risk groups better or worse quality risks (for instance because of its underwriting policy). The risk group appears only on the second level. In the above example, a general difference between the own company and the total of the other companies could not be found on level two from the data ($\widehat{\tau}_2^2$ was zero).

Exercise 6.2

a) Frequency of normal claims

cred. estim. on level 2					cred. estim. on level 1				
use of car	μ_0 in ‰	B_h in ‰	cred. estim. $\alpha_h^{(2)}$	$\mu(\Phi_h)$	risk class	$\mu(\Phi_h)$	B_i in ‰	cred. estim. $\alpha_i^{(1)}$	$\mu(\Theta_i)$
A	117.3	86.8	94.2%	88.5	A1	88.5	60.3	98.8%	60.3
					A2	88.5	74.9	99.5%	74.9
					A3	88.5	80.5	93.2%	80.5
					A4	88.5	91.5	95.6%	91.5
					A5	88.5	126.5	98.9%	126.5
B	117.3	148.2	92.9%	146.0	B1	146.0	134.6	54.6%	134.6
					B2	146.0	174.4	95.6%	174.4
					B3	146.0	98.2	60.1%	98.2
					B4	146.0	143.1	84.9%	143.1
					B5	146.0	165.3	98.6%	165.3

Frequency of large claims

cred. estim. on level 2					cred. estim. on level 1				
use of car	μ_0 in ‰	B_h in ‰	cred. estim. $\alpha_h^{(2)}$	$\mu(\Phi_h)$	risk class	$\mu(\Phi_h)$	B_i in ‰	cred. estim. $\alpha_i^{(1)}$	$\mu(\Theta_i)$
A	1.60	1.03	82.2%	1.13	A1	1.13	0.87	83.7%	0.91
					A2	1.13	0.97	92.5%	0.98
					A3	1.13	0.00	46.7%	0.60
					A4	1.13	1.07	58.2%	1.09
					A5	1.13	1.81	85.2%	1.71
B	1.60	2.27	69.7%	2.07	B1	2.07	0.00	7.2%	0.0
					B2	2.07	2.14	58.1%	0.0
					B3	2.07	0.00	8.8%	0.0
					B4	2.07	4.93	26.5%	0.0
					B5	2.07	1.95	82.1%	0.0

C.6 Exercises to Chapter 7

Exercise 7.1
Structural parameters:
$\widehat{\mu}_1 = 1.020, \quad \widehat{\mu}_2 = 0.932,$
$\widehat{\sigma}_1^2 = 33.2, \quad \widehat{\sigma}_2^2 = 32.3,$
$\widehat{T} = \begin{pmatrix} 0.279 & 0.203 \\ 0.203 & 0.198 \end{pmatrix}.$

Results:

risk group	categ.	cred.matrix A_i		μ	B	cred. est. of $\mu(\Theta_i)$
1	own	51.6%	44.7%	1.020	0.810	0.830
	others	11.5%	79.3%	0.932	0.750	0.763
2	own	33.2%	44.9%	1.020	0.940	0.961
	others	16.3%	54.4%	0.932	0.860	0.880
3	own	51.7%	36.7%	1.020	0.930	1.021
	others	21.9%	57.3%	0.932	1.060	0.986
4	own	31.7%	47.7%	1.020	1.140	1.038
	others	15.0%	57.6%	0.932	0.890	0.926
5	own	14.2%	74.6%	1.020	0.730	1.037
	others	4.3%	80.9%	0.932	1.010	0.983
6	own	33.3%	50.1%	1.020	1.040	0.991
	others	14.3%	62.4%	0.932	0.860	0.890
7	own	33.6%	50.1%	1.020	1.480	1.304
	others	14.3%	62.7%	0.932	1.190	1.160
8	own	29.3%	45.3%	1.020	1.100	0.979
	others	15.0%	52.8%	0.932	0.790	0.869
total	own				0.985	0.985
	others				0.905	0.905

The observations and the credibility estimates are displayed in Figure C.1.

Exercise 7.2
Structural parameters:
$\widehat{\mu}_1 = \widehat{\sigma}_1^2 = 1.092, \quad \widehat{\mu}_2 = \widehat{\sigma}_2^2 = 1.127,$
$\widehat{\sigma}_1^2 = 33.2, \quad \widehat{\sigma}_2^2 = 32.3$
$\widehat{T} = \begin{pmatrix} 0.192 & 0.193 \\ 0.193 & 0.223 \end{pmatrix}$
resulting credibility matrices and credibility estimates: see table “Results of Exercise 7.2”.

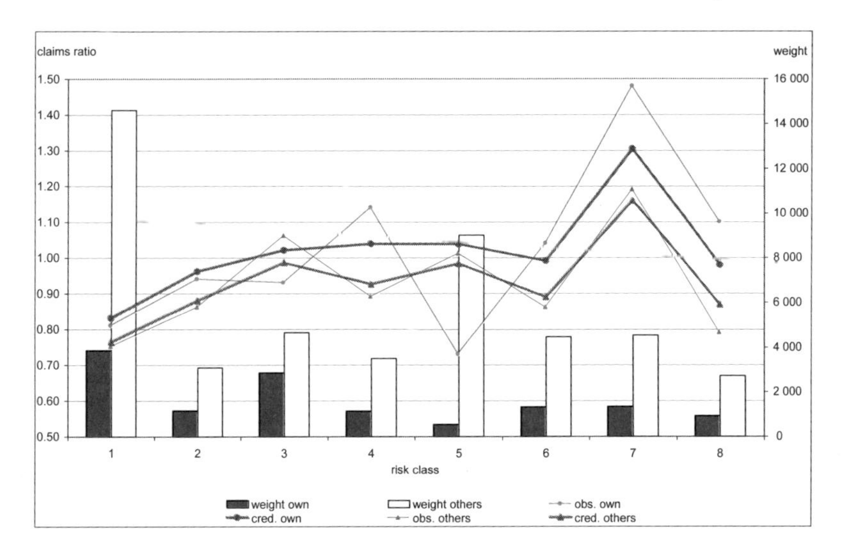

Fig. C.1. Observations and credibility estimates of Exercise 7.1

risk group	categ.	cred.matrix A_i		μ	F_i rel	cred. est. of $\lambda(\Theta_i)$ rel
A1	normal	97.5%	2.1%	1.09	0.60	0.60
	large	34.2%	57.6%	1.13	0.66	0.69
A2	normal	98.9%	0.9%	1.09	0.74	0.74
	large	28.0%	69.3%	1.13	0.74	0.76
A3	normal	88.1%	10.1%	1.09	0.80	0.72
	large	27.3%	33.2%	1.13	0.00	0.67
A4	normal	91.9%	7.0%	1.09	0.91	0.90
	large	31.4%	40.0%	1.13	0.81	0.94
A5	normal	97.8%	1.9%	1.09	1.25	1.25
	large	33.7%	59.0%	1.13	1.38	1.33
B1	normal	55.2%	32.5%	1.09	1.33	0.86
	large	5.3%	6.0%	1.13	0.00	1.07
B2	normal	91.9%	7.0%	1.09	1.73	1.71
	large	31.4%	39.9%	1.13	1.63	1.53
B3	normal	58.4%	31.2%	1.09	0.97	0.67
	large	6.5%	7.4%	1.13	0.00	1.04
B4	normal	77.7%	18.6%	1.09	1.42	1.83
	large	17.6%	20.3%	1.13	3.77	1.72
B5	normal	97.3%	2.4%	1.09	1.64	1.63
	large	34.6%	56.1%	1.13	1.49	1.52
total	normal					1.00
	large					1.00

Table: *Results of Exercise 7.2*

Exercise 7.5

a) Credibility matrix for $\nu = 5$

$$A = \begin{pmatrix} 16.5\% & 2.8\% & 2.8\% & -0.6\% \\ 2.8\% & 15.4\% & 1.6\% & 9.7\% \\ 2.8\% & 1.6\% & 15.4\% & 9.7\% \\ -0.6\% & 9.7\% & 9.7\% & 14.4\% \end{pmatrix}$$

b) The observations of region 2 and 3 are influenced from the characteristics of region 4 because of the positive correlations between region 4 and regions 2 and 3 (e.g. think of region 4 being a town and regions 2 and 3 agglomerations of this town; the observations of regions 2 and 3 are influenced by persons travelling from the agglomeration to the town). The negative value a_{14} can be interpreted as a correction of the region 4 effect inherent in the observations of regions 2 and 3.

c), d) and e)

(Inhomogeneous) credibility estimates of Θ:

Region:	1	2	3	4
Observation N_k/ν_k	0.20	0.60	1.20	2.00
$\widehat{\widehat{\Theta}}_{\text{multid.}}$	0.86	1.02	1.10	1.13
$\widehat{\widehat{\Theta}}_{\text{Bü–Str.}}$	0.87	0.93	1.03	1.17

Note the difference between the two methods: because of the high correlation (70%) between region 4 and regions 2 and 3 and the high observed value in region 4, the multidimensional estimates for regions 2 and 3 are substantially higher than the one-dimensional Bühlmann–Straub estimates.

Exercise 7.6

a) $\widetilde{\kappa} = (\text{Var}[\Theta])^{-1} = (0.25^2)^{-1} = 16;$
$v_\bullet = 49.5;$
$\alpha = v_\bullet / (v_\bullet + \widetilde{\kappa}) = 76\%;$
$\widetilde{F}_\bullet = N_\bullet / v_\bullet = 0.95;$
$\widehat{\widehat{\Theta}} = 1 + \alpha\left(\widetilde{F}_\bullet - 1\right) = 0.96.$

b) $\alpha_k = v_k / (v_k + \widetilde{\kappa})$; $\widetilde{F}_k = N_k / v_k$; $\widehat{\widehat{\Theta}}_k = 1 + \alpha_k \left(\widetilde{F}_k - 1 \right)$.

age class k	1	2	3	4	5	6	7	8	9	10
v_k	1.60	1.71	1.46	2.45	3.06	4.15	5.28	6.92	9.71	13.16
α_k	9.1%	9.7%	8.4%	13.3%	16.1%	20.6%	24.8%	30.2%	37.8%	45.1%
$\widetilde{F}_k$	0.00	0.00	0.00	0.41	0.65	0.72	1.14	1.16	1.13	1.22
$\widehat{\widehat{\Theta}}_k$	0.91	0.90	0.92	0.92	0.94	0.94	1.03	1.05	1.05	1.10

age group	n_k	N_k	N_k/n_k	Credibility estimate of Q_k: ρ= 0	ρ= 0.25	ρ= 0.5	ρ= 0.75	ρ= 1
16-20	1.60	0	0.00	0.91	0.88	0.84	0.79	0.96
21-25	1.71	0	0.00	0.90	0.86	0.81	0.77	0.96
26-30	1.46	0	0.00	0.92	0.87	0.82	0.78	0.96
31-35	2.45	1	0.41	0.92	0.88	0.84	0.81	0.96
36-40	3.06	2	0.65	0.94	0.91	0.88	0.86	0.96
41-45	4.15	3	0.72	0.94	0.94	0.93	0.91	0.96
46-50	5.28	6	1.14	1.03	1.03	1.02	1.00	0.96
51-55	6.92	8	1.16	1.05	1.06	1.07	1.05	0.96
56-60	9.71	11	1.13	1.05	1.07	1.09	1.08	0.96
61-65	13.16	16	1.22	1.10	1.10	1.11	1.10	0.96
Total	49.50	47	0.95					

c)

d) Note that the result for $\rho = 1$ is the same as the one found in a) and that the results for $\rho = 0$ are identical to the ones found in b).

In general, the correlation will be between the two extremes. Let us consider the case $\rho = 0.75$. Note the "pooling impact" of the multidimensional estimator in the lower age groups: the favourable observations in the lower age groups get more weight than if each of them is considered separately as in a).

The following table shows the credibility matrix for $\rho = 0.75$:

k	1	2	3	4	5	6	7	8	9	10
1	8.2%	5.9%	3.4%	3.7%	2.9%	2.4%	1.8%	1.3%	1.0%	0.7%
2	5.5%	8.3%	4.7%	5.1%	4.0%	3.3%	2.4%	1.8%	1.3%	1.0%
3	3.7%	5.5%	6.8%	7.3%	5.8%	4.7%	3.5%	2.5%	1.9%	1.4%
4	2.4%	3.6%	4.4%	10.5%	8.3%	6.8%	5.0%	3.6%	2.7%	2.0%
5	1.5%	2.3%	2.8%	6.6%	12.2%	10.0%	7.4%	5.4%	4.0%	3.0%
6	0.9%	1.4%	1.7%	4.0%	7.4%	15.2%	11.2%	8.2%	6.1%	4.5%
7	0.5%	0.8%	1.0%	2.3%	4.3%	8.8%	17.7%	12.9%	9.6%	7.1%
8	0.3%	0.4%	0.5%	1.3%	2.4%	4.9%	9.8%	21.1%	15.7%	11.7%
9	0.2%	0.2%	0.3%	0.7%	1.3%	2.6%	5.2%	11.2%	27.1%	20.3%
10	0.1%	0.1%	0.2%	0.4%	0.7%	1.4%	2.9%	6.2%	14.9%	37.6%

C.7 Exercises to Chapter 8

Exercise 8.1

a)

	intercept	slope
i)	107.8	4.0
ii)	107.8	10.0
iii)	130.0	4.0

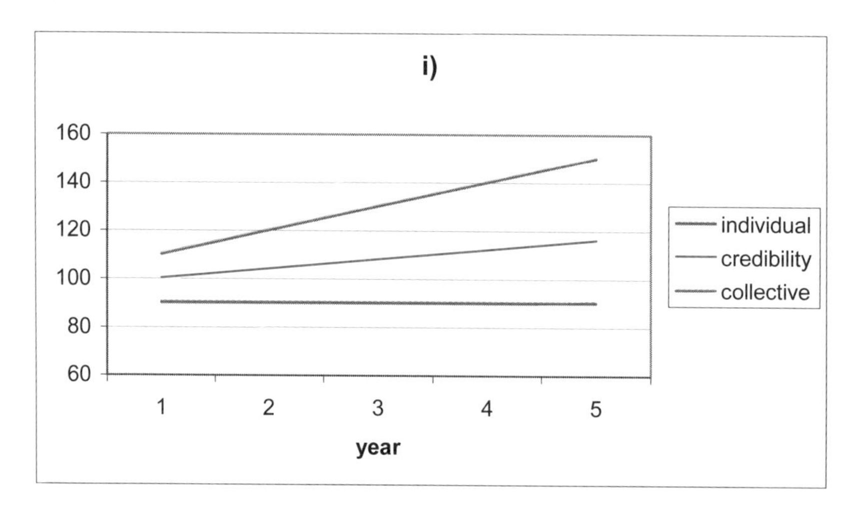

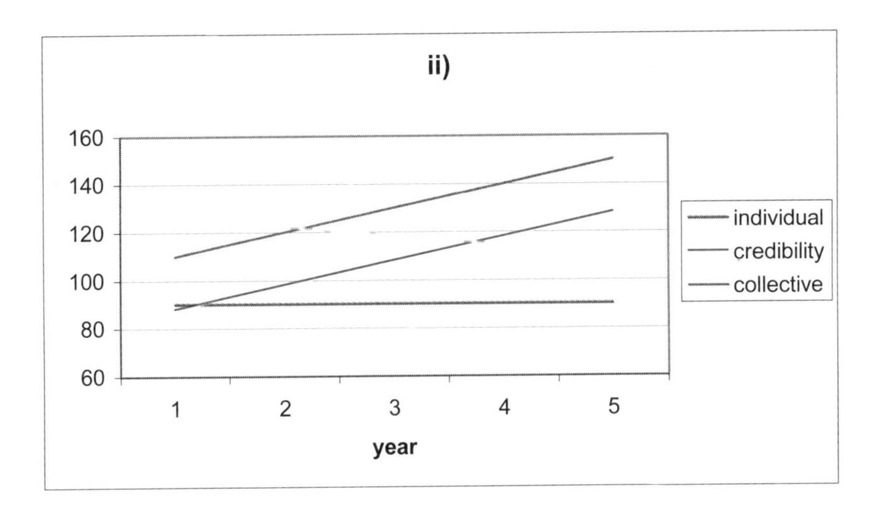

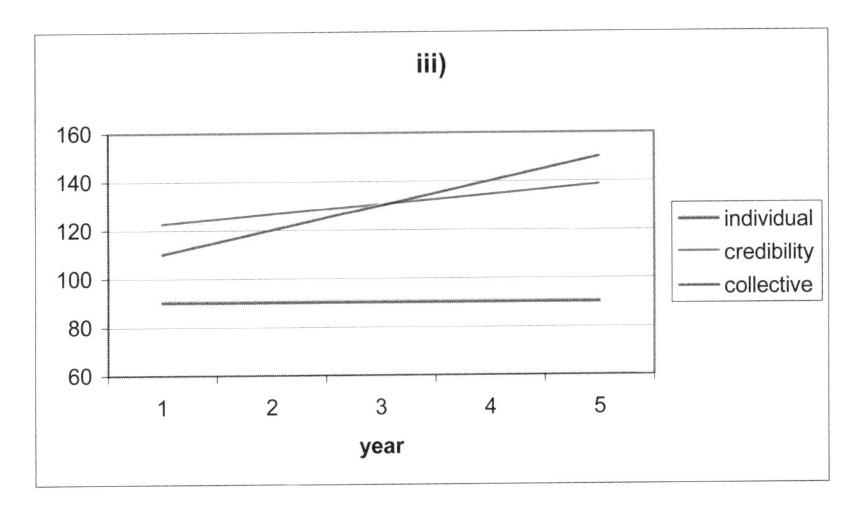

Comments:

i) Intercept and slope are between collective and individual regression line.

ii) Since τ_1^2 is near to zero, the slope is the same as in the collective regression line. Hence if we a priori assume, that the slope is the same for all risks considered, we just have to choose τ_1^2 very small (compared to σ^2).

iii) The slope is the same as in i), but because τ_0^2 is near to zero, the intercept coincides with the intercept of the collective regression line.

Exercise 8.2

a)

risk groups	1	2	3	4	global (j_0)
centres of gravity of time	3.50	3.33	3.26	3.30	3.40

Remark: the individual centres of gravity are very close to each other.

b)

	individual regression lines				collective regression line
risk groups	1	2	3	4	
intercept (at j_0)	6 425	6 113	6 856	4 996	6 098
slope	169	245	211	271	224

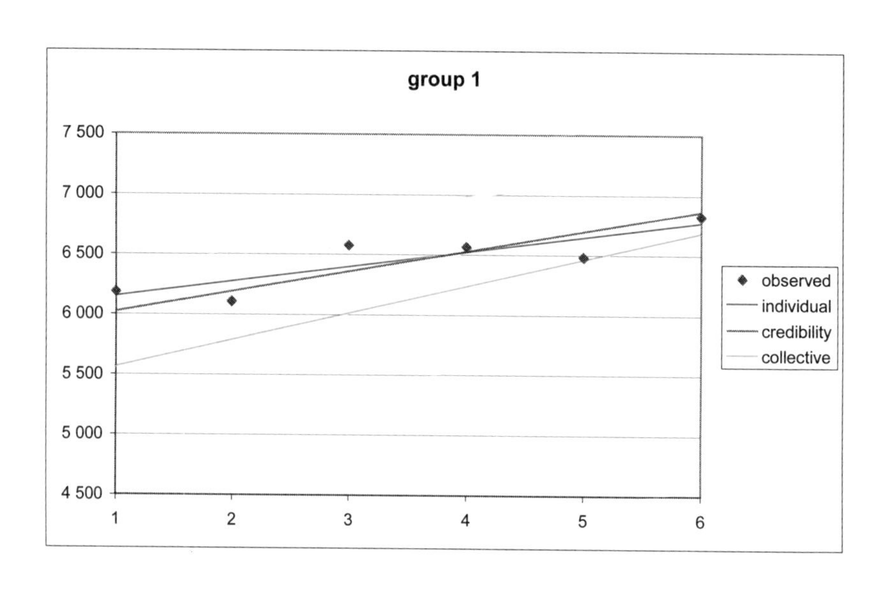

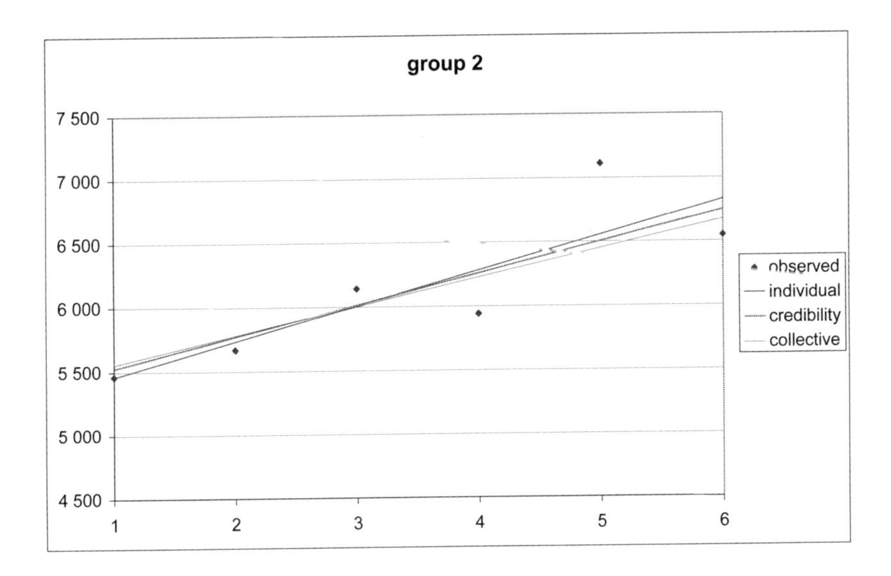
group 2
7 500
7 000
6 500
6 000
5 500
5 000
4 500
1
2
3
4
5
6
observed
individual
credibility
collective

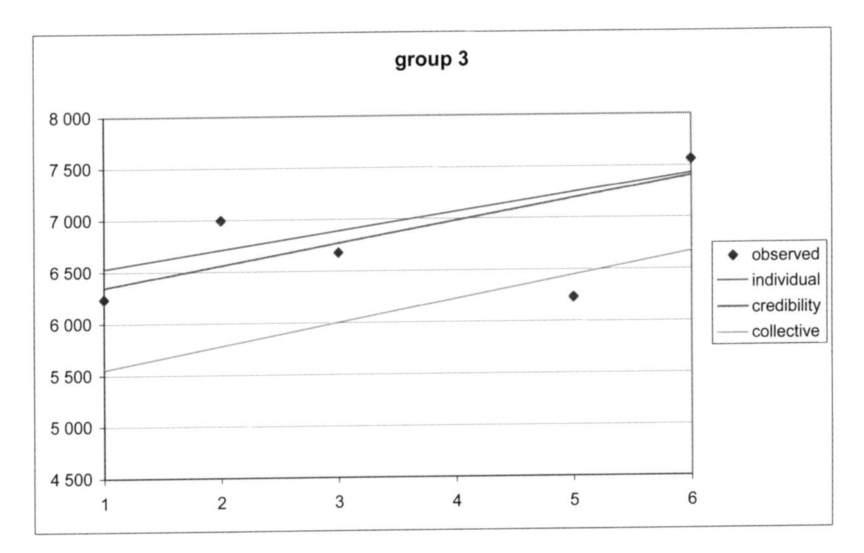
group 3
8 000
7 500
7 000
6 500
6 000
5 500
5 000
4 500
1
2
3
4
5
6
observed
individual
credibility
collective

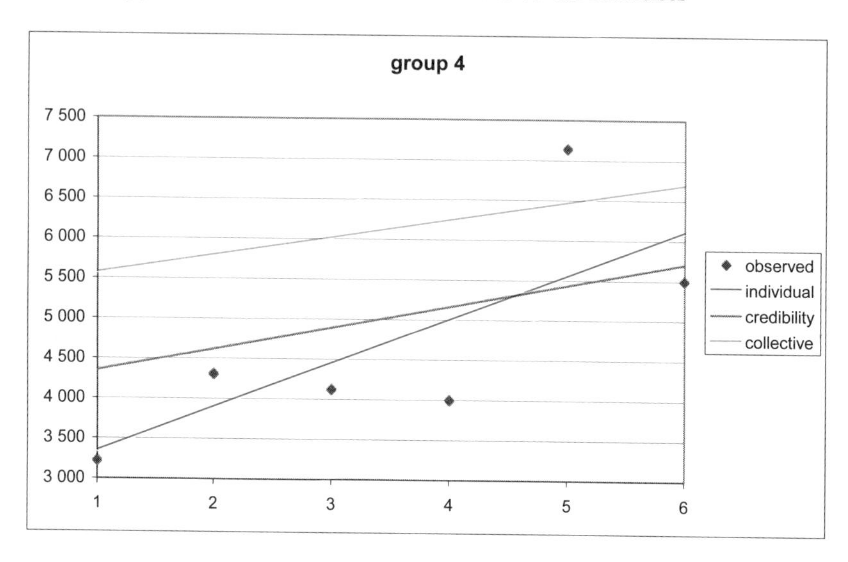

c)

risk groups	1	2	3	4
credibility estimate for $j = 8$	7 202	7 239	7 829	6 244

d) The same slope for all risk groups is identical to the assumption that τ_1^2 is equal to zero. The estimator of the slope of the collective regression line for $\tau_1^2 \to 0$ is equal to

$$\widehat{\beta}_1 = \sum_{j=1}^{4} \frac{w_{i\bullet}^*}{w_{\bullet\bullet}^*} B_{i1} = 204.$$

We then get the following results:

	credibility regression lines			
risk groups	1	2	3	4
intercept (at j_0)	6 425	6 113	6 856	4 996
slope	204	204	204	204

,

risk groups	1	2	3	4
credibility estimate for $j = 8$	7 362	7 050	7 793	5 933

References

[AL83] B. Abraham and J. Ledolter. *Statistical Methods for Forecasting.* John Wiley and Sons, New York, 1983.

[Bai45] A. L. Bailey. A generalized theory of credibility. *Proc. of Cas. Act. Soc.*, 32:13–20, 1945.

[Bai50] A. L. Bailey. Credibility procedures, Laplac's generalization of Bayes' Rule and the combination of collateral knowledge with observed data. *Proc. of Cas. Act. Soc.*, 37:7–23, 1950.

[BG97] H. Bühlmann and A. Gisler. Credibility in the regression case revisited. *ASTIN Bulletin*, 27:83–98, 1997.

[Bic64] F. Bichsel. Erfahrungstarifierung in der Motorfahrzeug-Haftpflicht-Versicherung. *Bulletin of Swiss Ass. of Act.*, pages 119–130, 1964.

[BJ87] H. Bühlmann and W. S. Jewell. Hierarchical credibility revisited. *Bulletin of Swiss Ass. of Act.*, pages 35–54, 1987.

[BS70] H. Bühlmann and E. Straub. Glaubwürdigkeit für Schadensätze. *Bulletin of Swiss Ass. of Act.*, pages 111–133, 1970.

[Büh64] H. Bühlmann. Optimale Prämienstufensysteme. *Bulletin of Swiss Ass. of Act.*, pages 193–214, 1964.

[Büh67] H. Bühlmann. Experience rating and credibility. *ASTIN Bulletin*, 4:199–207, 1967.

[Büh69] H. Bühlmann. Experience rating and credibility. *ASTIN Bulletin*, 5:157–165, 1969.

[Büh70] H. Bühlmann. *Mathematical Methods in Risk Theory.* Springer-Verlag, Berlin, 1970.

[Dan94] D. R. Dannenburg. Some results on the estimation of the credibility factor in the classical Bühlmann model. *Insurance: Mathematics and Economics*, 14:39–50, 1994.

[Dan96a] D. R. Dannenburg. An autoregressive credibility IBNR model. *Blätter der deutschen Gesellschaft für Versicherungsmathematik*, 22:235–248, 1996.

[Dan96b] D. R. Dannenburg. *Basic actuarial credibility models – Evaluations and extensions.* PhD thesis, Tinbergen Institute, Amsterdam, 1996.

[DeG70] M. H. DeGroot. *Optimal Statistical Decisions.* McGraw-Hill, New York, 1970.

[Del65] P. Delaporte. Tarification du risque individuel d'accidents d'automobiles par la prime modelée sur le risque. *ASTIN Bulletin*, 3:251–271, 1965.

[DF64] B. De Finetti. Sulla teoria della credibilità. *Giornale dell'Instituto Italiano degli Attuari*, 27:219–231, 1964.

[DG81] A. Dubey and A. Gisler. On parameter estimators in credibility. *Bulletin of Swiss Ass. of Act.*, pages 187–212, 1981.

[DKG96] D. R. Dannenburg, R. Kaas, and M. J. Goovaerts. *Practical Actuarial Credibility Models.* Institute of Actuarial Science and Econometrics, Amsterdam, 1996.

[DPDG76] N. De Pril, L. D'Hooge, and M. J. Goovaerts. A bibliography on credibility theory and its applications. *Journal of Computational and Applied Mathematics*, 2:55–62, 1976.

[DV76a] F. De Vylder. Geometrical credibility. *Scand. Act. J.*, pages 121–149, 1976.

[DV76b] F. De Vylder. Optimal semilinear credibility. *Bulletin of Swiss Ass. of Act.*, pages 27–40, 1976.

[DV77] F. De Vylder. Iterative credibility. *Bulletin of Swiss Ass. of Act.*, pages 25–33, 1977.

[DV78] F. De Vylder. Parameter estimation in credibility theory. *ASTIN Bulletin*, 10:99–112, 1978.

[DV81a] F. De Vylder. Practical credibility theory with emphasis on optimal parameter estimation. *ASTIN Bulletin*, 12:115–131, 1981.

[DV81b] F. De Vylder. Regression model with scalar credibility weights. *Bulletin of Swiss Ass. of Act.*, pages 27–39, 1981.

[DVB79] F. De Vylder and Y. Ballegeer. A numerical illustration of optimal semilinear credibility. *ASTIN Bulletin*, 10:131–148, 1979.

[DVC94] F. De Vylder and H. Cossette. Dependent contracts in Bühlmann's credibility model. *Bulletin of Swiss Ass. of Act.*, pages 117–147, 1994.

[DVG91] F. De Vylder and M. Goovaerts. Estimation of the heterogenity parameter in the Bühlmann-Straub credibility theory model. *Insurance: Mathematics and Economics*, 11:233–238, 1991.

[DY79] P. Diaconis and D. Ylvisaker. Conjugate priors for exponential families. *Annals of Statistics*, 7:269–281, 1979.

[Fer73] T. S. Ferguson. A Bayesian analysis of some nonparametric problems. *Annals of Statistics*, 1:209–230, 1973.

[Ger79] H. U. Gerber. *An Introduction to Mathematical Risk Theory.* Huebner Foundation Monograph no 8, Richard D. Irwin Inc., Homewood, Illinois, 1979.

[Ger95] H. U. Gerber. A teacher's remark on exact credibility. *ASTIN Bulletin*, 25:189–192, 1995.

[GH87] M. J. Goovaerts and W. J. Hoogstad. Credibility theory. *Survey of Acturial Studies, Nationale-Nederlanden*, 4, 1987.

[Gis80] A. Gisler. Optimum trimming of data in the credibility model. *Bulletin of Swiss Ass. of Act.*, pages 313–325, 1980.

[Gis89] A. Gisler. Optimales Stutzen von Daten in der Credibility-Theorie. In *Schriftenreihe Angewandte Versicherungsmathematik*, volume 22, pages 124–150. Verlag Versicherungswirtschaft, Karlsruhe, 1989.

[Gis90] A. Gisler. Credibility-theory made easy. *Bulletin of Swiss Ass. of Act.*, pages 75–100, 1990.

[Gis96] A. Gisler. Bonus-malus and tariff segmentation. ASTIN Colloquium, 1996.

[GJ75a] H. U. Gerber and D. A. Jones. Credibility formulae with geometric weights. *Transaction of the Society of Actuaries*, 27:39–52, 1975.

[GJ75b] H. U. Gerber and D. A. Jones. Credibility formulas of the updating type. In P. M. Kahn, editor, *Credibility: Theory and Applications*, Academic Press, New York, 1975.

[GR93] A. Gisler and P. Reinhard. Robust credibility. *ASTIN Bulletin*, 23:117–143, 1993.

[Hac75] C. A. Hachemeister. Credibility for regression models with application to trend. In P. M. Kahn, editor, *Credibility: Theory and Applications*, Academic Press, New York, 1975.

[Her99] T. N. Herzog. *Introduction to Credibility Theory.* ACTEX Publications, Winsted, CT, USA, 1999.

[Hes91] O. Hesselager. Prediction of outstanding claims, a hierachical credibility approach. *Scand. Act. J.*, pages 77–90, 1991.

[Hic75] J. C. Hickman. Introduction and historical overview of credibility. In P. M. Kahn, editor, *Credibility: Theory and Applications*, Academic Press, New York, 1975.

[HP98] M. R. Hardy and H. H. Panjer. A Credibility Approach to Mortality Risk. *ASTIN Bulletin*, 28:269–283, 1998.

[Jew73] W. S. Jewell. Multidimensional credibility. Report ORC Berkeley, 1973.

[Jew74a] W. S. Jewell. Credibility means are exact Bayesian for exponential families. *ASTIN Bulletin*, 8:77–90, 1974.

[Jew74b] W. S. Jewell. Exact multidimensional credibility. *Bulletin of Swiss Ass. of Act.*, pages 193–214, 1974.

[Jew75a] W. S. Jewell. Model variations in credibility theory. In P. M. Kahn, editor, *Credibility: Theory and Applications*, Academic Press, New York, 1975.

[Jew75b] W. S. Jewell. Regularity conditions for exact credibility. *ASTIN Bulletin*, 8:336–341, 1975.

[Jew75c] W. S. Jewell. The use of collateral data in credibility theory: a hierarchical model. *Giornale dell'Instituto Italiano degli Attuari*, 38:1–16, 1975.

[Jew76a] W. S. Jewell. Bayesian regression and credibility theory. Report ORC Berkeley, 1976.

[Jew76b] W. S. Jewell. A survey of credibility theory. Report ORC Berkeley, 1976.

[JZ83] P. de Jong and B. Zehnwirth. Credibility theory and the Kalman filter. *Insurance: Mathematics and Economics*, 2:281–286, 1983.

[Kal60] R. E. Kalman. A new approach to linear filtering and prediction problems. *Transactions of ASME–Journal of Basic Engineering*, 82:35–45, 1960.

[Kef29] R. Keffer. An experience rating formula. *Transaction of the Society of Actuaries*, 15:223–235, 1929.

[KF70] A. N. Kolmogorov and V. S. Fomin. *Introductory Real Analysis.* Dover Publications, New York, 1970.

[Klu92] S. A. Klugman. *Bayesian Statistics in Actuarial Science: With Emphasis on Credibility.* Kluwer, 1992.

[Kre82] E. Kremer. Credibility for some evolutionary models. *Scand. Act. J.*, pages 129–142, 1982.

[Kün92] H. R. Künsch. Robust methods for credibility. *ASTIN Bulletin*, 22:33–49, 1992.

[LC62] L. H. Longley-Cook. An introduction to credibility theory. *Proc. of Cas. Act. Soc.*, 49:194–221, 1962.

[Leh86] E. L. Lehmann. *Testing Statistical Hypothesis.* Wiley Series in Probability and Mathematical Statistics: Probability and Mathematical Statistics, New York, 1986.

[Lun40] O. Lundberg. On random processes and their applications to sickness and accident statistics. In *University of Stockholm Thesis*, Alquist and Wiksells, Uppsala, 1940.

[May64] A. L. Mayerson. A Bayesian view of credibility. *Proc. of Cas. Act. Soc.*, 51:85–104, 1964.

[Meh75] R. K. Mehra. Credibility theory and Kalman filtering with extensions. *Report RM 75-64, International Institute for Applied Systems Analysis, Schloss Laxenburg, Austria*, 1975.

[MN89] P. McCullagh and J. A. Nelder. *Generalized Linear Models.* Chapman and Hall, Cambridge, 2nd edition, 1989.

[Mow14] A. H. Mowbray. How extensive a payroll exposure is necessary to give a dependable pure premium. *Proc. of Cas. Act. Soc.*, 1:24–30, 1914.

[Mow27] A. H. Mowbray. Experience rating of risks for workmen's compensation insurance in the United States. *Transaction of the 8th Int. Congr. of Act.*, 1:324–335, 1927.

[Neu87] W. Neuhaus. Early warning. *Scand. Act. J.*, pages 128–156, 1987.

[Nor79] R. Norberg. The credibility approach to experience rating. *Scand. Act. J.*, 4:181–221, 1979.

[Nor80] R. Norberg. Empirical Bayes credibility. *Scand. Act. J.*, 4:177–194, 1980.

[Nor82] R. Norberg. On optimal parameter estimation in credibility. *Insurance: Mathematics and Economics*, 1:73–90, 1982.

[Nor86] R. Norberg. Hierarchical credibility: Analysis of a random effect linear model with nested claissification. *Scand. Act. J.*, pages 204–222, 1986.

[Nor89] R. Norberg. Experience rating in group life insurance. *Scand. Act. J.*, pages 194–224, 1989.

[Nor04] R. Norberg. Credibility theory. In J. L. Teugels and B. Sundt, editors, *Encyclopedia of Actuarial Science*, Wiley, Chichester, UK, 2004.

[Per32] F. S. Perryman. Some notes on credibility. *Proc. of Cas. Act. Soc.*, 19:65–84, 1932.

[Rob55] H. Robbins. An empirical Bayes approach to statistics. In *Berkeley Symposium on Mathematical Statistics and Probability*, University of California Press, Berkeley, 1955.

[Ryt90] M. Rytgaard. Estimation in the Pareto distribution. *ASTIN Bulletin*, 20:201–216, 1990.

[SAS93] SAS. The GENMOD procedure. Technical Report P-243, SAS/STAT Software, 1993.

[Sch04] R. Schnieper. Robust Bayesian experience rating. *ASTIN Bulletin*, 34:125–150, 2004.

[SL03] G. A. F. Seber and A. J. Lee. *Linear Regression Analysis.* Wiley, New York, 2003.

[SS04] E. S. W. Shiu and F. Y. Sing. Credibility theory and geometry. *Journal of Actuarial Practice*, 11:197–216, 2004.

[Sun79a] B. Sundt. A hierarchical regression credibility model. *Scand. Act. J.*, pages 107–114, 1979.

[Sun79b] B. Sundt. An insurance model with collective seasonal random factors. *Bulletin of Swiss Ass. of Act.*, pages 57–64, 1979.

[Sun79c] B. Sundt. On choice of statistics in credibility estimation. *Scand. Act. J.*, pages 115–123, 1979.

[Sun80] B. Sundt. A multi-level hierarchical credibility regression model. *Scand. Act. J.*, 1:25–32, 1980.

[Sun81] B. Sundt. Recursive credibility estimation. *Scand. Act. J.*, 1:3–22, 1981.

[Sun82] B. Sundt. Invariantly recursive credibility estimation. *Insurance: Mathematics and Economics*, 1:185–196, 1982.

[Sun83] B. Sundt. Finite credibility formulae in evoluationary models. *Scand. Act. J.*, pages 106–116, 1983.

[Sun84] B. Sundt. *An Introduction to Non-life Insurance Mathematics.* Verlag Versicherungswirtschaft, Karlsruhe, 1984.

[Sun97] B. Sundt. Book review: Practical actuarial credibility models by D. R. Dannenburg, R. Kaas and M. J. Goovaerts. *Bulletin of Swiss Ass. of Act.*, pages 89–93, 1997.

[Sun98] B. Sundt. Homogeneous credibility estimators. *Bulletin of Swiss Ass. of Act.*, pages 193–209, 1998.

[Tay74] G. C. Taylor. Experience rating with credibility adjustment of the manual premium. *ASTIN Bulletin*, 7:323–336, 1974.

[Tay75] G. C. Taylor. Credibility for time-heterogeneous loss ratios. In P. M. Kahn, editor, *Credibility: Theory and Applications*, Academic Press, New York, 1975.

[Tay79] G. C. Taylor. Credibility analysis of a general hierarchical model. *Scand. Act. J.*, pages 1–12, 1979.

[Wei05] S. Weisberg. *Applied Linear Regression.* Wiley, New York, 2005.

[Wen73] H. Wenger. Eine Tarifierungsmethode im Feuer-Industriegeschäft. *Bulletin of Swiss Ass. of Act.*, 1:95–111, 1973.

[Whi18] A.W. Whitney. The theory of experience rating. *Proc. of Cas. Act. Soc.*, 4:274–292, 1918.

[Wi86] Special issue on credibility theory. In G. W. De Wit, editor, *Insurance Abstracts and Reviews*, Elsevier Science Publishers, Amsterdam, 1986.

[Zeh77] B. Zehnwirth. The mean credibility formula is a Bayes rule. *Scand. Act. J.*, pages 212–216, 1977.

[Zeh79a] B. Zehnwirth. Credibility and the Dirichlet process. *Scand. Act. J.*, pages 13–23, 1979.

[Zeh79b] B. Zehnwirth. A hierarchical model for the estimation of claim rates in a motor car insurance portfolio. *Scand. Act. J.*, 2/3:75–82, 1979.

[Zeh84] B. Zehnwirth. Credibility: Estimation of structural parameters. In F. de Vylder M. Goovaerts and J. Haezendonck, editors, *Premium Calculation in Insurance*, Reidel, 1984.

[Zeh85] B. Zehnwirth. Linear filtering and recursive credibility estimation. *ASTIN Bulletin*, 15:19–38, 1985.

Index

Universitext

Aguilar, M.; Gitler, S.; Prieto, C.: Algebraic Topology from a Homotopical Viewpoint

Aksoy, A.; Khamsi, M. A.: Methods in Fixed Point Theory

Alevras, D.; Padberg M. W.: Linear Optimization and Extensions

Andersson, M.: Topics in Complex Analysis

Aoki, M.: State Space Modeling of Time Series

Arnold, V. I.: Lectures on Partial Differential Equations

Audin, M.: Geometry

Aupetit, B.: A Primer on Spectral Theory

Bachem, A.; Kern, W.: Linear Programming Duality

Bachmann, G.; Narici, L.; Beckenstein, E.: Fourier and Wavelet Analysis

Badescu, L.: Algebraic Surfaces

Balakrishnan, R.; Ranganathan, K.: A Textbook of Graph Theory

Balser, W.: Formal Power Series and Linear Systems of Meromorphic Ordinary Differential Equations

Bapat, R.B.: Linear Algebra and Linear Models

Benedetti, R.; Petronio, C.: Lectures on Hyperbolic Geometry

Benth, F. E.: Option Theory with Stochastic Analysis

Berberian, S. K.: Fundamentals of Real Analysis

Berger, M.: Geometry I, and II

Bliedtner, J.; Hansen, W.: Potential Theory

Blowey, J. F.; Coleman, J. P.; Craig, A. W. (Eds.): Theory and Numerics of Differential Equations

Blowey, J.; Craig, A.: Frontiers in Numerical Analysis. Durham 2004

Blyth, T. S.: Lattices and Ordered Algebraic Structures

Börger, E.; Grädel, E.; Gurevich, Y.: The Classical Decision Problem

Böttcher, A; Silbermann, B.: Introduction to Large Truncated Toeplitz Matrices

Boltyanski, V.; Martini, H.; Soltan, P. S.: Excursions into Combinatorial Geometry

Boltyanskii, V. G.; Efremovich, V. A.: Intuitive Combinatorial Topology

Bonnans, J. F.; Gilbert, J. C.; Lemaréchal, C.; Sagastizábal, C. A.: Numerical Optimization

Booss, B.; Bleecker, D. D.: Topology and Analysis

Borkar, V. S.: Probability Theory

Brunt B. van: The Calculus of Variations

Bühlmann, H.; Gisler, A.: A Course in Credibility Theory and its Applications

Carleson, L.; Gamelin, T. W.: Complex Dynamics

Cecil, T. E.: Lie Sphere Geometry: With Applications of Submanifolds

Chae, S. B.: Lebesgue Integration

Chandrasekharan, K.: Classical Fourier Transform

Charlap, L. S.: Bieberbach Groups and Flat Manifolds

Chern, S.: Complex Manifolds without Potential Theory

Chorin, A. J.; Marsden, J. E.: Mathematical Introduction to Fluid Mechanics

Cohn, H.: A Classical Invitation to Algebraic Numbers and Class Fields

Curtis, M. L.: Abstract Linear Algebra

Curtis, M. L.: Matrix Groups

Cyganowski, S.; Kloeden, P.; Ombach, J.: From Elementary Probability to Stochastic Differential Equations with MAPLE

Dalen, D. van: Logic and Structure

Das, A.: The Special Theory of Relativity: A Mathematical Exposition

Debarre, O.: Higher-Dimensional Algebraic Geometry

Deitmar, A.: A First Course in Harmonic Analysis

Demazure, M.: Bifurcations and Catastrophes

Devlin, K. J.: Fundamentals of Contemporary Set Theory

DiBenedetto, E.: Degenerate Parabolic Equations

Diener, F.; Diener, M.(Eds.): Nonstandard Analysis in Practice

Dimca, A.: Sheaves in Topology

Dimca, A.: Singularities and Topology of Hypersurfaces

DoCarmo, M. P.: Differential Forms and Applications

Duistermaat, J. J.; Kolk, J. A. C.: Lie Groups

Edwards, R. E.: A Formal Background to Higher Mathematics Ia, and Ib

Edwards, R. E.: A Formal Background to Higher Mathematics IIa, and IIb

Emery, M.: Stochastic Calculus in Manifolds

Emmanouil, I.: Idempotent Matrices over Complex Group Algebras

Endler, O.: Valuation Theory

Erez, B.: Galois Modules in Arithmetic

Everest, G.; Ward, T.: Heights of Polynomials and Entropy in Algebraic Dynamics

Farenick, D. R.: Algebras of Linear Transformations

Foulds, L. R.: Graph Theory Applications

Franke, J.; Härdle, W.; Hafner, C. M.: Statistics of Financial Markets: An Introduction

Frauenthal, J. C.: Mathematical Modeling in Epidemiology

Freitag, E.; Busam, R.: Complex Analysis

Friedman, R.: Algebraic Surfaces and Holomorphic Vector Bundles

Fuks, D. B.; Rokhlin, V. A.: Beginner's Course in Topology

Fuhrmann, P. A.: A Polynomial Approach to Linear Algebra

Gallot, S.; Hulin, D.; Lafontaine, J.: Riemannian Geometry

Gardiner, C. F.: A First Course in Group Theory

Gårding, L.; Tambour, T.: Algebra for Computer Science

Godbillon, C.: Dynamical Systems on Surfaces

Godement, R.: Analysis I, and II

Godement, R.: Analysis II

Goldblatt, R.: Orthogonality and Spacetime Geometry

Gouvêa, F. Q.: p-Adic Numbers

Gross, M. et al.: Calabi-Yau Manifolds and Related Geometries

Gustafson, K. E.; Rao, D. K. M.: Numerical Range. The Field of Values of Linear Operators and Matrices

Gustafson, S. J.; Sigal, I. M.: Mathematical Concepts of Quantum Mechanics

Hahn, A. J.: Quadratic Algebras, Clifford Algebras, and Arithmetic Witt Groups

Hájek, P.; Havránek, T.: Mechanizing Hypothesis Formation

Heinonen, J.: Lectures on Analysis on Metric Spaces

Hlawka, E.; Schoißengeier, J.; Taschner, R.: Geometric and Analytic Number Theory

Holmgren, R. A.: A First Course in Discrete Dynamical Systems

Howe, R., Tan, E. Ch.: Non-Abelian Harmonic Analysis

Howes, N. R.: Modern Analysis and Topology

Hsieh, P.-F.; Sibuya, Y. (Eds.): Basic Theory of Ordinary Differential Equations

Humi, M., Miller, W.: Second Course in Ordinary Differential Equations for Scientists and Engineers

Hurwitz, A.; Kritikos, N.: Lectures on Number Theory

Huybrechts, D.: Complex Geometry: An Introduction

Isaev, A.: Introduction to Mathematical Methods in Bioinformatics

Istas, J.: Mathematical Modeling for the Life Sciences

Iversen, B.: Cohomology of Sheaves

Jacod, J.; Protter, P.: Probability Essentials

Jennings, G. A.: Modern Geometry with Applications

Jones, A.; Morris, S. A.; Pearson, K. R.: Abstract Algebra and Famous Inpossibilities

Jost, J.: Compact Riemann Surfaces

Jost, J.: Dynamical Systems. Examples of Complex Behaviour

Jost, J.: Postmodern Analysis

Jost, J.: Riemannian Geometry and Geometric Analysis

Kac, V.; Cheung, P.: Quantum Calculus

Kannan, R.; Krueger, C. K.: Advanced Analysis on the Real Line

Kelly, P.; Matthews, G.: The Non-Euclidean Hyperbolic Plane

Kempf, G.: Complex Abelian Varieties and Theta Functions

Kitchens, B. P.: Symbolic Dynamics

Kloeden, P.; Ombach, J.; Cyganowski, S.: From Elementary Probability to Stochastic Differential Equations with MAPLE

Kloeden, P. E.; Platen; E.; Schurz, H.: Numerical Solution of SDE Through Computer Experiments

Kostrikin, A. I.: Introduction to Algebra

Krasnoselskii, M. A.; Pokrovskii, A. V.: Systems with Hysteresis

Kurzweil, H.; Stellmacher, B.: The Theory of Finite Groups. An Introduction

Lang, S.: Introduction to Differentiable Manifolds

Luecking, D. H., Rubel, L. A.: Complex Analysis. A Functional Analysis Approach

Ma, Zhi-Ming; Roeckner, M.: Introduction to the Theory of (non-symmetric) Dirichlet Forms

Mac Lane, S.; Moerdijk, I.: Sheaves in Geometry and Logic

Marcus, D. A.: Number Fields

Martinez, A.: An Introduction to Semiclassical and Microlocal Analysis

Matoušek, J.: Using the Borsuk-Ulam Theorem

Matsuki, K.: Introduction to the Mori Program

Mazzola, G.; Milmeister G.; Weissman J.: Comprehensive Mathematics for Computer Scientists 1

Mazzola, G.; Milmeister G.; Weissman J.: Comprehensive Mathematics for Computer Scientists 2

Mc Carthy, P. J.: Introduction to Arithmetical Functions

McCrimmon, K.: A Taste of Jordan Algebras

Meyer, R. M.: Essential Mathematics for Applied Field

Meyer-Nieberg, P.: Banach Lattices

Mikosch, T.: Non Life Insurance Mathematics

Mines, R.; Richman, F.; Ruitenburg, W.: A Course in Constructive Algebra

Moise, E. E.: Introductory Problem Courses in Analysis and Topology

Montesinos-Amilibia, J. M.: Classical Tessellations and Three Manifolds

Morris, P.: Introduction to Game Theory

Nikulin, V. V.; Shafarevich, I. R.: Geometries and Groups

Oden, J. J.; Reddy, J. N.: Variational Methods in Theoretical Mechanics

Øksendal, B.: Stochastic Differential Equations

Øksendal, B.; Sulem, A.: Applied Stochastic Control of Jump Diffusions

Poizat, B.: A Course in Model Theory

Polster, B.: A Geometrical Picture Book

Porter, J. R.; Woods, R. G.: Extensions and Absolutes of Hausdorff Spaces

Radjavi, H.; Rosenthal, P.: Simultaneous Triangularization

Ramsay, A.; Richtmeyer, R. D.: Introduction to Hyperbolic Geometry

Rees, E. G.: Notes on Geometry

Reisel, R. B.: Elementary Theory of Metric Spaces

Rey, W. J. J.: Introduction to Robust and Quasi-Robust Statistical Methods

Ribenboim, P.: Classical Theory of Algebraic Numbers

Rickart, C. E.: Natural Function Algebras

Roger G.: Analysis II

Rotman, J. J.: Galois Theory

Rubel, L. A.: Entire and Meromorphic Functions

Ruiz-Tolosa, J. R.; Castillo E.: From Vectors to Tensors

Runde, V.: A Taste of Topology

Rybakowski, K. P.: The Homotopy Index and Partial Differential Equations

Sagan, H.: Space-Filling Curves

Samelson, H.: Notes on Lie Algebras

Schiff, J. L.: Normal Families

Sengupta, J. K.: Optimal Decisions under Uncertainty

Séroul, R.: Programming for Mathematicians

Seydel, R.: Tools for Computational Finance

Shafarevich, I. R.: Discourses on Algebra

Shapiro, J. H.: Composition Operators and Classical Function Theory

Simonnet, M.: Measures and Probabilities

Smith, K. E.; Kahanpää, L.; Kekäläinen, P.; Traves, W.: An Invitation to Algebraic Geometry

Smith, K. T.: Power Series from a Computational Point of View

Smoryński, C.: Logical Number Theory I. An Introduction

Stichtenoth, H.: Algebraic Function Fields and Codes

Stillwell, J.: Geometry of Surfaces

Stroock, D. W.: An Introduction to the Theory of Large Deviations

Sunder, V. S.: An Invitation to von Neumann Algebras

Tamme, G.: Introduction to Étale Cohomology

Tondeur, P.: Foliations on Riemannian Manifolds

Toth, G.: Finite Möbius Groups, Minimal Immersions of Spheres, and Moduli

Verhulst, F.: Nonlinear Differential Equations and Dynamical Systems

Wong, M. W.: Weyl Transforms

Xambó-Descamps, S.: Block Error-Correcting Codes

Zaanen, A.C.: Continuity, Integration and Fourier Theory

Zhang, F.: Matrix Theory

Zong, C.: Sphere Packings

Zong, C.: Strange Phenomena in Convex and Discrete Geometry

Zorich, V. A.: Mathematical Analysis I

Zorich, V. A.: Mathematical Analysis II